Alexander A. Katz (Ed.)

Topological Algebras and their Applications

De Gruyter Proceedings in Mathematics

Topological Algebras and their Applications

Proceedings of the 8th International Conference on Topological Algebras and their Applications, 2014

Edited by
Alexander A. Katz

DE GRUYTER

Mathematics Subject Classification 2010
35-02, 65-02, 65C30, 65C05, 65N35, 65N75, 65N80

Editor
Prof. Alexander A. Katz, PhD
St. John's University
St. John's College of LAS
Department of Mathematics and Computer Sciences
8000 Utopia Parkway
SJH-334-G
Jamaica, NY 11439
USA
katza@stjohns.edu

ISBN 978-3-11-041433-2
e-ISBN (PDF) 978-3-11-041355-7
e-ISBN (EPUB) 978-3-11-041442-4

Library of Congress Control Number: 2018941019.

Bibliographic information published by the Deutsche Nationalbibliothek
The Deutsche Nationalbibliothek lists this publication in the Deutsche Nationalbibliografie; detailed bibliographic data are available on the Internet at http://dnb.dnb.de.

Printing and binding: CPI books GmbH, Leck
♾ Printed on acid-free paper
Printed in Germany

www.degruyter.com

In memory of Anastasios Mallios

Contents

Foreword

The current volume is the Proceedings of the 8th International Conference of Topological Algebras and Their Applications (ICTAA-2014) which was held on May 26-30, 2014 in Playa de Villas de Mar Beach, Dominican Republic. It is composed of delivered and invited papers of the conference.

This series of conferences started in 1999 in Tartu, Estonia and were subsequently held in Rabat, Moroco (2000), Oulu, Finland (2001), Oaxaca, Mexico (2002), Bedlewo, Poland (2003), Athens, Greece (2005) and Tartu, Estonia (2008 and 2013).

The topics of the conference include all areas of mathematics, connected with (preferably general) topological algebras and their applications, including all kinds of topological-algebraic structures as topological linear spaces, topological rings, topological modules, topological groups and semigroups; bornological-algebraic structures such as bornological linear spaces, bornological algebras, bornological groups, bornological rings and modules; algebraic and topological K-theory; topological module bundles, sheaves and others.

The conference and the present volume are dedicated to the memory of Professor Anastasios Mallios, 1932-2014 (Athens, Greece), who passed away just months before the conference.

The editor is grateful to the patience of De Gruyter's team who made this publication possible.

Many special thanks go to Dr. Mati Abel and Dr. Mart Abel (Tartu, Estonia) who were helping me with the list of potential conference participants and to Dr. Igor Nikolaev (St. John's University, NY, USA) who was helping me with formatting papers to the needed LaTeX template.

Alexander A. Katz, Editor

https://doi.org/10.1515/9783110413557-201

Mart Abel

Some results on spectral properties of unital algebras and on the algebra of linear operators on a unital algebra

Abstract: There are several books and papers written about the properties of the spectrum and the spectral radius of elements in Banach algebras (see [2], [3], [4] and [6]) or in Q-algebras ([5]). In [1] we already started to generalise some results connected with the spectral properties to the case of general topological algebras. The present paper is a continuation of the work started in [1].

Keywords: Spectrum, spectral radius, topological algebra, algebra of linear operators

MSC 2000: Primary 46H05; Secondary 46L10, 16-XX

1 Introduction and purely algebraic results

This paper got its motivation form the book [2] of B. Aupetit, where several results about the spectral properties of elements in Banach algebras were collected. While studying these proofs, it appeared that many of them could be generalised in more algebraic way, without using the properties of the norm. Herewith, we present the results with proofs for in the most general form we were able to prove them at the moment.

Let $\mathbb{K}$ denote either the field $\mathbb{R}$ of all real numbers or the field $\mathbb{C}$ of all complex numbers. All algebras that we will consider, will be associative ones. Let A be an algebra over $\mathbb{K}$. We denote by $\mathrm{Rad}(A)$ the Jacobson radical of A. It is known that the Jacobson radical of A can be represented as the intersection of all maximal left ideals of A or as the intersection of all maximal right ideals of A. An algebra is called semi-simple, if its Jacobson radical contains only the zero element θ_A of A, i.e., $\mathrm{Rad}(A) = \{\theta_A\}$.

Let X be a linear space over $\mathbb{K}$. Then it is known that the set $L(X)$ of all linear operators on X is an algebra with respect to the pointwise addition, pointwise scalar multiplication and multiplication as composition. In [1], we had some assumptions for the algebra $L(X)$, because we wanted the linear space X to be such that the algebra is semi-simple. In [2], p. 34, there is Theorem 3.1.4, which claims that $L(X)$ is a semi-simple algebra for every Banach space X. We start this paper by generalising the result of Aupetit for the case of an arbitrary linear space X.

Lemma 1.1. *Let X be a linear space over $\mathbb{K}$. Then $L(X)$ is a semi-simple algebra.*

Mart Abel, Institute of Mathematics, University of Tartu, 2 Liivi Str., Room 615, 50409 Tartu, Estonia; Institute of Mathematics and Natural Sciences, University of Tallinn, 29 Narva Str., Room A-416, 10120 Tallinn, Estonia; E-mail: mart.abel@tlu.ee, mart.abel@ut.ee

https://doi.org/10.1515/9783110413557-001

Proof. Fix any $x \in X \setminus \{\theta_X\}$ and define $I_x := \{T : T \in L(X), T(x) = \theta_X\}$. Then I_x is a left ideal of $L(X)$. We will show that I_x is a maximal left ideal of $L(X)$. For that, suppose that there exists a left ideal J of $L(X)$ such that $I_x \subset J$ and $I_x \neq J$. Then there exists $S \in J \setminus I_x$. Hence, $S(x) \neq \theta_X$. Now, $S(x) \subset J(x) = \{T(x) : T \in J\}$. Thus, $J(x) \neq \{\theta_X\}$. Since every $T \in J$ is a linear operator on X, then $J(x)$ is a linear subspace of X. Notice, that $S(J(x)) \subset J(x)$ for every $S \in L(X)$, because $ST \in J$ for every $T \in J$.

Suppose that $J(x) \neq X$. Then there exists $y \in X \setminus J(x)$. Since $J(x) \neq \{\theta_X\}$, there exists $z \in X$ such that $z \in J(x) \setminus \{\theta_X\}$. Hence, the set $\{z\}$ is linearly independent in X. Every linearly independent set can be completed to the (usual, in case X is finite-dimensional, or Hamel, if X is infinite-dimensional linear space) basis $\{z\} \cup \{e_i : i \in I\}$ of X. Hence, every $w \in X$ can be uniquely represented in the form

$$w = \lambda z + \sum_{i \in I} \lambda_i e_i \text{ with } \lambda, \lambda_i \in \mathbb{K} \text{ for all } i \in I.$$

Define

$$W(w) = W\Big(\lambda z + \sum_{i \in I} \lambda_i e_i\Big) := \lambda y$$

for every $w \in X$. Clearly, W is a linear map. Since every linear map is uniquely defined by its action on the basis, then we have constructed a linear map $W \in L(X)$ such that $W(z) = y$. So, we have $y = W(z) \in W(J(x)) \subset J(x)$, which is a contradiction with the choice of y. Hence, $J(x) = X$.

So, for each $x \in X = J(x)$, there is $U \in J$ such that $U(x) = x$. Take any $T \in L(X)$. Then $(T - TU)(x) = T(x) - T(U(x)) = T(x) - T(x) = \theta_X$, which implies that $T - TU \in I_x$ for any $T \in L(X)$. Since J is a left ideal of $L(X)$, then from $U \in J$ follows that $TU \in J$ for every $T \in L(X)$. We get now, that $T \in I_x + TU \subset I_x + J \subset J$ for every $T \in L(X)$. Hence, $J = L(X)$ and we have obtained that I_x is a maximal left ideal of $L(X)$. Since the Jacobson Radical of $L(X)$ is the intersection of all maximal left ideals of $L(X)$, then

$$\operatorname{Rad}(L(X)) \subset \bigcap_{x \in X \setminus \{\theta_X\}} I_x = \{\theta_{L(X)}\},$$

because $\theta_{L(X)}$, defined by $\theta_{L(X)}(w) = \theta_X$ for all $w \in X$, is the only linear map, which belongs to the intersection of all sets I_x with $x \in X \setminus \{\theta_X\}$. Hence, $\operatorname{Rad}(L(X)) = \{\theta_{L(X)}\}$ and $L(X)$ is a semi-simple algebra. □

From now on, all linear spaces and all algebras are considered to be over $\mathbb{C}$.

Let A be a unital algebra and a an arbitrary element of A. By e_A we will denote the unital element of A. We will use the denotions $\operatorname{Inv}(A)$ for the set of all invertible elements of A, $\sigma(a) = \{\lambda \in \mathbb{C} : a - \lambda e_A \notin \operatorname{Inv}(A)\}$ for the spectrum of an element $a \in A$ and $\rho(a) = \sup\{|\lambda| : \lambda \in \sigma(a)\}$ for the spectral radius of an element $a \in A$. Denote by $B(0, r)$ the set of all numbers $\lambda \in \mathbb{C}$ for which $|\lambda| < r$.

We will continue by generalising Theorem 3.2.4 from [2], p. 36.

Lemma 1.2. *Let A be an algebra with unit e_A and let $x, y \in A$ with $\rho(x) < \infty$ and $\rho(y) < \infty$. If $xy = e_A$ and $yx \neq e_A$, then there exists an open neighbourhood of zero in $\mathbb{C}$, which is contained both in $\sigma(x)$ and in $\sigma(y)$.*

Proof. From the assumption, we see that both x and y are not invertible in A. Let $p = yx \neq e_A$.

Take any $\lambda \in \mathbb{C}$. Notice that

$$(x - \lambda e_A)y = xy - \lambda y = e_A - \lambda y \quad \text{and} \quad y(x - \lambda e_A) = p - \lambda y \neq e_A - \lambda y.$$

Suppose that $\rho(y) \neq 0$ and $|\lambda| < \frac{1}{\rho(y)}$. Then either $\lambda = 0$ or $\frac{1}{|\lambda|} > \rho(y)$ and $y - \frac{1}{\lambda}e_A$ is invertible in A. Hence, in both cases, $e_A - \lambda y$ is invertible in A and $(x - \lambda e_A)y(e_A - \lambda y)^{-1} = e_A$. We see that $(x - \lambda e_A)$ is invertible in A if and only if $y(e_A - \lambda y)^{-1}(x - \lambda e_A) = e_A$.

Notice that from $y(e_A - \lambda y) = y - \lambda y^2 = (e_A - \lambda y)y$ follows that

$$\begin{aligned}(e_A - \lambda y)^{-1}y &= (e_A - \lambda y)^{-1}y(e_A - \lambda y)(e_A - \lambda y)^{-1} = \\ &= (e_A - \lambda y)^{-1}(e_A - \lambda y)y(e_A - \lambda y)^{-1} = y(e_A - \lambda y)^{-1}.\end{aligned}$$

Hence,

$$\begin{aligned}y(e_A - \lambda y)^{-1}(x - \lambda e_A) &= (e_A - \lambda y)^{-1}y(x - \lambda e_A) = \\ &= (e_A - \lambda y)^{-1}(p - \lambda y) \neq (e_A - \lambda y)^{-1}(e_A - \lambda y) = e_A.\end{aligned}$$

Therefore, $x - \lambda e_A$ is not invertible in A, which means that $\lambda \in \sigma(x)$. Thus, we have proved in case $\rho(y) \neq 0$ that from $|\lambda| < \frac{1}{\rho(y)}$ follows that $\lambda \in \sigma(x)$, or, equivalently, $B\left(0, \frac{1}{\rho(y)}\right) \subset \sigma(x)$.

In case $\rho(y) = 0$, we obtain that $y - \lambda e_A$ is invertible in A, whenever $\lambda \neq 0$. Similarily as before, we obtain that $x - \lambda e_A$ is invertible in A if and only if $y(e_A - \lambda y)^{-1}(x - \lambda e_A) = e_A$, but the last expression is not equal to e_A. Hence, $x - \lambda e_A$ is not invertible for any $\lambda \neq 0$ and $\sigma(x) = \mathbb{C}$, because x is also not invertible in A. Thus, $B(0, \mu) \subseteq \sigma(x)$ for any $\mu > 0$.

Analogically, in case $\rho(x) \neq 0$, starting from $|\lambda| < \frac{1}{\rho(x)}$, we obtain that $\lambda \in \sigma(y)$, or, equivalently, $B\left(0, \frac{1}{\rho(x)}\right) \subset \sigma(y)$ and in case $\rho(x) = 0$ we obtain $B(0, \mu) \subset \sigma(y)$ for any $\mu > 0$.

In case $\rho(x) \neq 0$ and $\rho(y) \neq 0$, take

$$r := \min\left\{\frac{1}{\rho(x)}, \frac{1}{\rho(y)}\right\}.$$

In case $\rho(x) = 0 = \rho(y)$, take $r = 1$. In case $\rho(x) = 0 \neq \rho(y)$ take $r = \frac{1}{\rho(y)}$ and in case $\rho(x) \neq 0 = \rho(y)$ take $r = \frac{1}{\rho(x)}$. Then we see that always there exists $r > 0$ such that $B(0, r) \subset \sigma(x)$ and $B(0, r) \subset \sigma(y)$. Hence, $B(0, r)$ is the open neighbourhood of 0, which is contained in both $\sigma(x)$ and in $\sigma(y)$. □

2 Results connected with spectrum and spectral radius of elements in topological algebras

We would like to remind that by a topological algebra we mean a topological vector space, which is equipped with the separately continuous multiplication making it also an algebra. A unital algebra A is a Q-algebra, if the set of invertible elements of A is open in A. This set is open, if for every invertible element $a \in A$ there exists a neighbourhood U of a in A such that every $x \in U$ is invertible.

We start this section by generalising Theorem 3.2.1 and Theorem 3.2.3 from [2], pp. 35–36.

Proposition 2.1. *Let A be a unital topological algebra in which there exists a neighbourhood O of zero such that*

$$\sum_{k=0}^{\infty} y^k \in A$$

for every $y \in O$. Then
(a) *$e_A - y \in \operatorname{Inv}(A)$ for every $y \in O$;*
(b) *A is a Q-algebra;*
(c) *$\sigma(x)$ is closed in $\mathbb{C}$ for every $x \in A$.*

Proof. (a) Let $y \in O$. Then

$$\sum_{k=0}^{\infty} y^k \in A \quad \text{and} \quad (e_A - y)\sum_{k=0}^{\infty} y^k = \left(\sum_{k=0}^{\infty} y^k\right)(e_A - y) = e_A,$$

which means that $e_A - y$ is invertible in A.

(b) Let $a \in A$ be an invertible element in A. Since the multiplication is separately continuous, there exists a neighbourhood V of zero in A such that $a^{-1}V \subset O$. Take $U = a - V = \{a - v : v \in V\}$. Then U is a neighbourhood of a in A.

Choose any $x \in U$. Then $a - x \in V$ and $a^{-1}(a - x) \in O$, which means that the element $e_A - a^{-1}(a - x)$ is invertible in A, by part (a). Hence,

$$a(e_A - a^{-1}(a - x)) = a - a + x = x$$

is also invertible in A, as a product of two invertible elements. Because it holds for any $x \in U$, then an invertible element a has a neighbourhood U such that every $x \in U$ is invertible. Since a was chosen arbitrarily from the set of invertible elements of A, then the set of invertible elements of A is open. Hence, A is a Q-algebra.

(c) Take any $x \in A$. If $\sigma(x) = \mathbb{C}$, then $\sigma(x)$ is closed in $\mathbb{C}$.

Suppose that $\sigma(x) \neq \mathbb{C}$. Then there exists $\lambda_0 \in \mathbb{C} \setminus \sigma(x)$. Now, $a = x - \lambda_0 e_A$ is invertible in A.

Take neighbourhood V of zero and neighbourhood U of a in A as in the part (b) of the proof. Then $a^{-1}V \subset O$ and $U = a - V$.

Choose a balanced neighbourhood W of zero in A such that $W \subset V$. Since the multiplication with complex numbers is continuous in A, then there exists $\epsilon > 0$ such that $\mu e_A \subset W$ for every $\mu \in \mathbb{C}$ with $|\mu| < \epsilon$. Take any $\lambda \in \mathbb{C}$ with $|\lambda - \lambda_0| < \epsilon$. Then $(x - \lambda e_A) - (x - \lambda_0 e_A) = (\lambda_0 - \lambda)e_A \in W$ and

$$x - \lambda e_A \in (x - \lambda_0 e_A) + W \subset a - W \subset a - V = U.$$

From the part (b) of the proof, we see that $x - \lambda e_A$ is invertible in A and $\lambda \in \mathbb{C} \setminus \sigma(x)$. Thus, every $\lambda_0 \in \mathbb{C} \setminus \sigma(x)$ has an open neighbourhood $Z(\lambda_0) = \{\lambda \in \mathbb{C} : |\lambda - \lambda_0| < \epsilon\}$ belonging to $\mathbb{C} \setminus \sigma(x)$. Therefore, $\mathbb{C} \setminus \sigma(x)$ is an open subset of $\mathbb{C}$ and $\sigma(x)$ is a closed subset of $\mathbb{C}$. □

In [6], p. 200, there is a Proposition 2.11, which claims the following.

Lemma 2.2. *If every element in an algebra has bounded spectrum, then every element has closed (hence, compact) spectrum.*

With the help of Lemma 2.2, we are able to generalize Theorem 3.3.5 from [2], p. 46 in part (a) and Theorem 3.4.1 from [2], p. 48 in part (b) of the following Lemma. The part (a) is also a generalisation of a result obtained in the proof of Proposition 2.3 from [1].

Lemma 2.3. *Let A be a unital algebra and $x, y \in A$. Then*
(a) $d(\{\lambda\}, \sigma(x)) = \inf\{|\mu - \lambda| : \mu \in \sigma(x)\} = \frac{1}{\rho((x-\lambda e_A)^{-1})}$ *for every* $\lambda \in \mathbb{C} \setminus \sigma(x)$;
(b) *if* $\rho((y - x)(x - \lambda e_A)^{-1}) \leqslant \rho(y - x)\rho((x - \lambda e_A)^{-1})$ *for every* $\lambda \in \sigma(y) \setminus \sigma(x)$, *then* $d(\sigma(y), \sigma(x)) = \inf\{|\lambda - \mu| : \lambda \in \sigma(y), \mu \in \sigma(x)\} \leqslant \rho(y - x)$.

Proof. (a) Let $\lambda \in \mathbb{C} \setminus \sigma(x)$. Then there exists $(x - \lambda e_A)^{-1} \in A$. By the Spectral Mapping Theorem (see [6], Theorem 2.1.10, p. 199), we know that

$$\sigma((x - \lambda e_A)^{-1}) = \left\{\frac{1}{\nu} : \nu \in \sigma(x - \lambda e_A)\right\}.$$

It is also clear that $\sigma(x - \lambda e_A) = \{\mu - \lambda : \mu \in \sigma(x)\}$. Therefore,

$$\rho((x - \lambda e_A)^{-1}) = \sup\left\{\left|\frac{1}{\nu}\right| : \nu \in \sigma(x - \lambda e_A)\right\} = \frac{1}{\inf\{|\nu| : \nu \in \sigma(x - \lambda e_A)\}} =$$

$$= \frac{1}{\inf\{|\mu - \lambda| : \mu \in \sigma(x)\}} = \frac{1}{d(\{\lambda\}, \sigma(x))}.$$

Hence, the claim follows.

(b) Suppose that there exists $\lambda \in \sigma(y)$ such that $d(\{\lambda\}, \sigma(x)) > \rho(y - x) \geqslant 0$. Then $y - \lambda e_A$ is not invertible in A. But $\lambda \notin \sigma(x)$ gives us that $x - \lambda e_A$ is invertible in A.

Using the formula obtained in part (a) of this proof, we see that $\frac{1}{\rho((x-\lambda e_A)^{-1})} = d(\{\lambda\}, \sigma(x)) > \rho(y - x)$. Hence, by our assumption, we obtain

$$\rho((y - x)(x - \lambda e_A)^{-1}) \leqslant \rho(y - x)\rho((x - \lambda e_A)^{-1}) < 1.$$

But then $-1 \notin \sigma((y-x)(x-\lambda e_A)^{-1})$ and $(y-x)(x-\lambda e_A)^{-1} + e_A$ is invertible in A. Notice, that $y - \lambda e_A = [(y-x)(x-\lambda e_A)^{-1} + e_A](x - \lambda e_A)$. Since $y - \lambda e_A$ is a product of two invertible elements, it must be also invertible, which is a contradiction with the assumption that $\lambda \in \sigma(y)$. Therefore, $d(\sigma(y), \sigma(x)) \leqslant \rho(y-x)$. ☐

Remark 2.4. *It is easy to show that if* $xy = yx$*, then also* $(y-x)(x-\lambda e_A)^{-1} = (x-\lambda e_A)^{-1}(y-x)$. *In [6], Corollary 3.1.6, p. 311, it is shown that if* A *is a spectral algebra (for the definition of spectral algebra, see [6], p. 233) and* $a, b \in A$ *are commuting, then* $\rho(ab) \leqslant \rho(a)\rho(b)$. *So, one could demand instead of condition (b) of Lemma 2.3 that* A *should be a spectral algebra with* x *and* y *commuting, but this would restrict us to spectral algebras and commuting elements. What we actually use in the proof is the inequality for the spectral radiuses, which could be true for also noncommuting elements.*

Let U be a subset of $\mathbb{C}$ and $\epsilon > 0$. In the following, we will need some subsets of $\mathbb{C}$ in the form

$$S_{(U,\epsilon)} := \{\lambda + \mu : \lambda \in U, \mu \in \mathbb{C}, |\mu| \leqslant \epsilon\}.$$

Let us remind that the spectrum function σ is called upper semicontinuous, if for every $a \in A$ and an open set U containing $\sigma(a)$, there exists a neighbourhood $O(a, U)$ of a such that for every $b \in O(a, U)$ holds $\sigma(b) \subseteq U$.

Next, we will generalise Theorem 3.4.2 from [2], p. 50 and find a sufficient condition for a topological algebra to have upper semicontinuous spectrum function.

Proposition 2.5. *Let* A *be a unital topological algebra in which the spectrum of each element is closed and the following condition holds:*

(P) for every $a \in A$ *and* $\epsilon > 0$ *there exists a neighbourhood* $O(a, \epsilon)$ *of* a *in* A *such that for every* $b \in O(a, \epsilon)$ *we have* $\rho(b-a) < \epsilon$ *and for every* $\lambda \in \sigma(b) \setminus \sigma(a)$ *we have* $\rho((b-a)(a-\lambda e_A)^{-1}) \leqslant \rho(b-a)\rho((a-\lambda e_A)^{-1})$

Then the spectrum function is upper semicontinuous.

Proof. Let $a \in A$ and let U be an open set containing $\sigma(a)$. In case $U = \mathbb{C}$, then we can choose $O(a, U) = A$.

Suppose that $U \neq \mathbb{C}$. Since U is an open set in $\mathbb{C}$, then there exists $\epsilon > 0$ such that $S_{(\sigma(a),\delta)} \subseteq U$ for every δ with $0 \leqslant \delta < \epsilon$.

By Lemma 2.3, part (b), we know that $d(\sigma(b), \sigma(a)) \leqslant \rho(b-a)$ for every $a, b \in A$. Hence, $\sigma(b) \subset S(\sigma(a), \rho(b-a))$ for all $a, b \in A$.

By the property (P), we can find such $O(a, \epsilon)$ that $\rho(b-a) < \epsilon$ for every $b \in O(a, \epsilon)$.

Take $O(a, U) = O(a, \epsilon)$. Then, for every $b \in O(a, U)$, we have

$$\sigma(b) \subset S_{(\sigma(a),\rho(b-a))} \subseteq U,$$

because $0 \leqslant \rho(b-a) < \epsilon$. Thus, the spectral radius is upper semicontinuous. ☐

In [5], Theorem 9, p. 59, it is shown that a unital topological algebra is a Q-algebra if and only if the spectrum function is upper semicontinuous. Using this result, we can state also a sufficient condition for a unital topological algebra to be a Q-algebra.

Corollary 2.6. *Let A be a unital topological algebra, for which the condition (P) of Proposition 2.5 holds and in which the spectrum of every element is closed. Then A is a Q-algebra.*

A neighbourhood U of zero in algebra A is called *bounded* if for every neighbourhood O of zero in A there exists $M_O > 0$ such that $U \subset M_O O$.

We will finish this section with the generalisation of Theorem 3.2.11 from [2], p. 41. Unfortunately, we could not avoid the joint continuity of the multiplication in the following result.

Theorem 2.7. *Let A be a unital topological algebra with jointly continuous multiplication in which there exists a neighbourhood O of zero such that*

$$\sum_{k=0}^{\infty} y^k \in A$$

for every $y \in O$. If $(x_n)_{n\in\mathbb{N}}$ is a sequence of invertible elements of A converging to some non-invertible element x of A, then the following properties hold:

(a) *there exists a neighbourhood U of zero in A such that for every $M > 0$ and for every $N_0 \in \mathbb{N}$ there exists $n > N_0$ for which $x_n^{-1} \notin MU$;*

(b) *if U is a bounded neighbourhood of zero in A, then for every $M > 0$ and every $N_0 \in \mathbb{N}$ there exists $n > N_0$ for which $x_n^{-1} \notin MU$.*

Proof. (a) Suppose that the property (a) does not hold. It means that for every neighbourhood U of zero in A, there exists $M_U > 0$ and $N_U \in \mathbb{N}$ such that $x_n^{-1} \in M_U U$ for every $n > N_U$.

Since the multiplication is jointly continuous in A, then there exists a neighbourhood V of zero in A such that $VV \subseteq O$. Now, there exists $M_V > 0$ and $N_V \in \mathbb{N}$ such that $x_n^{-1} \in M_V V$ for every $n > N_V$. Notice, that $\frac{1}{M_V} V$ is also a neighbourhood of zero in A. Since the sequence $(x_n)_{n\in\mathbb{N}}$ converges to x, then there exists $N_1 \in \mathbb{N}$ such that $x_n - x \in \frac{1}{M_V} V$ for every $n > N_1$. Take $N := \max\{N_V, N_1\}$ and fix any $n_0 > N$. Then

$$x_{n_0}^{-1}(x_{n_0} - x) \in (M_V V)\left(\frac{1}{M_V} V\right) \subseteq VV \subseteq O.$$

Therefore, element $e_A - x_{n_0}^{-1}(x_{n_0} - x)$ is invertible in A, by part (a) of the Proposition 2.1. Thus,

$$x = x_{n_0}[e_A - x_{n_0}^{-1}(x_{n_0} - x)]$$

is invertible in A as a product of two invertible elements. But this is a contradiction with the assumption. Hence, the property (a) holds.

(b) Suppose that U is a bounded neighbourhood of zero in A such that the property (b) does not hold. It means that there exists $M_U > 0$ and $N_U \in \mathbb{N}$ such that $x_n^{-1} \in M_U U$ for every $n > N_U$.

Since the multiplication is jointly continuous in A, there exists a neighbourhood V of zero in A such that $VV \subseteq O$. Since U is bounded, then there exists $M_V > 0$ such that $U \subset M_V V$. Notice, that $\frac{1}{M_U M_V} V$ is also a neighbourhood of zero in A. Since the sequence $(x_n)_{n \in \mathbb{N}}$ converges to x, then there exists $N_1 \in \mathbb{N}$ such that $x_n - x \in \frac{1}{M_U M_V} V$ for every $n > N_1$. Take $N := \max\{N_U, N_1\}$ and fix any $n_0 > N$. Then

$$x_{n_0}^{-1}(x_{n_0} - x) \in (M_U U)\left(\frac{1}{M_U M_V} V\right) \subseteq (M_U M_V V)\left(\frac{1}{M_U M_V} V\right) \subseteq VV \subseteq O.$$

Hence, element $e_A - x_{n_0}^{-1}(x_{n_0} - x)$ is invertible in A, by part (a) of the Proposition 2.1. Thus,

$$x = x_{n_0}[e_A - x_{n_0}^{-1}(x_{n_0} - x)]$$

is invertible in A as a product of two invertible elements. But this is a contradiction with the assumption. Therefore, the property (b) holds. □

3 Results for linear maps on unital algebras

Let $T : A \to B$ be a map between algebras A and B. Denote the unital element of B by e_B and the zero element of B by θ_B. We remind, that

$$\mathrm{Ker}(T) = \{a \in A : T(a) = \theta_B\}.$$

By saying that $T : A \to B$ is an *algebra homomorphism*, we mean that the following conditions are satisfied for every $a_1, a_2 \in A$ and for every $\lambda_1, \lambda_2 \in \mathbb{C}$:

a) $T(\lambda_1 a_1 + \lambda_2 a_2) = \lambda_1 T(a_1) + \lambda_2 T(a_2)$;
b) $T(a_1 a_2) = T(a_1)T(a_2)$.

In case A and B are unital algebras with units e_A and e_B, we call an algebra homomorphism $T : A \to B$ a *unital algebra homomorphism*, if it is an algebra homomorphism and the condition

c) $T(e_A) = e_B$

is also fulfilled.

We would like to start with some purely algebraic results.

Lemma 3.1. *Let A and B be unital algebras and $T : A \to B$ an algebra homomorphism with the property $T(e_A) \in \mathrm{Inv}(B)$. Then $T(e_A) = e_B$, i.e., T is a unital algebra homomorphism.*

Proof. Let $b \in \mathrm{Inv}(B)$ be an element such that $T(e_A) = b$. Then there exists $b^{-1} \in B$ and $b = T(e_A) = T(e_A e_A) = T(e_A)T(e_A) = b^2$. Hence, $e_B = bb^{-1} = b^2 b^{-1} = b$ and we obtain that $T(e_A) = e_B$. Thus, T is a unital algebra homomorphism. □

Now, we generalise a part of Theorem 3.3.8 from [2], p. 47.

Theorem 3.2. *Let A and B be unital algebras and $T : A \to B$ an algebra homomorphism for which $T(e_A) \in \mathrm{Inv}(B)$. Then*
(a)

$$\sigma(T(x)) \subseteq \bigcap_{y \in \mathrm{Ker}(T)} \sigma(x+y) \quad \textit{for every} \quad x \in A.$$

Moreover, if T is onto and satisfies the property $T^{-1}(e_B) = \{e_A\}$, then
(b) *$T(\mathrm{Inv}(A)) = \mathrm{Inv}(B)$ and the restriction of T to $\mathrm{Inv}(A)$ is a bijection between $\mathrm{Inv}(A)$ and $\mathrm{Inv}(B)$;*
(c)

$$\sigma(T(x)) = \bigcap_{y \in \mathrm{Ker}(T)} \sigma(x+y) \quad \textit{for every} \quad x \in A.$$

Proof. By Lemma 3.1, we know that actually T is a unital algebra homomorphism. Hence, $T(e_A) = e_B$. It is clear that $T(\mathrm{Inv}(A)) \subseteq \mathrm{Inv}(B)$, because for every $a \in \mathrm{Inv}(A)$ we have

$$T(a)T(a^{-1}) = T(aa^{-1}) = T(e_A) = e_B = T(e_A) = T(a^{-1}a) = T(a^{-1})T(a),$$

which means that $T(a) \in \mathrm{Inv}(B)$.

(a) Suppose, that

$$\lambda \notin \bigcap_{y \in \mathrm{Ker}(T)} \sigma(x+y).$$

Then there exists $y_0 \in \mathrm{Ker}(T)$ such that $\lambda \notin \sigma(x+y_0)$. But this means that $x+y_0-\lambda e_A \in \mathrm{Inv}(A)$ and $T(x)-\lambda e_B = T(x+y_0-\lambda e_A) \in T(\mathrm{Inv}(A)) \subseteq \mathrm{Inv}(B)$. This gives us $\lambda \notin \sigma(T(x))$. Hence,

$$\sigma(T(x)) \subseteq \bigcap_{y \in \mathrm{Ker}(T)} \sigma(x+y)$$

for every $x \in A$.

(b) Take any $b \in \mathrm{Inv}(B)$. Then there exists $b^{-1} \in \mathrm{Inv}(B)$. Now, there are $a, c \in A$ such that $T(a) = b$ and $T(c) = b^{-1}$, because T is an onto map. Notice, that

$$T(ac) = T(a)T(c) = bb^{-1} = e_B = b^{-1}b = T(c)T(a) = T(ca).$$

Hence, $ac, ca \in T^{-1}(e_B) = \{e_A\}$. Therefore, $ac = ca = e_A$ and $a, c \in \mathrm{Inv}(A)$. So, $T(\mathrm{Inv}(A)) \supseteq \mathrm{Inv}(B)$ and we obtain $T(\mathrm{Inv}(A)) = \mathrm{Inv}(B)$.

We already proved that the restriction of T to $\mathrm{Inv}(A)$ is onto $\mathrm{Inv}(B)$. Now we show that it is also one-to-one. For that, take any $a_1, a_2 \in \mathrm{Inv}(A)$ and suppose that $T(a_1) = T(a_2)$. Then there exist $a_1^{-1}, a_2^{-1} \in \mathrm{Inv}(A)$ such that $a_1 a_2^{-1} \in \mathrm{Inv}(A)$ and

$$T(a_2)T(a_2^{-1}) = T(a_2 a_2^{-1}) = T(e_A) = e_B = T(e_A) = T(a_1 a_1^{-1}) = T(a_1)T(a_1^{-1}).$$

But now,

$$T(a_1 a_2^{-1}) = T(a_1)T(a_2^{-1}) = T(a_2)T(a_2^{-1}) = e_B$$

gives us $a_1 a_2^{-1} \in T^{-1}(e_B) = \{e_A\}$. Hence, $a_1 = (a_1 a_2^{-1})a_2 = e_A a_2 = a_2$. Therefore, the restriction of T to $\mathrm{Inv}(A)$ is one-to-one map. We conclude that the restriction of T to $\mathrm{Inv}(A)$ is a bijection.

(c) Take any $x \in A$ and any $\lambda \notin \sigma(T(x))$. Then

$$T(x) - \lambda e_B \in \mathrm{Inv}(B) = T(\mathrm{Inv}(A)),$$

which means that there exists $z \in \mathrm{Inv}(A)$ such that $T(x) - \lambda e_B = T(z)$. Take $y = z - x + \lambda e_A$. Then $z = x + y - \lambda e_A \in \mathrm{Inv}(A)$ gives us $\lambda \notin \sigma(x+y)$. Moreover, $T(y) = T(z - x + \lambda e_A) = T(z) - T(x) + \lambda e_B = \theta_B$, which means that $y \in \mathrm{Ker}(T)$. Hence, we obtain that

$$\bigcap_{y \in \mathrm{Ker}(T)} \sigma(x + y) \subseteq \sigma(T(x)),$$

as well. So, the condition (c) holds. □

Let A be a unital topological Hausdorff algebra. Denote by $E(A)$ the set of elements x of A, for which the series

$$\sum_{k=0}^{\infty} \frac{x^k}{k!}$$

converges to an element of A, i.e.,

$$E(A) := \left\{ x \in A : e^x := \sum_{k=0}^{\infty} \frac{x^k}{k!} \in A \right\}.$$

We will use the following sets:

$$\exp(A) = \{e^x : x \in E(A)\},$$

$$\epsilon(x) = \{\lambda \in \mathbb{C} : x - \lambda e_A \notin \exp(A)\}.$$

Now, we are ready to generalise the rest of Theorem 3.3.8 from [2], p. 47. It will be also a generalisation of Theorem 2.1.14, part (c) from [6], p. 208.

Theorem 3.3. *Let A and B be unital topological Hausdorff algebras and $T : A \to B$ an algebra homomorphism, which satisfies the properties $T(e_A) \in \mathrm{Inv}(B)$ and $T(e^a) = e^{T(a)}$ for every $a \in A$. Then*

(a) $T(\exp(A)) \subseteq \exp(B)$;

(b)

$$\epsilon(T(x)) \subseteq \bigcap_{y \in \mathrm{Ker}(T)} \epsilon(x+y) \quad \textit{for every} \quad x \in A.$$

Moreover, if T is onto, then

(c) $T(\exp(A)) = \exp(B)$;

(d)

$$\epsilon(T(x)) = \bigcap_{y \in \mathrm{Ker}(T)} \epsilon(x+y) \quad \textit{for every} \quad x \in A.$$

Proof. Notice, that T is a unital algebra homomorphism, by Lemma 3.1.

(a) Take any $x \in E(A)$. Since T is an algebra homomorphism with the given property, then $T(e^x) = e^{T(x)}$, which means that $T(\exp(A)) \subseteq \exp(B)$.

(b) Suppose, that

$$\lambda \notin \bigcap_{y \in \mathrm{Ker}(T)} \epsilon(x+y).$$

Then there exists $y_0 \in \mathrm{Ker}(T)$ such that $\lambda \notin \epsilon(x+y_0)$. But this means that $x+y_0-\lambda e_A \in \exp(A)$ and $T(x)-\lambda e_B = T(x+y_0-\lambda e_A) \in T(\exp(A)) \subseteq \exp(B)$. This gives us $\lambda \notin \epsilon(T(x))$. Hence,

$$\epsilon(T(x)) \subseteq \bigcap_{y \in \mathrm{Ker}(T)} \epsilon(x+y)$$

for every $x \in A$.

(c) Take any $y \in E(B)$. As T is onto, then there exists $x \in E(A)$ such that $T(x) = y$. Since T has the mentioned property, $T(e^x) = e^y$. Hence, $\exp(B) \subseteq T(\exp(A))$. Consequently, $T(\exp(A)) = \exp(B)$.

(d) Take any $x \in A$ and any $\lambda \notin \epsilon(T(x))$. Then

$$T(x) - \lambda e_B \in \exp(B) = T(\exp(A)),$$

which means that there exists $z \in \exp(A)$ such that $T(x) - \lambda e_B = T(z)$. Take $y := z - x + \lambda e_A$. Then $z = x + y - \lambda e_A \in \exp(A)$ gives us $\lambda \notin \epsilon(x+y)$. Moreover, $T(y) = T(z-x+\lambda e_A) = T(z) - T(x) + \lambda e_B = \theta_B$, which means that $y \in \mathrm{Ker}(T)$. Hence, we obtain that

$$\bigcap_{y \in \mathrm{Ker}(T)} \epsilon(x+y) \subseteq \epsilon(T(x)),$$

as well. So, we conclude that the condition (d) holds. □

Remark 3.4. *If T is a continuous algebra homomorphism with the property* $T(e_A) \in \mathrm{Inv}(B)$, *then the condition* $T(e^a) = e^{T(a)}$ *is fulfilled for every* $a \in A$.

There is a following Theorem by N. Jacobson (see [6], Lemma 3.1.2, p. 33).

Theorem 3.5. *Let A be a unital algebra, $a, b \in A$ and $\lambda \in \mathbb{C} \setminus \{0\}$. Then $ab - \lambda e_A \in \operatorname{Inv}(A)$ if and only if $ba - \lambda e_A \in \operatorname{Inv}(A)$.*

The following result follows from the Spectral Mapping Theorem (see [6], Proposition 2.2.1, pp. 208–209).

Theorem 3.6. *Let A be an algebra.*
(a) *Any $a \in A$ and any $\lambda \in \mathbb{C}$ satisfy*
$\rho(\lambda a) = \lambda\rho(a)$; $\rho(a^n) = (\rho(a))^n$; *if A is unital, then $\rho(e_A) = 1$.*
(b) *Any $a, b \in A$ satisfy $\rho(ab) = \rho(ba)$ unless one product has spectrum $\{0\}$ and the other has empty spectrum.*
(c) *Let A and B be algebras and let $\phi : A \to B$ be an algebra homomorphism. Then any element $a \in A$ with nonempty spectrum satisfies*

$$\rho_B(\phi(a)) \leqslant \inf\{\rho_A(a + b) : b \in \operatorname{Ker}(\phi)\} \leqslant \rho_A(a).$$

(d) *If B is a subalgebra of A, then any $b \in B$ with nonempty spectrum in B satisfies*

$$\rho_A(b) \leqslant \rho_B(b).$$

(e) *If A is not unital, then any $\lambda + a \in A^1 = \{\mu + a : \mu \in \mathbb{C}, a \in A\}$ (unitization of A) satisfies*

$$\rho(\lambda + a) \leqslant |\lambda| + \rho(a) \leqslant 3\rho(\lambda + a).$$

We say that an element $a \in A$ in a topological algebra A is a *topological divisor of zero*, if there exists a family $(a_\iota)_{\iota\in I}$ of elements of A such that both $(aa_\iota)_{\iota\in I}$ and $(a_\iota a)_{\iota\in I}$ converge to θ_A but $(a_\iota)_{\iota\in I}$ does not converge to θ_A. It is easy to see that, if we can find such $(a_\iota)_{\iota\in I}$ for which $\rho(a_\iota) = 1$ for every $\iota \in I$, then $(a_\iota)_{\iota\in I}$ does not converge to θ_A, because $\rho(\theta_A) = 0$.

Let

$$\tau(a) = \{\mu \in \mathbb{C} : a - \mu e_A \text{ is a topological divisor of zero}\}$$

and denote by ∂U a topological boundary of any subset U of $\mathbb{C}$.

Notice, that the condition $-E(A) \subseteq E(A)$ is equivalent with the condition $\exp(A) \subseteq \operatorname{Inv}(A)$ and, since $\theta_A \in E(A)$, then the condition $E(A) + E(A) = E(A)$ is equivalent with the the condition $E(A) + E(A) \subseteq E(A)$, which is equivalent with the condition $ab \in \exp(A)$ for every $a, b \in \exp(A)$. All of those conditions are automatically fulfilled for a Banach algebra A.

Now, we are ready to generalise Theorem 1 from [4], p. 115.

Theorem 3.7. *Let A be a unital topological algebra.*
(a) *Then the inclusion $\tau(a) \subseteq \sigma(a)$ holds.*
(b) *Suppose that $-E(A) \subseteq E(A)$. Then the inclusion $\sigma(a) \subseteq \epsilon(a)$ holds.*
(c) *Suppose that the following conditions are fulfilled:*

(i) *there exists* $k > 0$ *such that from* $a \in A$ *and* $\rho(a) < k$ *follows that* $e_A - a \in \exp(A)$;
(ii) $\rho(a) < \infty$ *for every* $a \in \mathrm{Inv}(A)$;
(iii) $\rho(a) = 0$ *if and only if* $a = \theta_A$;
(iv) $-E(A) \subseteq E(A)$;
(v) $E(A) + E(A) = E(A)$.

Then we have the inclusion $\partial\epsilon(a) \subseteq \tau(a)$ *for every* $a \in A$ *with* $\epsilon(a)$ *closed.*

Proof. (a) Take any $a \in A$ and any $s \in \tau(a)$. Then there exists a family $(a_\iota)_{\iota\in I}$ of elements of A such that $((a - se_A)a_\iota)_{\iota\in I}$ converges to θ_A but $(a_\iota)_{\iota\in I}$ does not converge to θ_A. Suppose that $s \notin \sigma(a)$. Then there exists $(a - se_A)^{-1} \in A$. Hence, $(a_\iota)_{\iota\in I} = ((a-se_A)^{-1}(a-se_A)a_\iota)_{\iota\in I}$ converges to $(a-se_A)^{-1}\theta_A = \theta_A$, which contradicts the assumption that $(a_\iota)_{\iota\in I}$ does not converge to θ_A. Hence, $s \in \sigma(a)$ and $\tau(a) \subseteq \sigma(a)$ holds.

(b) Take any $a \in A$. Since $-E(A) \subseteq E(A)$, then $\exp(A) \subseteq \mathrm{Inv}(A)$. Hence, it is clear that $\sigma(a) \subseteq \epsilon(a)$ holds.

(c) To show the last inclusion, choose any $a \in A$ for which $\epsilon(a)$ is closed and an arbitrary $s \in \partial\epsilon(a) \subset \epsilon(a)$. Then there exists a sequence (s_n) of elements of $\mathbb{C}\setminus\epsilon(a)$ such that (s_n) converges to s. Now, $a - se_A \notin \exp(A)$, but, because of the condition (iv), $a - s_ne_A \in \exp(A) \subseteq \mathrm{Inv}(A)$ for every $n \in \mathbb{N}$. Hence, for every $n \in \mathbb{N}$, we have $(a - s_ne_A)^{-1} \in A$.

Notice, that for every $n \in \mathbb{N}$, we have

$$(a - se_A)(a - s_ne_A)^{-1} = [(a - s_ne_A) + (s_ne_A - se_A)](a - s_ne_A)^{-1} =$$
$$= e_A + (s_n - s)(a - s_ne_A)^{-1} = e_A - (s - s_n)(a - s_ne_A)^{-1}.$$

Suppose that for some $n_0 \in \mathbb{N}$ we have $\rho((s-s_{n_0})(a-s_{n_0}e_A)^{-1}) < k$. Then $(a-se_A)(a-s_{n_0}e_A)^{-1} \in \exp(A)$, by the condition (i). Condition (v) gives us

$$(a - se_A) = (a - se_A)(a - s_{n_0}e_A)^{-1}(a - s_{n_0}e_A) \in \exp(A),$$

which is a contradiction with $s \in \epsilon(a)$. Hence, we have

$$|s_n - s|\, \rho((a - s_ne_A)^{-1}) = |s - s_n|\, \rho((a - s_ne_A)^{-1}) = \rho((s - s_n)(a - s_ne_A)^{-1}) \geqslant k$$

for every $n \in \mathbb{N}$. Thus, we have $\rho((a - s_ne_A)^{-1}) > 0$ for every $n \in \mathbb{N}$.

Take

$$b_n := \frac{(a - s_ne_A)^{-1}}{\rho((a - s_ne_A)^{-1})}$$

for every $n \in \mathbb{N}$. Then $\rho(b_n) = 1$ for every $n \in \mathbb{N}$. Therefore, $\sigma(b_n) \neq \emptyset$ and $\sigma(b_n) \neq \{0\}$ for every $n \in \mathbb{N}$. Hence, the conditions of Theorem 3.6, part (b), are fulfilled and we have $\rho((a - se_A)b_n) = \rho(b_n(a - se_A))$ for every $n \in \mathbb{N}$. Consequently,

$$\rho((a - se_A)b_n) = \rho(b_n(a - se_A)) = \frac{\rho((a - se_A)(a - s_ne_A)^{-1})}{\rho((a - s_ne_A)^{-1})} =$$

$$= \frac{\rho(e_A + (s_n - s)(a - s_n e_A)^{-1})}{\rho((a - s_n e_A)^{-1})} = \frac{1 + |s_n - s|\, \rho((a - s_n e_A)^{-1})}{\rho((a - s_n e_A)^{-1})} \leqslant$$

$$\leqslant \frac{|s_n - s|}{k} + |s_n - s| = \frac{k+1}{k} |s_n - s|$$

for every $n \in \mathbb{N}$. Since s_n converges to s and k is a constant, then $\rho((a - se_A)b_n)$ converges to 0. By the condition (iii), we obtain that the sequence $((a - se_A)b_n)$ converges to θ_A, which means that $a - se_A$ is a topological divisor of zero, hence, $s \in \tau(a)$. Therefore, $\partial(a) \subseteq \tau(a)$ for every $a \in A$ with $\epsilon(a)$ closed. □

The next result deals with the openness and closedness of sets $\exp(A)$ and $\epsilon(a)$.

Lemma 3.8. *Let A be a unital topological algebra, for which the following conditions are fulfilled:*

(i) *there exists a neighbourhood O of zero in A such that from $a \in O$ follows that $e_A - a \in \exp(A)$;*

(ii) $-E(A) \subseteq E(A)$;

(iii) $E(A) + E(A) = E(A)$.

Then

(a) $\exp(A)$ *is open in* A;

(b) $\epsilon(a)$ *is closed in* $\mathbb{C}$ *for every* $a \in A$.

Proof. (a) Take an arbitrary $b \in \exp(A)$. Then the condition (ii) gives us that $b \in \mathrm{Inv}(A)$. Hence, there exist $b^{-1}, x \in A$ such that $b = e^x$. Since the multiplication in A is separately continuous, there exists a balanced neighbourhood U of zero in A such that $b^{-1}U \subseteq O$.

Notice, that $b + U$ is a neighbourhood of b in A. Take any $c \in b + U$. Then $c - b \in U$ and $e_A - b^{-1}c = b^{-1}(b - c) \in b^{-1}U \subseteq O$. Hence,

$$b^{-1}c = e_A - (e_A - b^{-1}c) \in \exp(A),$$

which means that there exists $y \in A$ such that $b^{-1}c = e^y$. Thus, by condition (iii), from $x, y \in \mathrm{Exp}(A)$ follows that $x + y \in \mathrm{Exp}(A)$. Hence,

$$c = b(b^{-1}c) = e^x e^y = e^{x+y} \in \exp(A).$$

Therefore, $b + U \in \exp(A)$. Since b was an arbitrary element of $\exp(A)$, then $\exp(A)$ is open in A.

(b) Take any $a \in A$ and consider the map $f : \mathbb{C} \to A$ defined by $f(\lambda) := a - \lambda e_A$ for every $\lambda \in \mathbb{C}$. Then it is easy to see that f is a continuous map. Moreover, $f^{-1}(\exp(A)) = \mathbb{C} \setminus \epsilon(a)$. Since $\exp(A)$ is open, then $\epsilon(a)$ is closed in A. □

We remind that by a character of an algebra A we mean a linear and multiplicative map $\chi : A \to \mathbb{C}$. The following result will be needed in the proof of the Theorem 3.10.

Lemma 3.9. *Let A be topological algebra and χ a character of A. Then* $\mathrm{Ker}(\chi)$ *is a two-sided ideal of A, which is maximal as left ideal, maximal as right ideal and maximal as two-sided ideal.*

Proof. It is clear that $\mathrm{Ker}(\chi)$ is a two-sided ideal of A. Suppose that there exists left (right or two-sided) ideal J of A such that $\mathrm{Ker}(\chi) \subset J$ and $J \setminus \mathrm{Ker}(\chi) \neq \emptyset$. Take any $a \in A$ and some $b \in J \setminus \mathrm{Ker}(\chi)$. Then $\chi(b) \neq 0$ and

$$c := a - \frac{\chi(a)}{\chi(b)} b \in \mathrm{Ker}(\chi) \subset J.$$

Now, we get that $a = c + \frac{\chi(a)}{\chi(b)} b \in J$, implying $J = A$. Hence, $\mathrm{Ker}(\chi)$ is maximal as left ideal, maximal as right ideal and maximal as two-sided ideal. □

Let us remind that A is *a Gelfand-Mazur algebra*, if the quotient algebra A/I is topologically isomorphic to $\mathbb{C}$ for every closed two-sided ideal I of A, which is maximal as left or right ideal.

Now we are ready to generalise Theorem 4.1.2 from [2], p. 70.

Theorem 3.10. *Let A be a unital algebra. Then*
(a) *for every $a \in A$ we have $\{\chi(a) : \chi$ is a character of $A\} \subseteq \sigma(a)$.*
Suppose that A is a unital Gelfand-Mazur algebra. Then
(b) $\chi \mapsto \mathrm{Ker}(\chi)$ *defines a bijection from the set of all continuous characters of A onto the set of all closed two-sided ideals of A, which are maximal as left or right ideals;*
(c) *if for every $b \in A \setminus \mathrm{Inv}(A)$ there exists a closed two-sided ideal M which is maximal as left or right ideal such that $bA \subseteq M$, then for every $a \in A$ holds also $\{\chi(a) : \chi$ is a character of $A\} = \sigma(a)$.*

Proof. (a) Fix an arbitrary $a \in A$. Take any character χ of A and suppose that there exists an inverse $b \in A$ of $a - \chi(a)e_A$. Then $(a - \chi(a)e_A)b = e_A$ and

$$0 = (\chi(a) - \chi(a)\chi(e_A))\chi(b) = \chi((a - \chi(a)e_A)b) = \chi(e_A) = 1,$$

which is a contradiction. Hence, $a - \chi(a)$ is not invertible and $\chi(a) \in \sigma(a)$. Since it holds for every character χ of A, then $\{\chi(a) : \chi$ is a character of $A\} \subseteq \sigma(a)$ for any $a \in A$.

(b) If χ is continuous, then $\mathrm{Ker}(\chi)$ is closed. It is now clear that the map $\chi \mapsto \mathrm{Ker}(\chi)$ is into, by Lemma 3.9.

Take any closed two-sided ideal I of A, which is maximal as a left or right ideal. Since A is a Gelfand-Mazur algebra, then there exists a topological isomorphism $\phi : A/I \to \mathbb{C}$. Let $\pi : A \to A/I$ be the canonical homomorphism. Take $\chi := \phi \circ \pi$. Then χ is a character of A and $\mathrm{Ker}(\chi) = I$. Moreover, since the maps ϕ and π are continuous, then χ is a continuous character of A. Hence, our map $\chi \mapsto \mathrm{Ker}(\chi)$ is onto the set of all closed maximal two-sided ideals of A.

Take any two characters χ_1, χ_2 of A. Then $\chi_1(e_A) = 1 = \chi_2(e_A)$. Suppose that $\mathrm{Ker}(\chi_1) = \mathrm{Ker}(\chi_2)$. Take any $a \in A$ and let $\lambda := \chi_1(a)$. Then $\chi_1(a - \lambda e_A) = 0$ implies $a - \lambda e_A \in \mathrm{Ker}(\chi_1) = \mathrm{Ker}(\chi_2)$. Hence, $\chi_2(a - \lambda e_A) = 0$ and $\chi_2(a) = \lambda = \chi_1(a)$. Thus, $\chi_1 = \chi_2$ and the map $\chi \mapsto \mathrm{Ker}(\chi)$ is one-to-one, therefore also a bijection.

(c) Take any $\lambda \in \sigma(a)$. Then $a - \lambda e_A \notin \mathrm{Inv}(A)$. By the assumption, there exists closed two-sided ideal M of A such that M is maximal as left or as right ideal and $(a - \lambda e_A)A \subseteq M$.

By part (b), there exists a (continuous) character χ of A such that $M = \mathrm{Ker}(\chi)$. Therefore, $0 = \chi((a - \lambda e_A)e_A) = \chi(a) - \lambda$. Thus, $\lambda = \chi(a)$. So, we have obtained that $\sigma(a) \subseteq \{\chi(a) : \chi \text{ is a character of } A\}$. Together with part (a), we have that $\{\chi(a) : \chi \text{ is a character of } A\} = \sigma(a)$. □

Let A be a unital algebra over $\mathbb{K}$ and X a linear space over the same field $\mathbb{K}$. A representation of an algebra A on X is a unital algebra homomorphism from A into the algebra $L(X)$. Remember, that a representation π of an algebra A on a linear space X is called irreducible, if the following conditions hold:

a) there exists $a \in A$ such that $\pi(a) \neq \theta_{L(X)}$;
b) $\{\theta_X\}$ and X are only such linear subspaces Y of X, which *are invariant under* π, i.e., for which $\pi(a)Y \subseteq Y$ for every $a \in A$.

In [6], p. 443, there is a generalisation of the Schur's Lemma in the following form.

Lemma 3.11. *Let π be an irreducible representation of an algebra A on a linear space X. Then the set*

$$C = \{T : T \in L(X), T\pi(x) = \pi(x)T \text{ for all } x \in A\}$$

is a division algebra.

In the following result, which generalises Lemma 4.2.3 from [2], p. 82, we need a bit more, we need that C is isomorphic to the field of complex numbers.

Proposition 3.12. *Let A be a unital algebra, X a linear space and $\pi : A \to L(X)$ an irreducible representation of A on X. If the set*

$$C = \{T : T \in L(X), T\pi(x) = \pi(x)T \text{ for all } x \in A\}$$

is isomorphic to $\mathbb{C}$ and ξ_1, ξ_2 are linearly independent elements of X, then there exists $a \in A$ such that $[\pi(a)](\xi_1) = \theta_X$ and $[\pi(a)](\xi_2) \neq \theta_X$.

Proof. Let ξ_1 and ξ_2 be linearly independent elements of X. Take

$$L_i := \{x : x \in A, [\pi(x)](\xi_i) = \theta_X\} \quad \text{for} \quad i \in \{1, 2\}.$$

Then both L_1 and L_2 are left ideals of A. Since π is irreducible, then L_1 and L_2 are also maximal ideals of A. Suppose that $[\pi(x)](\xi_1) = \theta_X$ implies $[\pi(x)](\xi_2) = \theta_X$ for all $x \in A$. Then $L_1 \subseteq L_2$. Since L_1 and L_2 are maximal ideals, we have $L_1 = L_2$ and $\overline{A} = A/L_1 = A/L_2$. For $i \in \{1, 2\}$, define maps $T_i : \overline{A} \to X$ by $T_i(\overline{a}) := [\pi(a)](\xi_i)$ for all $\overline{a} \in \overline{A}$. Then it is easy to check that the maps T_1 and T_2 are bijections and linear maps. Hence, there exists a linear map $D := T_2 T_1^{-1} \in L(X)$, which is not constantly zero.

Take any $\xi \in X$. Since ξ_1 and ξ_2 are linearly independent, then $\xi_1 \neq \theta_X$. Consider the set $F := \{[\pi(x)](\xi_1) : x \in A\}$. The set F is a linear subset of X, which is invariant under π. Since $\xi_1 = [\pi(e_A)](\xi_1) \in F$, then $F \neq \{\theta_X\}$. As π is irreducible, then $F = X$. Hence, there exists $b \in A$ such that $\xi = [\pi(b)](\xi_1)$.

Now, for every $a \in A$, we have

$$(\pi(a)D)(\xi) = [\pi(a)](D(\xi)) = [\pi(a)]([T_2 T_1^{-1}](\xi)) =$$
$$= [\pi(a)](T_2(T_1^{-1}([\pi(b)](\xi_1)))) = [\pi(a)](T_2(\overline{b})) = [\pi(a)]([\pi(b)](\xi_2)) =$$
$$= [\pi(ab)](\xi_2) = T_2(\overline{ab}) = T_2(T_1^{-1}([\pi(ab)](\xi_1))) = (T_2 T_1^{-1})([\pi(a)\pi(b)](\xi_1)) =$$
$$= D([\pi(a)](\xi)) = (D\pi(a))(\xi).$$

Hence, $\pi(a)D = D\pi(a)$ for every $a \in A$ and $D \in C$. As C is isomorphic with $\mathbb{C}$ and D is not a zero map, then there exists a complex number $\lambda \neq 0$ such that $D = \lambda I$, where $I \in L(X)$ is the map defined by $I(x) = x$ for every $x \in X$. So, from $T_2 T_1^{-1} = \lambda I$ follows $T_2 = \lambda T_1$. Hence, taking $a = e_A$, we have

$$\xi_2 = [\pi(e_A)](\xi_2) = T_2(\overline{e_A}) = \lambda T_1(\overline{e_A}) = \lambda[\pi(e_A)](\xi_1) = \lambda \xi_1,$$

which is a contradiction with the assumption that ξ_1 and ξ_2 are linearly independent in X. Hence, there must exist $a \in A$ such that $[\pi(a)](\xi_1) = \theta_X$ but $[\pi(a)](\xi_2) \neq \theta_X$. □

From this Proposition, we obtain our last result of this paper, which generalises Lemma 4.2.4 from [2], p. 83.

Corollary 3.13. *Let A be a unital algebra, X a linear space and $\pi : A \to L(X)$ an irreducible representation of A on X. Take any integer $n \geqslant 2$. If the set*

$$C = \{T : T \in L(X),\ T\pi(x) = \pi(x)T \text{ for all } x \in A\}$$

is isomorphic with $\mathbb{C}$ and $\xi_1, \ldots, \xi_n$ are linearly independent elements of X, then there exists $a \in A$ such that $[\pi(a)](\xi_i) = \theta_X$ for $i \leqslant n-1$ and $[\pi(a)](\xi_n) \neq \theta_X$.

Proof. The proof is identical with the proof of Lemma 4.2.4 in [2], using only Proposition 3.12 instead of Lemma 4.2.3. □

Acknowledgment: The research was supported by institutional research funding IUT20-57 of the Estonian Ministry of Education and Research.

Bibliography

[1] Abel, M. *On algebraic properties of the spectrum and spectral radius of elements in a unital algebra, Function Spaces in Analysis*, Contemp. Math. 645 (2015), 1–12.

[2] Aupetit, B. *A Primer on Spectral Theory.* Universitext. Springer-Verlag, New York, 1991.

[3] Burlando, L. *Continuity of spectrum and spectral radius in Banach algebras, Functional analysis and operator theory, (Warsaw, 1992)*, Banach Center Publ., 30, Polish Acad. Sci., Warsaw, 1994, 53–100.

[4] Harte, R. *The Exponential Spectrum in Banach Algebras, Proc. Amer. Math. Soc.* 58 (1976), 114–118.

[5] Honary, T. G. Najafi Tavani, M., *Upper semicontinuity of the spectrum function and automatic continuity in topological Q-algebras, Note Mat.* 28 (2008), no. 2, 57–62.

[6] Palmer, T. W. *Banach algebras and the general theory of *-algebras. Vol. I. Algebras and Banach Algebras.* Encyclopedia of Mathematics and its Applications, 49. Cambridge University Press, Cambridge, 1994.

Mart Abel, Mati Abel, and Paul Tammo

Descriptions of all closed maximal one-sided ideals in topological algebras

Abstract: Let $f: A \to B$ be a homomorphism of topological algebras, I a closed maximal regular left (right) ideal of A and J a closed maximal regular left (right) ideal of B. We study under which conditions $f(I)$ or $f^{-1}(J)$ is a closed maximal regular left (right) ideal and the relationship between closed maximal regular left (right) ideals in A and B. Using these results, descriptions of all closed maximal regular one-sided ideals in topological algebras by primitive ideals and by extendible ideals are given.

Keywords: Topological algebras, closed maximal one-sided ideals in topological algebras, closed primitive ideals, extendible ideals

MSC: 46H05, 46H10

1 Introduction

Let $\mathbf{K}$ denote one of the fields $\mathbf{R}$ or $\mathbf{C}$ of real or complex numbers. An associative algebra A is a *topological algebra over* $\mathbf{K}$ (shortly, a *topological algebra*) if A is a topological linear space over $\mathbf{K}$ (with respect to the same linear operations) and the multiplication in A is separately continuous. We denote by θ_A the zero element of A.

We assume that an ideal means a proper ideal. A left (right or two-sided) ideal I of A is *regular* (sometimes also called *modular*) if there exists an element $u \in A$ such that $a - au \in I$ ($a - ua \in I$ or $a - au, a - ua \in I$, respectively) for all $a \in A$. Such element u is called a *right* (*left* or *two-sided*, respectively) *regular unit for* I. Note that $u \notin I$. If $u \in I$, then for every $a \in A$ we have $au \in I$ and therefore $a = (a - au) + au \in I$ which means that $I = A$, a contradiction. The same holds for any left or two-sided regular unit.

Let M be a maximal regular left ideal of A with a right regular unit u and P be the *primitive ideal, defined by* M, that is, $P = \{p \in A : pA \subset M\}$. If $p \in P$ and $a \in A$, then $pA \subset M$ and $aM \subset M$. Therefore, $apA \subset M$, which means that $ap \in P$. Similarily, $pa \in P$. Thus, P is a two-sided ideal of A. Also, as M is regular, $p = (p - pu) + pu \in M$ for each $p \in P$ means that $P \subset M$. Moreover, if M is closed and $(a_\lambda)_{\lambda \in \Lambda} \subset P$ is a net which converges to a, then $a_\lambda A \subset M$ for all $\lambda \in \Lambda$ and $(a_\lambda b)_{\lambda \in \Lambda} \to ab \in M$ for any

Mart Abel, School of Digital Technologies, Tallinn University, 29 Narva Str., Room A-416, 10120 Tallinn, Estonia and Institute of Mathematics, University of Tartu, 2 Liivi Str., Room 615, 50409 Tartu, Estonia; E-mail: mart.abel@tlu.ee and mart.abel@ut.ee
Mati Abel, 2 J. Liivi Str., Room 614, Institute of Mathematics, University of Tartu, 50409 Tartu, Estonia; E-mail: mati.abel@ut.ee
Paul Tammo, 2 J. Liivi Str., Room 428a, Institute of Mathematics, University of Tartu, 50409 Tartu, Estonia; E-mail: paul.tammo@ut.ee

https://doi.org/10.1515/9783110413557-002

$b \in A$, since M is closed. It means that $aA \in M$ and therefore $a \in P$. Hence, P is closed if M is closed.

Similarily, $P = \{p \in A : Ap \subset M\}$ is a closed two-sided ideal if M is a closed maximal regular right ideal of A.

We denote by $M_l(A)$ ($M_r(A)$) the set of all maximal left (respectively, right) ideals of A and by $m_l(A)$ ($m_r(A)$ or $m_t(A)$) the set of all closed maximal regular left (respectively, right or two-sided) ideals of A. In addition, $\pi_l(A)$ ($\pi_r(A)$) is the set of all closed primitive ideals of A, defined by closed maximal regular left (respectively, right) ideals. For any closed ideal $I \subset A$,

$$h_k^A(I) = \{M \in m_k(A) : I \subset M\},$$

where $k = l$, $k = r$ or $k = t$.

The closure of $B \subset A$ in a topological space A is denoted by $cl_A B$. For every $S \subset A$, let

$$I(S) = cl_A\left\{\sum_{k=1}^{n} a_k s_k + s_k b_k + c_k s_k d_k + \lambda_k s_k : n \in \mathbf{N}, a_k, b_k, c_k, d_k \in A, s_k \in S, \lambda_k \in \mathbf{K} \text{ for all } k\right\}.$$

Then $I(S)$ is a minimal closed two-sided ideal of A which contains S. If M is an ideal and $I(M) \neq A$, then the ideal M is called an *extendible ideal of A*.

All closed maximal regular one-sided ideals in some classes of Gelfand-Mazur algebras are described in [6, 11, 12] and [12]. In the present paper, all closed maximal regular one-sided ideals in topological algebras are described by primitive ideals and by extendible ideals.

2 Algebraic preliminaries

The following results describe the properties of image and preimage of a homomorphism of Abelian groups. Since every algebra is also an Abelian group with respect to the addition, we can apply these results later on to algebras as well.

Lemma 2.1. *Let A and B be (additive) Abelian groups, M and N subgroups of A, P a subset of A and $f : A \to B$ a homomorphism. Then*
(a) $f^{-1}(f(P)) = \ker f + P$,
(b) $f(P) = f(A)$ *if and only if* $\ker f + P = A$,
(c) $f^{-1}(f(M)) = M$ *if and only if* $\ker f \subset M$,
(d) *if* $\ker f \subset M$ *and* $\ker f \subset N$, *then* $f(M) = f(N)$ *if and only if* $M = N$.

Proof. (a) If $x \in f^{-1}(f(P))$, then there is $p \in P$ such that $f(x) = f(p)$. Hence, $x - p \in \ker f$ and therefore $x = (x - p) + p \in \ker f + P$. Thus, $f^{-1}(f(P)) \subset \ker f + P$. Let now

$y \in \ker f + P$. Then there are $k \in \ker f$ and $p \in P$ such that $y = k + p$. Since $f(y) = f(k + p) = f(p)$, then $y \in f^{-1}(f(P))$ and $\ker f + P \subset f^{-1}(f(P))$.

(b) If $f(P) = f(A)$, then $A \subset f^{-1}(f(A)) = f^{-1}(f(P)) = \ker f + P \subset A$, by (a), and thus, $\ker f + P = A$. If $\ker f + P = A$, then

$$f(A) = f(\ker f + P) = f(f^{-1}(f(P))) = f(P) \cap f(A) = f(P),$$

by (a).

(c) On one hand, $f^{-1}(f(M)) = \ker f + M$, by (a). On the other hand, $\ker f + M = M$ if and only if $\ker f \subset M$.

(d) If $M = N$, then $f(M) = f(N)$. Let now $f(M) = f(N)$ and $m \in M$. Then there is $n \in N$ such that $f(m) = f(n)$ which means that $m - n \in \ker f \subset N$. Now, $m = (m-n)+n \in N$ and we have shown that $M \subset N$. Similarily, $N \subset M$ and therefore, $M = N$. □

The next lemma describes the properties of the quotient map which is also a group homomorpism.

Lemma 2.2. *Let A be an Abelian group, I, J, M and N subgroups of A, P a subset of A, Q a nonempty subset of A, and R a subset of A/I. Let $\kappa_I : A \to A/I$ be the quotient map. Then*

(a) $\kappa_I^{-1}(\kappa_I(P)) = I + P$,

(b) $\kappa_I(P) = A/I$ *if and only if* $I + P = A$,

(c) $\kappa_I^{-1}(\kappa_I(M)) = M$ *if and only if* $I \subset M$,

(d) *if* $I \subset M$ *and* $I \subset N$, *then* $\kappa_I(M) = \kappa_I(N)$ *if and only if* $M = N$,

(e) *if* $\kappa_I(P) = \kappa_J(Q)$, *then* $I = J$,

(f) $\kappa_I(\kappa_I^{-1}(R)) = R$.

Proof. Assertions (a), (b), (c) and (d) follow from parts (a), (b), (c) and (d) of Lemma 2.1, respectively.

To prove (e), we first note that P is nonempty since Q is nonempty and $\kappa_I(P) = \kappa_J(Q)$. Let $p \in P$ and $q \in Q$ be such that $p + I = q + J$. Then $I = (q - p) + J$. Since $\theta_A \in J$, then $q-p \in I$ and also $p-q \in I$. Therefore, $I = (p-q)+I = (p-q)+(q-p)+J = J$.

Assertion (f) holds since κ is a surjection. □

The following lemma is similar to Lemma 1.2.16 in [10]. Remind that in our case an ideal is a proper ideal. In case A and B are algebras, by an epimorphism $f : A \to B$ we mean a surjective homomorphism.

Lemma 2.3. *Let A and B be algebras and $f : A \to B$ an epimorphism.*

(a) *For any left (right or two-sided) ideal I of A, $f(I)$ is either a left (respectively, right or two-sided) ideal of B or $f(I) = B$.*

(b) *If I is a regular left (right or two-sided) ideal of A with a right (respectively, left or two-sided) regular unit u, then $f(I)$ is either a regular left (respectively, right or two-*

sided) ideal of B with a right (respectively, left or two-sided) regular unit $f(u)$ or $f(I) = B$.

Proof. (a) Let I be a left ideal of A. If $f(I) = B$, then (a) holds. If $f(I) \neq B$, let $j \in f(I)$ and $b \in B$. Since f is an epimorphism, there exist $i \in I$ and $a \in A$ such that $f(i) = j$ and $f(a) = b$. Now,

$$bj = f(a)f(i) = f(ai) \in f(I)$$

and we have shown that $Bf(I) \subset f(I)$. Similarily it can be shown that $\mathbf{K}f(I) \subset f(I)$ and $f(I)+f(I) \subset f(I)$. Thus, $f(I)$ is a left ideal of B. The proof for right and two-sided cases is similar.

(b) By (a), $f(I)$ is a left (right or two-sided) ideal of B or $f(I) = B$. If $f(I) = B$, then (b) holds. If $f(I) \neq B$, let u be a right regular unit for I. The regularity of $f(I)$ follows from the equality $f(x)f(u) - f(x) = f(xu - x)$ for all $x \in A$, when I is a left regular ideal and from the similar equality when I is a right regular ideal. □

Now we turn our attention to the properties of the preimage of a left (right or two-sided) ideal. Parts (a) and (b) of Lemma 2.4 are similar to parts (i) and (ii) of Lemma 1.2.17. in [10], respectively.

Lemma 2.4. *Let A and B be algebras and $f: A \to B$ a homomorphism.*

(a) *For any left (right or two-sided) ideal J of B, $f^{-1}(J)$ is a left (respectively, right or two-sided) ideal of A or $f^{-1}(J) = A$.*

(b) *If J is a regular left (right or two-sided) ideal of B with a right (respectively, left or two-sided) regular unit $u \in f(A)$, then $f^{-1}(J)$ is a regular left (respectively, right or two-sided) ideal of A with any $v \in f^{-1}(u)$ as a right (respectively, left or two-sided) regular unit of $f^{-1}(J)$.*

(c) *If $I \subset A$ is a left (right or two-sided) ideal and* $\ker f \subset I$, *then $f(I) \neq B$.*

Proof. (a) If $f^{-1}(J) = A$, then (a) holds. If $f^{-1}(J) \neq A$, let $a, b \in f^{-1}(J)$, $c \in A$ and $\lambda \in \mathbf{K}$. Then $f(a + b) = f(a) + f(b) \in J$, $f(ca) = f(c)f(a) \in J$, and $f(\lambda a) = \lambda f(a) \in J$. Therefore, $a + b, ca, \lambda a \in f^{-1}(J)$. So, we have shown that $f^{-1}(J)$ is a left ideal of A. The proof is similar for right and two-sided cases.

(b) By (a), $f^{-1}(J)$ is a left ideal of A or $f^{-1}(J) = A$. Since $u \in f(A)$, $f^{-1}(u) \neq \emptyset$. Let $v \in f^{-1}(u)$ and $a \in A$. Then $f(a - av) = f(a) - f(a)u \in J$ and therefore $a - av \in f^{-1}(J)$. Moreover, since $u \in B \setminus J$, we have $v \in A \setminus f^{-1}(J)$, that is, $f^{-1}(J) \neq A$. We conclude that $f^{-1}(J)$ is a regular left ideal of A with a right regular unit v. The proof is similar for right and two-sided cases.

(c) Since I is a left (right or two-sided) ideal and $\ker f \subset I$, we have $\ker f + I = I \neq A$. Now, by Lemma 2.1 part (b), $f(I) \neq f(A) \subset B$. □

The next result remains true also in case of groups, subgroups and group homomorphisms.

Lemma 2.5. *Let A and B be algebras, $M \subset A$ a maximal left (right or two-sided) ideal, and $f: A \to B$ an epimorphism. Then the following statements are equivalent*
(a) $\ker f \subset M$,
(b) $\ker f + M \neq A$,
(c) $f(M) \neq B$,
(d) $f(M)$ *is a maximal left (respectively, right or two-sided) ideal of B.*

Proof. (a) $\Rightarrow$ (b)If $\ker f \subset M$, then $\ker f + M = M \neq A$, since M is an ideal.
(b) $\Rightarrow$ (c)If $\ker f + M \neq A$, then, by part (b) of Lemma 2.1, $f(M) \neq f(A) = B$, since f is an epimorphism.
(c) $\Rightarrow$ (d)If $f(M) \neq B$, then, by part (a) of Lemma 2.3, $f(M)$ is a left (respectively, right or two-sided) ideal of B. To show the maximality of $f(M)$, let $f(M) \subset W$ for a left (respectively, right or two-sided) ideal W of B. Now we have

$$M \subset f^{-1}(f(M)) \subset f^{-1}(W).$$

Since W is a left (respectively, right or two-sided) ideal of B, then, by Lemma 2.4 part (a), $f^{-1}(W)$ is also a left (respectively, right or two-sided) ideal of A or $f^{-1}(W) = A$. Since f is a surjection, $f(f^{-1}(W)) = W$. Now, by the maximality of M, either $f^{-1}(W) = A$, which gives us a contradiction $W = B$, or $f^{-1}(W) = M$, which gives us $f(M) = W$. Thus, $f(M)$ is a maximal left (respectively, right or two-sided) ideal of B.
(d) $\Rightarrow$ (a)Since $f(M)$ is an ideal of B, then $f(M) \neq B$ and $f^{-1}(f(M)) \supset M$ is an ideal of A, by Lemma 2.4 part (a). By the maximality of M, $f^{-1}(f(M)) = M$. Therefore, $\ker f = f^{-1}(\theta_B) = f^{-1}(f(\theta_A)) \subset f^{-1}(f(M)) = M$. □

Corollary 2.6. *Let A and B be algebras, $M \subset A$ a maximal regular left (right or two-sided) ideal, and $f: A \to B$ an epimorphism. Then $f(M) = B$ or $f(M)$ is a maximal regular left (right or two-sided) ideal of B.*

Proof. If $f(M) = B$, then the assertion holds. If $f(M) \neq B$, then, by statements (c) and (d) of Lemma 2.5, $f(M)$ is a maximal left (right or two-sided) ideal of B, and, by Lemma 2.3 part (b), $f(M)$ is regular. □

The following is essentially part (iii) of Lemma 1.2.17. in [10].

Lemma 2.7. *Let A and B be algebras, $f: A \to B$ an epimorphism and $\mathcal{M}$ a maximal regular left (right or two-sided) ideal of B. Then $f^{-1}(\mathcal{M})$ is a maximal regular left (respectively, right or two-sided) ideal of A.*

Proof. By Lemma 2.4 part (b), $f^{-1}(\mathcal{M})$ is a regular left ideal of A. Since f is an epimorphism, $f(f^{-1}(\mathcal{M})) = \mathcal{M}$. Let I be a left ideal of A such that $f^{-1}(\mathcal{M}) \subset I$. Then $\mathcal{M} \subset f(I)$ and, as $\mathcal{M}$ is maximal, $\mathcal{M} = f(I)$ or $f(I) = B$. Since $\ker f = f^{-1}(\theta_B) \subset f^{-1}(\mathcal{M}) \subset I$, then, by Lemma 2.4 part (c), $f(I) \neq B$. Now,

$I \subset f^{-1}(f(I)) = f^{-1}(\mathcal{M})$. Consequently, $f^{-1}(\mathcal{M})$ is a maximal regular left ideal of A. The proof is similar for right and two-sided cases. □

3 Topological preliminaries

Our aim in this section is to enrich the previous results with topology. We start with a simple observation that the closure of a non-dense ideal is an ideal.

Lemma 3.1. *Let A be a topological algebra and $I \subset A$ a left (right or two-sided) non-dense ideal. Then $cl_A I$ is a left (respectively, right or two-sided) ideal of A.*

Proof. Let I be a left ideal of A and $a \in cl_A I$. Then there is a net $(a_\lambda)_{\lambda\in\Lambda} \subset I$ such that $(a_\lambda)_{\lambda\in\Lambda} \to a$. Let $b \in A$. As the multiplication is separately continuous, $(ba_\lambda)_{\lambda\in\Lambda} \to ba$. Since $ba_\lambda \in I$ for all λ, then $ba \in cl_A I$, that is, $A cl_A I \subset cl_A I$. Similarily, $\mathbf{K} cl_A I \subset cl_A I$.

Let now $a, b \in cl_A I$. Then there are nets $(a_\lambda)_{\lambda\in\Lambda} \subset I$ and $(b_\mu)_{\mu\in M} \subset I$ such that $(a_\lambda)_{\lambda\in\Lambda} \to a$ and $(b_\mu)_{\mu\in M} \to b$. Fix $\mu \in M$. Since the addition in A is continuous, $(a_\lambda + b_\mu)_{\lambda\in\Lambda} \to a + b_\mu$ and therefore $a + b_\mu \in cl_A I$ for all $\mu \in M$. Also, $(a + b_\mu)_{\mu\in M} \to a + b \in cl_A I$. We conclude that $cl_A I$ is a left ideal of A.

The proof for right and two-sided cases is similar. □

Lemma 3.2. (a) *Let A and B be topological algebras, $I \subset A$ a closed left (right or two-sided) ideal, $f: A \to B$ an epimorphism and $cl_B f(I) \neq B$. Then $cl_B f(I)$ is a closed left (respectively, right or two-sided) ideal of B.*

(b) *Let $f: A \to B$ be a continuous open epimorphism, I a closed regular left (right or two-sided) ideal of A and $\ker f \subset I$. Then $cl_B f(I) \neq B$.*

(c) *Let $f: A \to B$ be a continuous open epimorphism, M a closed maximal regular left (right or two-sided) ideal of A and $\ker f \subset M$. Then $f(M)$ is a closed maximal regular left (respectively, right or two-sided) ideal of B.*

Proof. (a) By Lemma 2.3 part (a), $f(I)$ is a left (right or two-sided) ideal of B. Since $f(I)$ is non-dence by the assumption, then, by Lemma 3.1, $cl_B f(I)$ is a closed left (respectively, right or two-sided) ideal of B.

(b) Let u be a right (respectively, left or two-sided) regular unit for I and let us assume that $cl_B f(I) = B$. Then $f(u) \in B$ and there is a net $(m_\lambda)_{\lambda\in\Lambda} \subset I$ such that $(f(m_\lambda))_{\lambda\in\Lambda} \to f(u)$, hence $(f(m_\lambda - u))_{\lambda\in\Lambda} \to \theta_B$. Let O_A be a neighbourhood of zero in A. Then $O_B = f(O_A)$ is a neighbourhood of zero in B, because f is an open map, and there is a $\lambda_0 \in \Lambda$ such that $f(m_\lambda - u) \in O_B$ if $\lambda \succ \lambda_0$. Let $\lambda_1 \succ \lambda_0$. Now, $f(m_{\lambda_1} - u) \in f(O_A)$,

$$m_{\lambda_1} - u \in f^{-1}(f(m_{\lambda_1} - u)) \subset f^{-1}(f(O_A)) = O_A + \ker f \subset O_A + I,$$

by Lemma 2.1 (a), and

$$u = m_{\lambda_1} - (m_{\lambda_1} - u) \in I + O_A + I = O_A + I,$$

which means that $u \in cl_A I = I$ (see, for example, [13], Statement 1.1), a contradiction. Therefore, $cl_B f(I) \neq B$.

(c) By (b), $cl_B f(M) \neq B$, hence, by (a), $cl_B f(M)$ is a closed left (respectively, right or two-sided) ideal of B. By Lemma 2.3 part (b), $f(M)$ is a regular ideal and, by statements (a) and (d) of Lemma 2.5, $f(M)$ is maximal. Now, from $f(M) \subset cl_B f(M) \neq B$ follows that $f(M) = cl_B f(M)$. Thus, $f(M)$ is a closed regular left (respectively, right or two-sided) ideal of B. □

Corollary 3.3. *Let A be a topological algebra, I a two-sided ideal of A, and $\kappa_I : A \to A/I$ the quotient map.*

(a) *If $M \in h_k^A(I)$, then $\kappa_I(M) \in m_k(A/I)$, where $k = l$, $k = r$ or $k = t$.*

(b) *If $\mathcal{M} \in m_k(A/I)$, then $\kappa_I^{-1}(\mathcal{M}) \in h_k^A(I)$, where $k = l$, $k = r$ or $k = t$.*

Proof. (a) The quotient map κ_I is a continuous open epimorphism. Since $I \subset M$ and $\ker \kappa_I = I$, then $\ker \kappa_I \subset M$ and, by Lemma 3.2 part (c), $\kappa_I(M)$ is a closed maximal regular left (respectively, right or two-sided) ideal of A/I.

(b) By Lemma 2.7, $\kappa_I^{-1}(\mathcal{M})$ is a maximal regular left (respectively, right or two-sided) ideal of A. Since κ_I is continuous, $\kappa_I^{-1}(\mathcal{M})$ is closed. Also, $I = \kappa_I^{-1}(\theta_{A/I}) \subset \kappa_I^{-1}(\mathcal{M})$. □

4 Main results

First, we describe the relationship between closed maximal regular left (right or two-sided) ideals in topological algebras A and B provided that there is a continuous open epimorphism $f : A \to B$.

Theorem 4.1. *Let A and B be topological algebras, $f : A \to B$ a continuous open epimorphism. Then the map $\varphi : h_k^A(\ker f) \to m_k(B)$, where $k = l$, $k = r$ or $k = t$, defined by $\varphi(M) = f(M)$ for every $M \in h_k^A(\ker f)$, is a bijection.*

Proof. We prove the theorem in case $k = l$. The proof for right and two-sided case is similar.

To show that the map φ is correctly defined, let $M \in h_l^A(\ker f)$. Since f is a continuous open epimorphism, M a closed regular left ideal and $\ker f \subset M$, then, by Lemma 3.2 part (c), $f(M)$ is a closed maximal regular left ideal of B, that is, $\varphi(M) = f(M) \in m_l(B)$.

If $\mathcal{M} \in m_l(B)$, then, by Lemma 2.7, $f^{-1}(\mathcal{M})$ is a maximal regular left ideal of A and $\ker f = f^{-1}(\theta_B) \subset f^{-1}(\mathcal{M})$. Since f is continuous, $f^{-1}(\mathcal{M})$ is closed. Therefore,

$f^{-1}(\mathcal{M}) \in h_l^A(\ker f)$ and $\varphi(f^{-1}(\mathcal{M})) = f(f^{-1}(\mathcal{M})) = f(A) \cap \mathcal{M} = \mathcal{M}$. Therefore, φ is a surjection.

Let now $M_1, M_2 \in h_l^A(\ker f)$ and $f(M_1) = f(M_2)$. Then, by Lemma 2.1 part (d), $M_1 = M_2$. It means that φ is an injection. □

Corollary 4.2. *Let A be a topological algebra, I a two-sided ideal of A and $\kappa_I : A \to A/I$ the quotient map. Then the map $\varphi : h_k^A(I) \to m_k(A/I)$, where $k = l, k = r$ or $k = t$, defined by $\varphi(M) = \kappa_I(M)$ for every $M \in h_k^A(I)$, is a bijection.*

Proof. Since the quotient map is an open epimorphism, the assertion follows from Theorem 4.1. □

Now we give a description of all closed maximal regular one-sided ideals of a topological algebra using primitive ideals.

Theorem 4.3. *Let A be a topological algebra and $M \in m_l(A)$ ($M \in m_r(A)$). Then*

$$M = \kappa_P^{-1}(\mathcal{M})$$

for some $P \in \pi_l(A)$ ($P \in \pi_r(A)$, respectively) and $\mathcal{M} \in m_l(A/P)$ ($\mathcal{M} \in m_r(A/P)$, respectively), where $\kappa_P : A \to A/P$ is the quotient map.

Proof. Let P be a primitive ideal of A, defined by a closed maximal regular left ideal M of A and $\mathcal{M} = \kappa_P(M)$. Then $P \in \pi_l(A)$. Since M is regular, then $P \subset M$. The quotient map $\kappa_P : A \to A/P$ is a continuous open epimorphism and $\ker \kappa_P = P$. Hence, by Lemma 3.2 part (c), $\mathcal{M} \in m_l(A/P)$ and by Lemma 2.2 part (c), $M = \kappa_P^{-1}(\kappa_P(M))$, that is, $M = \kappa_P^{-1}(\mathcal{M})$.

The proof for right-sided case is similar. □

Note that the sets $m_k(A/P)$ ($P \in \pi_k(A)$, $k = l$ or $k = r$) in the next theorem are disjoint. Indeed, let $P_1, P_2 \in \pi_k(A)$ and suppose that there is

$$\mathcal{M} \in m_k(A/P_1) \cap m_k(A/P_2).$$

Then $\mathcal{M} \subset (A/P_1) \cap (A/P_2)$ and there are $a_1, a_2 \in A$ such that $a_1 + P_1 = a_2 + P_2$. Therefore, $(a_1 - a_2) + P_1 = P_2$. Since $\theta_A \in P_2$, we get that $-(a_1 - a_2) \in P_1$ and thus, $(a_1 - a_2) + P_1 = P_1$. It follows that $P_1 = P_2$.

Theorem 4.4. *Let A be a topological algebra and the map*

$$\Lambda : \bigcup_{P \in \pi_k(A)} m_k(A/P) \to m_k(A)$$

(here $k = l$ or $k = r$) be defined by

$$\Lambda(\mathcal{M}) = \kappa_P^{-1}(\mathcal{M})$$

for every $\mathcal{M} \in m_k(A/P)$ and $P \in \pi_k(A)$, where $\kappa_P\colon A \to A/P$ is the quotient map. Then Λ is a surjection and for every fixed P an injection. Hence, the restriction $\Lambda|_{m_k(A/P)}$ is a bijection for every fixed P.

Proof. Let

$$\mathcal{M} \in \bigcup_{P \in \pi_k(A)} m_k(A/P).$$

Then $\mathcal{M} \in m_k(A/P)$ for some $P \in \pi_k(A)$. Hence $\Lambda(\mathcal{M}) = \kappa_P^{-1}(\mathcal{M})$ is a closed maximal left ideal of A, by Corollary 3.3 part (b). Since the quotient map is a homomorphism, $\Lambda(\mathcal{M})$ is regular by Lemma 2.4 part (b), consequently $\Lambda(\mathcal{M}) \in m_k(A)$. Therefore, Λ is correctly defined.

To show that Λ is an onto map, let $M \in m_k(A)$ and P be the closed primitive ideal of A, defined by M. Then $P \subset M$, $\kappa_P(M) \in m_k(A/P)$ by Lemma 3.2 part (c) and $\Lambda(\kappa_P(M)) = \kappa_P^{-1}(\kappa_P(M)) = M$ by Lemma 2.2 part (c). Thus, Λ is an onto map.

Let now $P \in \pi_k(A)$ and $\mathcal{M}_1, \mathcal{M}_2 \in m_k(A/P)$. If $\Lambda(\mathcal{M}_1) = \Lambda(\mathcal{M}_2)$, that is, $\kappa_P^{-1}(\mathcal{M}_1) = \kappa_P^{-1}(\mathcal{M}_2)$, then $\mathcal{M}_1 = \kappa_P(\kappa_P^{-1}(\mathcal{M}_1)) = \kappa_P(\kappa_P^{-1}(\mathcal{M}_2)) = \mathcal{M}_2$ since κ_P is a surjection. Hence, Λ is an injection for every fixed P. □

We give another description of all closed maximal regular left or right ideals of a topological algebra using extendible ideals.

Theorem 4.5. *Let A be a topological algebra, $M \in m_k(A)$ ($k = l$ or $k = r$) and E a subalgebra of A such that $I(M \cap E) \subset M$. Then*

$$M = \kappa_{I(M\cap E)}^{-1}(\mathcal{M})$$

for some $\mathcal{M} \in m_k(A/I(M \cap E))$, where $\kappa_{I(M\cap E)}\colon A \to A/I(M \cap E)$ is the quotient map.

Proof. Let $M \in m_k(A)$ and $\mathcal{M} = \kappa_{I(M\cap E)}(M)$. Then $\mathcal{M} \in m_k(A/I(M \cap E))$, by Lemma 3.2 part (c), because $\kappa_{I(M\cap E)}$ is a continuous open epimorphism. Now, $M \subset \kappa_{I(M\cap E)}^{-1}(\kappa_{I(M\cap E)}(M)) = \kappa_{I(M\cap E)}^{-1}(\mathcal{M})$. Since $\kappa_{I(M\cap E)}^{-1}(\mathcal{M})$ is a left (right) ideal of A, by Lemma 2.7, and M is maximal, then $M = \kappa_{I(M\cap E)}^{-1}(\mathcal{M})$. □

Remark 4.6. *In case of two-sided ideals, Theorem 4.1 is proved in [14], Theorem 2.6.6, and in case when the subalgebra E of A is a subalgebra of the centre of A, Theorem 4.5 is proved in [6, 11, 12] and [12].*

Acknowledgment: The research is partially supported by institutional research funding IUT20-57 of the Estonian Ministry of Education and Research.

Bibliography

[1] Abel, M. *Description of closed maximal ideals in Gelfand-Mazur algebras*. General topological algebras (Tartu, 1999), 7–13, Math. Stud. (Tartu), 1, Est. Math. Soc., Tartu, 2001.

[2] Abel, M. *Structure of Gelfand-Mazur algebras*, Dissertation, University of Tartu. Dissertationes Mathematicae Universitatis Tartuensis, 31. Tartu University Press, Tartu, 2003.

[3] Abel, M. *Description of closed maximal regular one-sided ideals in Gelfand-Mazur algebras without a unit.* Acta Univ. Oulu. Ser. A Sci. Rerum Natur. No. 408 (2004), 9–24.

[4] Balachandran, V. K. Topological Algebras, *North-Holland Mathematics Studies*, 185. North-Holland Publishing Co., Amsterdam, 2000.

[5] Panova, O. *Real Gelfand-Mazur algebras. Dissertation*, University of Tartu, Tartu, 2006. Dissertationes Mathematicae Universitatis Tartuensis, 48. Tartu University Press, Tartu, 2006.

[6] Rickart, Ch. E. *General theory of Banach algebras.* The University Series in Higher Mathematics, D. van Nostrand Co., Inc., Princeton, N.J.-Toronto-London-New York 1960.

[7] Schaefer, H. H. and Wolff, M. P. *Topological vector spaces*. Second edition. Graduate Texts in Mathematics, 3. Springer-Verlag, New York, 1999.

Fabio Bagarello* and Giorgia Bellomonte

On non self-adjoint operators defined by Riesz bases in Hilbert and rigged Hilbert spaces

Abstract: In this paper we discuss some results on non self-adjoint Hamiltonians with real discrete simple spectrum under the assumption that their eigenvectors form Riesz bases of a certain Hilbert space. Also, we exhibit a generalization of those results to the case of rigged Hilbert spaces, and we also consider the problem of the factorization of the aforementioned Hamiltonians in terms of generalized lowering and raising operators.

Keywords: Non self-adjoint Hamiltonians, Riesz bases, rigged Hilbert spaces

MSC: Primary 46N50; Secondary 47N50

1 Introduction

In recent years non self-adjoint Hamiltonians have attracted a lot of interest, mainly in quantum physics. The reason for this is that, since the seminal paper [9], people understood that self-adjointness of an Hamiltonian is a sufficient but not necessary condition to get real observable (in the physical sense) eigenvalues. After several preliminary attempts to check (numerically or perturbatively) that the spectrum of $H = p^2 + ix^3$ has only a punctual part, and that the eigenvalues are real, mathematical proofs appeared, see, for instance, [12, 13, 19]. Nowadays, it is a completely accepted fact that this kind of Hamiltonians really exists, and this spurred also the interest of the mathematical community, [6].

A particularly interesting, and reasonably simple, situation arise when one considers operators of the form $H = BA$, where A and B satisfy (on some dense domain) the commutation rule $[A, B] = AB - BA = frm[o]--$. In this case, [3], the eigenvalues and the eigenvectors of H can indeed be easily found, at least under some mild further conditions on A and B, and one can also look for the eigenstates of the adjoint of H, H^*, whose eigenvalues turn out to coincide with those of H. Notice that, if in particular $A = B^*$, then H reduces to the standard quantum harmonic oscillator Hamiltonian, [17], a very well-know example in literature.

One of the essential points, in the analysis of Hamiltonians satisfying $H \neq H^*$, is that the set of their eigenstates $\mathcal{F}_\phi = \{\phi_n,\ n \in \mathbb{N}_0\}$, with $\mathbb{N}_0 = \mathbb{N}\bigcup\{0\}$, if they exist,

***Corresponding author: Fabio Bagarello,** Dipartimento di Energia, Ingegneria dell'Informa-zione e Modelli Matematici, Scuola Politecnica, Università degli Studi di Palermo, Viale delle Scienze, I-90128 Palermo, and INFN, Università di Torino, Italy
Giorgia Bellomonte, Dipartimento di Matematica e Informatica, Università degli Studi di Palermo, Via Archirafi 34, I-90123 Palermo, Italy
https://doi.org/10.1515/9783110413557-003

is not necessarily an orthonormal (o.n.) basis for the Hilbert space $\mathcal{H}$ of the system. Hence we are forced to abandon the notion of o.n. basis, to move to something more general. The first step toward this generalization is provided by the concept of Riesz basis, [10, 14]. This means that there exist an o.n. basis $\mathcal{E} = \{e_n, n \in \mathbb{N}_0\}$ of $\mathcal{H}$ and an operator T bounded in $\mathcal{H}$, invertible and with bounded inverse T^{-1}, such that $T\phi_n = e_n$ for all $n \in \mathbb{N}_0$. Let us further define $\psi_n = T^* e_n$ and $\mathcal{F}_\psi = \{\psi_n, n \in \mathbb{N}_0\}$. It is clear that $\mathcal{F}_\psi$ is a Riesz basis, too. Moreover, the vectors of the two sets $\mathcal{F}_\phi$ and $\mathcal{F}_\psi$ are biorthogonal, i.e. $\langle \phi_n \mid \psi_m \rangle = \delta_{n,m}$, $\forall n, m$, and

$$f = \sum_{n=0}^{\infty} \langle f \mid \phi_n \rangle \, \psi_n = \sum_{n=0}^{\infty} \langle f \mid \psi_n \rangle \, \phi_n, \quad \forall f \in \mathcal{H}.$$

Furthermore, the operators S_ϕ and S_ψ defined by

$$S_\phi f = \sum_{n=0}^{\infty} \langle f \mid \phi_n \rangle \, \phi_n, \qquad S_\psi f = \sum_{n=0}^{\infty} \langle f \mid \psi_n \rangle \, \psi_n,$$

are bounded, everywhere defined in $\mathcal{H}$, positive and, therefore, self-adjoint. Also, they are one the inverse of the other: $S_\phi = (S_\psi)^{-1}$ and, finally, $S_\phi = (T^{-1})(T^{-1})^*$. From what we said before, one see that Riesz bases may appear as eigenvectors of nonself-adjoint operators. The simplest situation when this happens is the following: let H be a self-adjoint operator with (just) discrete spectrum defined on a subset $D(H)$ of the Hilbert space $\mathcal{H}$. Assume that each eigenvalue λ_n is simple. Then the corresponding eigenvectors $\{e_n\}$ constitute an o. n. basis of $\mathcal{H}$. If $\tilde{H}$ is another operator *similar* to H (i.e., there exists a bounded operator V with bounded inverse V^{-1} which intertwines $\tilde{H}$ and H, in the sense that $V : D(H) \to D(\tilde{H})$ and $\tilde{H}V\xi = VH\xi$, for every $\xi \in D(H)$), then the vectors $\{\varphi_n\} := Ve_n$ are eigenvectors of $\tilde{H}$ and constitute, by definition, a Riesz basis for $\mathcal{H}$.

In a recent paper, [5], A. Inoue and C. Trapani, together with one of us (FB), discussed the possibility of using Riesz bases as the main ingredient to construct non self-adjoint Hamiltonians (and related objects, e.g. lowering and raising operators associated to them) in a rigorous way. This reverses what usually happens in physics, where the Hamiltonian H is given and its eigenvectors must be deduced. Also, one has to check whether these eigenvectors produce or not a Riesz basis. In this case, its unique biorthogonal Riesz basis consists in the eigenvectors of H^*.

More recently, [8], C. Trapani together with one of the authors (GB), generalized the notion of Riesz bases to rigged Hilbert spaces $\mathcal{D}[t] \subset \mathcal{H}[\|\cdot\|] \subset \mathcal{D}^\times[t^\times]$. The idea, there, was derived from some physical applications where the intertwining operator T exists but at least one between T and T^{-1} is unbounded, as in the case of the aforementioned cubic Hamiltonian $X = p^2 + ix^3$, for which it has been proved that there is no intertwining operator bounded with bounded inverse which makes it similar to a self-adjoint operator [19]. A *strict generalized* Riesz basis is a *Schauder* basis $\{\xi_n\}$ for $\mathcal{D}[t]$ (i.e. a basis for which the *coefficient functionals* are continuous in $\mathcal{D}[t]$), such that there exists an invertible operator T continuous from $\mathcal{D}[t]$ to $\mathcal{H}[\|\cdot\|]$ such that $\{T\xi_n\}$ is an orthonormal basis for $\mathcal{H}$, and whose inverse T^{-1} is a *bounded* operator in $\mathcal{H}$. In addition to properties of strict generalized Riesz bases, in [8], following the line in [5], some operators defined by strict generalized Riesz basis have been studied and some intertwining relation has been proved. In this paper we go a little further, by defining some lowering and raising operators via strict generalized Riesz bases and their dual basis, showing the possibility of using them to factorize the Hamiltonians defined in [8]. Moreover, we discuss some examples which were not considered before.

The paper is organized as follows: in Section 2 we start reviewing the outcomes in [5], discussing some examples. In Section 3 we gather some preliminaries, the main definitions and results about strict generalized Riesz basis, and about the simplest operators that can be defined by them and their *dual bases*. Furthermore, we introduce and study the properties of some generalized lowering and raising operators by which it is possible to factorize Hamiltonians defined in [8]. Finally, we give a new example of strict generalized Riesz basis and of the operator T appearing in its definition. Section 4 contains our final comments and our future plans.

2 Some operators defined by Riesz bases

The first part of this section is devoted to a brief review of the results originally given in [5]. Let $\mathcal{F}_\phi$ be a Riesz basis in $\mathcal{H}$ and, as above, $\psi_n = T^* e_n$, $n \in \mathbb{N}_0$. We adopt the same notation as in Section 1. As remarked before, $\mathcal{F}_\psi$ is also a Riesz basis and it is biorthogonal to $\mathcal{F}_\phi$: $\langle \phi_n \mid \psi_m \rangle = \delta_{nm}$.

Throughout this section let $\boldsymbol{\alpha} = \{\alpha_n\}$ be any sequence of complex numbers. We define the operators

$$H^{\boldsymbol{\alpha}}_{\phi,\psi} = \sum_{n=0}^{\infty} \alpha_n \phi_n \otimes \overline{\psi}_n \tag{1}$$

and

$$H^{\boldsymbol{\alpha}}_{\psi,\phi} = \sum_{n=0}^{\infty} \alpha_n \psi_n \otimes \overline{\phi}_n \tag{2}$$

as follows[1]:

$$\begin{cases} D(H^{\boldsymbol{\alpha}}_{\phi,\psi}) = \{f \in \mathcal{H}; \sum_{n=0}^{\infty} \alpha_n \langle f \mid \psi_n \rangle \phi_n \text{ exists in } \mathcal{H}\} \\ \\ H^{\boldsymbol{\alpha}}_{\phi,\psi} f = \sum_{n=0}^{\infty} \alpha_n \langle f \mid \psi_n \rangle \phi_n, \; f \in D(H^{\boldsymbol{\alpha}}_{\phi,\psi}) \end{cases}$$

1 If $\phi, \psi \in \mathcal{H}$, we denote, as usual, by $\phi \otimes \overline{\psi}$ the linear operator defined by $(\phi \otimes \overline{\psi})(f) = \langle f \mid \psi \rangle \phi$, $f \in \mathcal{H}$.

and

$$\begin{cases} D(H^{\boldsymbol{\alpha}}_{\psi,\phi}) = \{f \in \mathcal{H}; \sum_{n=0}^{\infty} \alpha_n \langle f \mid \phi_n \rangle \psi_n \text{ exists in } \mathcal{H}\} \\ H^{\boldsymbol{\alpha}}_{\psi,\phi} f = \sum_{n=0}^{\infty} \alpha_n \langle f \mid \phi_n \rangle \psi_n, \ f \in D(H^{\boldsymbol{\alpha}}_{\phi,\psi}) \end{cases}.$$

Then, introducing $\mathcal{D}_\phi := \text{span}\{\phi_n\}$ and $\mathcal{D}_\psi := \text{span}\{\psi_n\}$, it is clear that $\mathcal{D}_\phi \subset D(H^{\boldsymbol{\alpha}}_{\phi,\psi})$, $\mathcal{D}_\psi \subset D(H^{\boldsymbol{\alpha}}_{\psi,\phi})$ and that $H^{\boldsymbol{\alpha}}_{\phi,\psi}\phi_k = \alpha_k\phi_k$, and $H^{\boldsymbol{\alpha}}_{\psi,\phi}\psi_k = \alpha_k\psi_k$, $k \in \mathbb{N}_0$ Hence, in particular, $H^{\boldsymbol{\alpha}}_{\phi,\psi}$ and $H^{\boldsymbol{\alpha}}_{\psi,\phi}$ are densely defined. In [5] the following proposition has been proved:

Proposition 2.1. *The following statements hold.*
(1) $D(H^{\boldsymbol{\alpha}}_{\phi,\psi}) = \{f \in \mathcal{H}; \sum_{n=0}^{\infty} |\alpha_n|^2 |\langle f \mid \psi_n \rangle|^2 < \infty\}$,
$D(H^{\boldsymbol{\alpha}}_{\psi,\phi}) = \{f \in \mathcal{H}; \sum_{n=0}^{\infty} |\alpha_n|^2 |\langle f \mid \phi_n \rangle|^2 < \infty\}$.
(2) $H^{\boldsymbol{\alpha}}_{\phi,\psi}$ *and* $H^{\boldsymbol{\alpha}}_{\psi,\phi}$ *are closed.*
(3) $\left(H^{\boldsymbol{\alpha}}_{\phi,\psi}\right)^* = H^{\overline{\boldsymbol{\alpha}}}_{\psi,\phi}$, *where* $\overline{\boldsymbol{\alpha}} = \{\overline{\alpha}_n\}$.
(4) $H^{\boldsymbol{\alpha}}_{\phi,\psi}$ *is bounded if and only if* $H^{\boldsymbol{\alpha}}_{\psi,\phi}$ *is bounded and if and only if* $\boldsymbol{\alpha}$ *is a bounded sequence. In particular* $H^{\mathbf{1}}_{\phi,\psi} = H^{\mathbf{1}}_{\psi,\phi} = frm[o]--$, *where* $\mathbf{1}$ *is the sequence constantly equal to* 1 *and* $frm[o]--$ *is the identity operator on* $\mathcal{H}$.

Extending what is usually done in the literature, where β_n (see below) is taken to be one for all n, in [5] the authors introduced a second sequence of complex numbers, $\boldsymbol{\beta} := \{\beta_n\}$, and two operators

$$S^{\boldsymbol{\beta}}_\phi = \sum_{n=0}^{\infty} \beta_n \phi_n \otimes \overline{\phi}_n$$

and

$$S^{\boldsymbol{\beta}}_\psi = \sum_{n=0}^{\infty} \beta_n \psi_n \otimes \overline{\psi}_n$$

as follows:

$$\begin{cases} D(S^{\boldsymbol{\beta}}_\phi) = \{f \in \mathcal{H}; \sum_{n=0}^{\infty} \beta_n \langle f \mid \phi_n \rangle \phi_n \text{ exists in } \mathcal{H}\} \\ S^{\boldsymbol{\beta}}_\phi f = \sum_{n=0}^{\infty} \beta_n \langle f \mid \phi_n \rangle \phi_n, \ f \in D(S^{\boldsymbol{\beta}}_\phi) \end{cases}$$

and

$$\begin{cases} D(S^{\boldsymbol{\beta}}_\psi) = \{f \in \mathcal{H}; \sum_{n=0}^{\infty} \beta_n \langle f \mid \psi_n \rangle \psi_n \text{ exists in } \mathcal{H}\} \\ S^{\boldsymbol{\beta}}_\psi f = \sum_{n=0}^{\infty} \beta_n \langle f \mid \psi_n \rangle \psi_n, \ f \in D(S^{\boldsymbol{\beta}}_\psi) \end{cases}.$$

It is clear that

$$\mathcal{D}_\psi \subset D(S_\phi^{\boldsymbol{\beta}}) \quad \text{and} \quad S_\phi^{\boldsymbol{\beta}} \psi_k = \beta_k \phi_k, \; k \in \mathbb{N}_0 \; ; \tag{3}$$

$$\mathcal{D}_\phi \subset D(S_\psi^{\boldsymbol{\beta}}) \quad \text{and} \quad S_\phi^{\boldsymbol{\beta}} \phi_k = \beta_k \psi_k, \; k \in \mathbb{N}_0 \tag{4}$$

Hence, in particular, $S_\phi^{\boldsymbol{\beta}}$ and $S_\psi^{\boldsymbol{\beta}}$ are densely defined, and, see [5]:

Proposition 2.2. *The following statements hold.*

(1) $D(S_\phi^{\boldsymbol{\beta}}) = \{f \in \mathcal{H}; \sum_{n=0}^{\infty} |\beta_n|^2 |\langle f \mid \phi_n \rangle|^2 < \infty\} = D(H_{\psi,\phi}^{\boldsymbol{\beta}})$,
$D(S_\psi^{\boldsymbol{\beta}}) = \{f \in \mathcal{H}; \sum_{n=0}^{\infty} |\beta_n|^2 |\langle f \mid \psi_n \rangle|^2 < \infty\} = D(H_{\phi,\psi}^{\boldsymbol{\beta}})$.

(2) $S_\phi^{\boldsymbol{\beta}}$ *and* $S_\psi^{\boldsymbol{\beta}}$ *are closed.*

(3) $\left(S_\phi^{\boldsymbol{\beta}}\right)^* = S_\phi^{\overline{\boldsymbol{\beta}}}$ *and* $\left(S_\psi^{\boldsymbol{\beta}}\right)^* = S_\psi^{\overline{\boldsymbol{\beta}}}$, *where* $\overline{\boldsymbol{\beta}} = \{\overline{\beta_n}\}$.

(4) If $\{\beta_n\} \subset \mathbb{R}$ *(respectively,* $\{\beta_n\} \subset \mathbb{R}^+$*) then* $S_\phi^{\boldsymbol{\beta}}$ *and* $S_\psi^{\boldsymbol{\beta}}$ *are self-adjoint (respectively, positive self-adjoint). Furthermore,* $S_\phi^{\boldsymbol{\beta}}$ *is bounded if and only if* $S_\psi^{\boldsymbol{\beta}}$ *is bounded and if and only if* $\boldsymbol{\beta}$ *is a bounded sequence.*

(5) If, $\boldsymbol{\beta} = \mathbf{1}$, *where, as before,* $\mathbf{1}$ *denotes the sequence constantly equal to* 1, *then* $S_\phi := S_\phi^{\mathbf{1}}$ *and* $S_\psi := S_\psi^{\mathbf{1}}$ *are bounded positive self-adjoint operators on* $\mathcal{H}$ *and they are inverses of each other, that is* $S_\phi = (S_\psi)^{-1}$, *and* $S_\phi = T^{-1}(T^{-1})^*$, $S_\psi = T^*T$.

(6) Moreover, if, $\boldsymbol{\beta} = \mathbf{1}$, *then*

$$\begin{cases} S_\psi H_{\phi,\psi}^{\boldsymbol{\alpha}} = H_{\psi,\phi}^{\boldsymbol{\alpha}} S_\psi = S_\psi^{\boldsymbol{\alpha}}, \\ S_\phi H_{\psi,\phi}^{\boldsymbol{\alpha}} = H_{\phi,\psi}^{\boldsymbol{\alpha}} S_\phi = S_\phi^{\boldsymbol{\alpha}}. \end{cases} \tag{5}$$

Remark 2.3. *The equations in (5) are particularly interesting for physical applications. These kind of equations are called* intertwining relations *and extend, with non trivial consequences, [15, 16, 18], the standard commutation rules: in particular, they ensure the existence of a deep relation between the eigenvectors of two operators (*$H_{\phi,\psi}^{\boldsymbol{\alpha}}$ *and* $H_{\psi,\phi}^{\boldsymbol{\alpha}}$ *in this case, with* S_ψ *and* S_ϕ *acting as intertwining operators) and the fact that their eigenvalues coincide.*

Going back to the general case, let $\boldsymbol{\alpha} = \{\alpha_n\}$ be a sequence of complex numbers. We define the operators $\mathsf{h}_{\phi,\psi}^{\boldsymbol{\alpha}}$ and $\mathsf{h}_{\psi,\phi}^{\boldsymbol{\alpha}}$ as follows:

$$\begin{cases} \mathsf{h}_{\phi,\psi}^{\boldsymbol{\alpha}} = S_\psi^{1/2} H_{\phi,\psi}^{\boldsymbol{\alpha}} S_\phi^{1/2}, \\ \mathsf{h}_{\psi,\phi}^{\boldsymbol{\alpha}} = S_\phi^{1/2} H_{\psi,\phi}^{\boldsymbol{\alpha}} S_\psi^{1/2}. \end{cases} \tag{6}$$

Then, [5], $D(\mathrm{h}^{\boldsymbol{\alpha}}_{\phi,\psi}) = \{S_\psi^{1/2}f; f \in D(H^{\boldsymbol{\alpha}}_{\phi,\psi})\}$, $D(\mathrm{h}^{\boldsymbol{\alpha}}_{\psi,\phi}) = \{S_\phi^{1/2}f; f \in D(H^{\boldsymbol{\alpha}}_{\psi,\phi})\}$ and these are both dense in $\mathcal{H}$. Moreover $(\mathrm{h}^{\boldsymbol{\alpha}}_{\phi,\psi})^* = \mathrm{h}^{\overline{\boldsymbol{\alpha}}}_{\psi,\phi}$. Finally, and very important, if $\{\alpha_n\} \subset \mathbb{R}$, then $\mathrm{h}^{\boldsymbol{\alpha}}_{\phi,\psi}$ is self-adjoint.

Equations in (6) can be rewritten as $S_\phi^{1/2}\mathrm{h}^{\boldsymbol{\alpha}}_{\phi,\psi} = H^{\boldsymbol{\alpha}}_{\phi,\psi}S_\phi^{1/2}$ and $S_\psi^{1/2}\mathrm{h}^{\boldsymbol{\alpha}}_{\psi,\phi} = H^{\boldsymbol{\alpha}}_{\psi,\phi}S_\psi^{1/2}$, which, similarly to the equalities in (5), are also intertwining relations between different operators having the same eigenvalues and related eigenvectors, [15, 16, 18]. However, contrarily to what happens for (5), these new intertwining relations involve both self-adjoint and non self-adjoint operators, which appears to be similar.

2.1 Generalized lowering and raising operators

A crucial role in quantum mechanics, is the possibility of factorizing an operator, and in particular the Hamiltonian H of a certain physical system $\mathcal{S}$, i.e. the possibility of introducing two operators X and Y such that, in a suitable sense, one can write $H = XY$. The first prototype example of such a situation is provided by the quantum harmonic oscillator, whose dynamics is driven by the Hamiltonian $H_0 = \frac{1}{2}(p^2 + x^2)$. Here x and p are the well known position and momentum operators acting on $\mathcal{H} = \mathcal{L}^2(R)$ and satisfying the commutation rule $[x, p] = xp - px = ifrm[o]--$, $frm[o]--$ being the identity operator on $\mathcal{H}$. In fact, except for an unessential addictive constant, we can rewrite $H_0 = a^*a$, where $a = \frac{1}{\sqrt{2}}(x - ip)$ and a^* is its hermitian conjugate. So we see that H_0 can be factorized in a very peculiar way, by means of a single operator. In general, if an Hamiltonian H can be factorized, the computation of its eigenvectors and eigenvalues is significantly simplified. Moreover, when a H admits such a representation, then all the machinery of super-symmetric quantum mechanics can be applied. This means, substantially, that another operator can be usually found whose eigenvectors can be identified, [11]. This, not surprisingly, is useful in quantum mechanics since it produces new *exactly solvable* Schrödinger equations.

Going back to our particular situation, we will now show that, at least formally, the operators $H^{\boldsymbol{\alpha}}_{\phi,\psi}$ and $H^{\boldsymbol{\alpha}}_{\psi,\phi}$ can in fact be factorized.

For that, following what we did in [6], we first introduce generalized lowering and raising operators as follows:

$$A^{\boldsymbol{\gamma}}_{\phi,\psi} = \sum_{n=1}^{\infty} \gamma_n \phi_{n-1} \otimes \overline{\psi_n}, \qquad A^{\boldsymbol{\gamma}}_{\psi,\phi} = \sum_{n=1}^{\infty} \gamma_n \psi_{n-1} \otimes \overline{\phi_n}$$

$$B^{\boldsymbol{\gamma}}_{\phi,\psi} = \sum_{n=0}^{\infty} \gamma_{n+1} \phi_{n+1} \otimes \overline{\psi_n}, \qquad B^{\boldsymbol{\gamma}}_{\psi,\phi} = \sum_{n=0}^{\infty} \gamma_{n+1} \psi_{n+1} \otimes \overline{\phi_n}.$$

These operators turn out to be densely defined closed operators in $\mathcal{H}$, and they are related as follow: $\left(A^{\boldsymbol{y}}_{\phi,\psi}\right)^* = B^{\overline{\boldsymbol{y}}}_{\psi,\phi}$ and $\left(A^{\boldsymbol{y}}_{\psi,\phi}\right)^* = B^{\overline{\boldsymbol{y}}}_{\phi,\psi}$. Here $\boldsymbol{y} = \{y_n\}$, and $y_n \in C$, in general. Moreover,

$$D(A^{\boldsymbol{y}}_{\phi,\psi}) = \left\{f \in \mathcal{H};\ \sum_{n=1}^{\infty} |y_n|^2 |\langle f \mid \psi_n\rangle|^2 < \infty\right\} = D(H^{\{\alpha_n\}}_{\phi,\psi}),$$

$$D(A^{\boldsymbol{y}}_{\psi,\phi}) = \left\{f \in \mathcal{H};\ \sum_{n=1}^{\infty} |y_n|^2 |\langle f \mid \phi_n\rangle|^2 < \infty\right\} = D(H^{\{\alpha_n\}}_{\psi,\phi}),$$

$$D(B^{\boldsymbol{y}}_{\phi,\psi}) = \left\{f \in \mathcal{H};\ \sum_{n=0}^{\infty} |y_{n+1}|^2 |\langle f \mid \psi_n\rangle|^2 < \infty\right\} = D(H^{\{\alpha_{n+1}\}}_{\phi,\psi}),$$

$$D(B^{\boldsymbol{y}}_{\psi,\phi}) = \left\{f \in \mathcal{H};\ \sum_{n=0}^{\infty} |y_{n+1}|^2 |\langle f \mid \phi_n\rangle|^2 < \infty\right\} = D(H^{\{\alpha_{n+1}\}}_{\psi,\phi}).$$

To simplify the notation, and without any loss of generality, we will assume here that $y_0 = 0$. This corresponds to a shift in the Hamiltonian, which, from a purely physical point of view, has no consequence at all. In [5], mainly in view of our factorization problem, a particular interest has been given to products of these operators. In particular, it has been proved that

$$D(A^{\boldsymbol{y}}_{\phi,\psi}B^{\boldsymbol{y}}_{\phi,\psi}) = \left\{f \in \mathcal{H};\ \sum_{n=0}^{\infty} |y_{n+1}|^2 |\langle f \mid \psi_n\rangle|^2 < \infty \text{ and } \sum_{n=0}^{\infty} |y_{n+1}|^4 |\langle f \mid \psi_n\rangle|^2 < \infty\right\}$$

and that $A^{\boldsymbol{y}}_{\phi,\psi}B^{\boldsymbol{y}}_{\phi,\psi} \subset \sum_{n=0}^{\infty} y_{n+1}^2 \phi_n \otimes \overline{\psi}_n = H^{\{y_{n+1}^2\}}_{\phi,\psi}$. Moreover,

$$D(B^{\boldsymbol{y}}_{\phi,\psi}A^{\boldsymbol{y}}_{\phi,\psi}) = \left\{f \in \mathcal{H};\ \sum_{n=0}^{\infty} |y_n|^2 |\langle f \mid \psi_n\rangle|^2 < \infty \text{ and } \sum_{n=0}^{\infty} |y_n|^4 |\langle f \mid \psi_n\rangle|^2 < \infty\right\}$$

and $B^{\boldsymbol{y}}_{\phi,\psi}A^{\boldsymbol{y}}_{\phi,\psi} \subset \sum_{n=0}^{\infty} y_n^2 \phi_n \otimes \overline{\psi}_n = H^{\{y_n^2\}}_{\phi,\psi}$. Also, suppose that $0 \equiv |y_0| \le |y_1| \le |y_2| \le \cdots$. Then:

$$D(B^{\boldsymbol{y}}_{\phi,\psi}A^{\boldsymbol{y}}_{\phi,\psi}) = \left\{f \in \mathcal{H};\ \sum_{n=0}^{\infty} |y_n|^4 |\langle f \mid \psi_n\rangle|^2 < \infty\right\} = D(H^{\{y_n^2\}}_{\phi,\psi});$$

$$D(A^{\boldsymbol{y}}_{\phi,\psi}B^{\boldsymbol{y}}_{\phi,\psi}) = \left\{f \in \mathcal{H};\ \sum_{n=0}^{\infty} |y_{n+1}|^4 |\langle f \mid \psi_n\rangle|^2 < \infty\right\} = D(H^{\{y_{n+1}^2\}}_{\phi,\psi});$$

$$B^{\boldsymbol{y}}_{\phi,\psi}A^{\boldsymbol{y}}_{\phi,\psi} \subset A^{\boldsymbol{y}}_{\phi,\psi}B^{\boldsymbol{y}}_{\phi,\psi};$$

$$B^{\boldsymbol{\gamma}}_{\phi,\psi}A^{\boldsymbol{\gamma}}_{\phi,\psi} = \sum_{n=0}^{\infty} \gamma_n^2 \phi_n \otimes \overline{\psi}_n = H^{\{\gamma_n^2\}}_{\phi,\psi};$$

$$A^{\boldsymbol{\gamma}}_{\phi,\psi}B^{\boldsymbol{\gamma}}_{\phi,\psi} = \sum_{n=0}^{\infty} \gamma_{n+1}^2 \phi_n \otimes \overline{\psi}_n = H^{\{\gamma_{n+1}^2\}}_{\phi,\psi}.$$

Finally,

$$A^{\boldsymbol{\gamma}}_{\phi,\psi}B^{\boldsymbol{\gamma}}_{\phi,\psi}f - B^{\boldsymbol{\gamma}}_{\phi,\psi}A^{\boldsymbol{\gamma}}_{\phi,\psi}f = H^{\{\gamma_{n+1}^2\}}_{\phi,\psi}f - H^{\{\gamma_n^2\}}_{\phi,\psi}f = \left(\sum_{n=0}^{\infty}(\gamma_{n+1}^2 - \gamma_n^2)\phi_n \otimes \overline{\psi}_n\right)f,$$

for every $f \in D(H^{\{\gamma_{n+1}^2\}}_{\phi,\psi})$.

It is clear that the lowering and raising operators considered above constitute a generalization of annihilation and creation operators of quantum mechanics and, as in that case, they can be used to factorize the original Hamiltonian. In particular, if the sequence $\boldsymbol{\alpha} = \{\alpha_n\}$ introduced in the previous section is such that $0 = \alpha_0 < \alpha_1 < \alpha_2 < \ldots$, and if we take $\gamma_n = \sqrt{\alpha_n}$ here, the sequence $\boldsymbol{\gamma}$ satisfies the assumptions above, and we find, for instance that $B^{\boldsymbol{\gamma}}_{\phi,\psi}A^{\boldsymbol{\gamma}}_{\phi,\psi} = \sum_{n=0}^{\infty} \alpha_n \phi_n \otimes \overline{\psi}_n = H^{\boldsymbol{\alpha}}_{\phi,\psi}$: hence $H^{\boldsymbol{\alpha}}_{\phi,\psi}$ can be factorized.

2.2 Examples

2.2.1 First example

This first example extends what has been done in [5]. Let $P = P^* = P^2$ be an orthogonal projection on $\mathcal{H}$. Let us put $V_{\alpha,\beta} = \alpha frm[o]-- + \beta P$, $\alpha, \beta \in C$. Clearly, $V_{\alpha,\beta}$ is bounded and, if α and $\alpha+\beta$ are both non zero, has bounded inverse. In fact, $V^{-1}_{\alpha,\beta} = V_{\frac{1}{\alpha},\frac{-\beta}{\alpha(\alpha+\beta)}}$. The operators $S_\phi = T_{\alpha,\beta}T^*_{\alpha,\beta}$ and $S_\psi = S_\phi^{-1}$ introduced in Proposition 2.2, with T replaced by $T_{\alpha,\beta} := V^{-1}_{\alpha,\beta}$, can now easily computed. In particular we get

$$S_\phi = |\alpha|^2 frm[o]-- + \left(\alpha\overline{\beta} + \overline{\alpha}\beta + |\beta|^2\right)P = T_{|\alpha|^2,\alpha\overline{\beta}+\overline{\alpha}\beta+|\beta|^2}.$$

The next ingredient of our construction is the positive square root of this operator, and its inverse. To simplify the example, we just focus here to the case in which $\alpha = \beta \neq 0$. Then, calling for simplicity $T = T_{\alpha,\alpha}$, we have:

$$T = \alpha(frm[o]-- + P), \qquad T^{-1} = \frac{1}{\alpha}\left(frm[o]-- - \frac{1}{2}P\right),$$

$$S_\phi = |\alpha|^2(frm[o]-- + 3P), \qquad S_\psi = \frac{1}{|\alpha|^2}\left(frm[o]-- - \frac{3}{4}P\right),$$

from which we recover the positive operators

$$S_\phi^{1/2} = |\alpha|(frm[o]-- + P), \qquad S_\psi^{1/2} = \frac{1}{|\alpha|}\left(frm[o]-- - \frac{1}{2}P\right).$$

We are now ready to discuss a simple *example in the example*. For that, let $I \subset R$ be a subset of the real line with measure one: $\mu(I) = 1$, and let $\chi_I(x)$ be the characteristic function on I: $\chi_I(x) = 1$ if $x \in I$ and $\chi_I(x) = 0$ otherwise. We now introduce an orthogonal projector operator P_I using $\chi_I(x)$:

$$P_I = \chi_I(x) \otimes \overline{\chi_I(x)} \quad \Longrightarrow \quad (P_I f)(x) = \langle f \mid \chi_I \rangle \chi_I(x) = \left(\int_I f(y)dy \right) \chi_I(x),$$

for all $f(x) \in \mathcal{H} = \mathcal{L}^2(R)$. Let now $\mathcal{E} = \{e_n(x),\ n \in \mathbb{N}_0\}$ be an orthonormal basis for $\mathcal{H}$. For instance, $\mathcal{E}$ could be the set of the (Hermite) eigenstates of the quantum harmonic oscillator, H_0. Then, as discussed in Section 2, if we put $\phi_n = Te_n$ and $\psi_n = (T^{-1})^* e_n$, $n \in \mathbb{N}_0$ the sets $\mathcal{F}_\phi = \{\phi_n,\ n \in \mathbb{N}_0\}$ and $\mathcal{F}_\psi = \{\psi_n,\ n \in \mathbb{N}_0\}$ are two biorthogonal Riesz bases. In particular we find

$$\phi_n(x) = \alpha \begin{cases} e_n(x), & \text{if } x \notin I, \\ e_n(x) + \int_I e_n(y)dy, & \text{if } x \in I, \end{cases}$$

and

$$\psi_n(x) = \frac{1}{\overline{\alpha}} \begin{cases} e_n(x), & \text{if } x \notin I, \\ e_n(x) - \frac{1}{2} \int_I e_n(y)dy, & \text{if } x \in I. \end{cases}$$

A direct computation shows that these functions, as expected, are indeed biorthogonal: $\langle \phi_n \mid \psi_m \rangle = \delta_{n,m}$. Moreover they are both (Riesz) bases, by construction. Hence, they can be used to define the Hamiltonian $H^{\boldsymbol{\alpha}}_{\phi,\psi} = \sum_{n=0}^{\infty} \alpha_n \phi_n \otimes \overline{\psi}_n$ as in the beginning of Section 2. Here $\boldsymbol{\alpha} = \{\alpha_n\}$ is a sequence of real numbers. In particular one can show, by direct computations, that

$$H^{\boldsymbol{\alpha}}_{\phi,\psi} = H^{\boldsymbol{\alpha}}_{e,e} + P_I H^{\boldsymbol{\alpha}}_{e,e} - \frac{1}{2} H^{\boldsymbol{\alpha}}_{e,e} P_I - \frac{1}{2} P_I H^{\boldsymbol{\alpha}}_{e,e} P_I,$$

where we have introduced the following (formally) self-adjoint Hamiltonian $H^{\boldsymbol{\alpha}}_{e,e} = \sum_{n=0}^{\infty} \alpha_n e_n \otimes \overline{e}_n$. A simple computation also shows that $H^{\boldsymbol{\alpha}}_{e,e} = \mathsf{h}^{\boldsymbol{\alpha}}_{\phi,\psi} = S_\psi^{1/2} H^{\boldsymbol{\alpha}}_{\phi,\psi} S_\phi^{1/2}$, so that, as stated after equation (6), we see that $H^{\boldsymbol{\alpha}}_{e,e}$ is really (and not just formally) self-adjoint.

Similar computations can be carried out in relation with the generalized lowering and raising operators. In particular, if $e_n(x)$ are the eigenstates of the quantum harmonic oscillators, and if we take $\alpha_n = n$, then $H^{\boldsymbol{\alpha}}_{e,e}$ is the well known Hamiltonian $H_0 = a^* a$ already introduced, where a and a^*, satisfying $[a, a^*] = aa^* - a^* a = frm[o]--$ (in a suitable sense), are the well known standard bosonic operators. With these choices, and taking $\gamma_n = \sqrt{n}$, we deduce that $A^{\boldsymbol{\gamma}}_{\phi,\psi} = a$ and $B^{\boldsymbol{\gamma}}_{\phi,\psi} = a^*$.

2.2.2 Second example

We consider here the Hilbert space $\mathcal{H} = \mathcal{L}^2(K)$, $K = [-\pi, \pi]$, and the following o.n. basis: $\mathcal{E} = \{e_n(x) = \frac{1}{\sqrt{2\pi}} e^{inx},\ n \in Z\}$. Next we introduce two multiplicative bounded operators, with bounded inverse: $T_1 = (1+x^2)^{-1}$ and $T_2 = \frac{1}{\sqrt{2\pi}}e^{-iDx}$, $D \in R$. It is clear that, while $T_1 = T_1^*$, T_2 is not self-adjoint. We have $\|T_1^{-1}\| \leq (1+\pi^2)$, $\|T_1\| \leq 1$, while $\|T_2\| = \|T_2^{-1}\| = \frac{1}{\sqrt{2\pi}}$. We can now use these operators to construct two families of biorthogonal Riesz bases out of $\mathcal{E}$. The first pair consists in the following functions:

$$\phi_n(x) = (T_1^{-1}e_n)(x) = \frac{1}{\sqrt{2\pi}}(1+x^2)e^{inx},$$

$$\psi_n(x) = (T_1^* e_n)(x) = \frac{1}{\sqrt{2\pi}(1+x^2)}e^{inx}.$$

Hence we can define, as we did in Section 2, $H^{\boldsymbol{\alpha}}_{\phi,\psi} = \sum_{n=0}^{\infty} \alpha_n \phi_n \otimes \overline{\psi}_n$ and we can see, using the above relations between ϕ_n, ψ_n and e_n, that

$$H^{\boldsymbol{\alpha}}_{\phi,\psi} = T^{-1} H^{\boldsymbol{\alpha}}_{e,e} T,$$

with the usual definition for $H^{\boldsymbol{\alpha}}_{e,e}$ and with T respectively T_1 and T_2. Now, since T_1 is positive, it is easy to check that $S_\phi^{1/2} = T_1$ and, consequently, that $S_\psi^{1/2} = T_1^{-1}$. The conclusion is that $h^{\boldsymbol{\alpha}}_{\phi,\psi} = H^{\boldsymbol{\alpha}}_{e,e}$, which, for real α_n, is self-adjoint.

Slightly different is the situation if we consider T_2. In this case, in fact, $S_\phi^{1/2} = \sqrt{2\pi} \neq T_2$, and therefore $S_\psi^{1/2} = \frac{1}{\sqrt{2\pi}} \neq T_2^{-1}$, and $h^{\boldsymbol{\alpha}}_{\phi,\psi} = S_\psi^{1/2} H^{\boldsymbol{\alpha}}_{\phi,\psi} S_\phi^{1/2} = H^{\boldsymbol{\alpha}}_{\phi,\psi}$, which is different from $H^{\boldsymbol{\alpha}}_{e,e}$.

We leave to the reader the analysis of the latter operators, since this is not particularly relevant for us here.

3 A generalization to rigged Hilbert spaces

3.1 Strict generalized Riesz bases in rigged Hilbert spaces

In this section we will summarize the main results given in [8], where a generalized notion of Riesz basis in a Hilbert space to the more general setting of rigged Hilbert spaces has been introduced. Before going forth we introduce some notation and recall some definition.

Let $\mathcal{D}$ be a dense subspace of $\mathcal{H}$. A locally convex topology t on $\mathcal{D}$ finer than the topology induced by the Hilbert norm defines, in standard fashion, a *rigged Hilbert space* (RHS)

$$\mathcal{D}[t] \hookrightarrow \mathcal{H} \hookrightarrow \mathcal{D}^\times[t^\times], \tag{1}$$

where $\mathcal{D}^\times$ is the vector space of all continuous conjugate linear functionals on $\mathcal{D}[t]$, i.e., the conjugate dual of $\mathcal{D}[t]$, endowed with the *strong dual topology* $t^\times = \beta(\mathcal{D}^\times, \mathcal{D})$ and $\hookrightarrow$ denotes a continuous embedding. Since the Hilbert space $\mathcal{H}$ can be identified with a subspace of $\mathcal{D}^\times[t^\times]$, we will systematically read (1) as a chain of topological inclusions: $\mathcal{D}[t] \subset \mathcal{H} \subset \mathcal{D}^\times[t^\times]$. These identifications imply that the sesquilinear form $B(\cdot,\cdot)$ that puts $\mathcal{D}$ and $\mathcal{D}^\times$ in duality is an extension of the inner product of $\mathcal{D}$; i.e. $B(\xi,\eta) = \langle \xi \mid \eta \rangle$, for every $\xi, \eta \in \mathcal{D}$ and we will adopt the symbol $\langle \cdot \mid \cdot \rangle$ for both of them.

Example 3.1. *A very familiar example of RHS in distribution theory is the triplet*

$$(\mathcal{S}(\mathbb{R}), \mathcal{L}^2(\mathbb{R}), \mathcal{S}'(\mathbb{R})),$$

constituted by the Schwartz space of rapidly decreasing C^∞-functions, the Lebesgue $\mathcal{L}^2$-space, and the space of tempered distributions.

Example 3.2. *Let $\mathcal{D}$ be a dense domain in Hilbert space $\mathcal{H}$. A typical example of rigged Hilbert space can be obtained as follows. Denote by $\mathcal{L}^\dagger(\mathcal{D})$ the *-algebra consisting of all closable operators A with $D(A) = \mathcal{D}$, which together with their adjoints, A^*, leave $\mathcal{D}$ invariant. The involution of $\mathcal{L}^\dagger(\mathcal{D})$ is defined by $A \mapsto A^\dagger$, where $A^\dagger = A^* \upharpoonright \mathcal{D}$. The *-algebra $\mathcal{L}^\dagger(\mathcal{D})$ defines in $\mathcal{D}$ the* graph topology *$t_\dagger$ by the family of seminorms*

$$\xi \in \mathcal{D} \to \|\xi\|_A := \|(frm[o]-- + A^*\overline{A})^{1/2}\xi\|, \quad A \in \mathcal{L}^\dagger(\mathcal{D}).$$

For every fixed $A \in \mathcal{L}^\dagger(\mathcal{D})$, the completion of $\mathcal{D}$ with respect to $\|\cdot\|_A$ is a Hilbert space denoted by $\mathcal{H}_A$. Since the topology $t_\dagger$ is finer than the topology induced on $\mathcal{D}$ by the Hilbert norm of $\mathcal{H}$, it defines in natural way a structure of rigged Hilbert space.

Let now $\mathcal{D}[t] \subset \mathcal{H} \subset \mathcal{D}^\times[t^\times]$ be a RHS and let $\mathfrak{L}(\mathcal{D}, \mathcal{D}^\times)$ denote the vector space of all continuous linear maps from $\mathcal{D}[t]$ into $\mathcal{D}^\times[t^\times]$. In $\mathfrak{L}(\mathcal{D}, \mathcal{D}^\times)$ an involution $X \mapsto X^\dagger$ can be introduced by the equality[2]

$$\langle X^\dagger \eta \mid \xi \rangle = \overline{\langle X\xi \mid \eta \rangle}, \quad \forall \xi, \eta \in \mathcal{D}.$$

Hence $\mathfrak{L}(\mathcal{D}, \mathcal{D}^\times)$ is a †-invariant vector space. Let $\mathcal{E}, \mathcal{F} \in \{\mathcal{D}, \mathcal{H}, \mathcal{D}^\times\}$ and $\mathfrak{L}(\mathcal{E}, \mathcal{F})$ the space of all continuous linear maps from $\mathcal{E}$ into $\mathcal{F}$. We put

$$\mathcal{C}(\mathcal{E}, \mathcal{F}) := \{X \in \mathfrak{L}(\mathcal{D}, \mathcal{D}^\times) : \exists Y \in \mathfrak{L}(\mathcal{E}, \mathcal{F}),\ Y\xi = X\xi, \forall \xi \in \mathcal{D}\}.$$

In particular, if $X \in C(\mathcal{D}, \mathcal{H})$ then its adjoint $X^\dagger \in \mathfrak{L}(\mathcal{D}, \mathcal{D}^\times)$ has an extension from $\mathcal{H}$ into $\mathcal{D}^\times$, which we denote by the same symbol.

Let us now recall some definitions given in [8]. Let $\mathcal{E}$ be a locally convex vector space with topology $t_\mathcal{E}$ and $\{\xi_n\}$ be a sequence of vectors of $\mathcal{E}$.

2 In the following we will prefer to maintain the notation with a vector in the dual $\mathcal{D}^\times$ in the first entry of the bilinear form $\langle \cdot \mid \cdot \rangle$ and a vector in the Hilbert space $\mathcal{H}$ or in the dense subspace $\mathcal{D}$ in the second one.

The sequence $\{\xi_n\}$ is a *basis* for $\mathcal{E}$ if, for every $\phi \in \mathcal{E}$, there exists a *unique* sequence $\{c_n\}$ of complex numbers such that

$$\phi = \sum_{n=0}^{\infty} c_n \xi_n, \tag{2}$$

where the series on the right hand side converges in $\mathcal{E}[t_{\mathcal{E}}]$.

If $\{\xi_n\}$ is a basis for $\mathcal{E}$, then $\{\xi_n\}$ is *ω-independent*; i.e., if $\sum_{n=0}^{\infty} c_n \xi_n = 0$, then $c_n = 0$, for every $n \in \mathbb{N}$. This in turn implies that the sequence $\{\xi_n\}$ consists of linearly independent vectors.

It is easy to see that every coefficient $c_n = c_n(\phi)$ in (2) depends linearly on $\phi \in \mathcal{E}$; hence, it can be regarded as a linear functional on $\mathcal{E}$. A basis $\{\xi_n\}$ of $\mathcal{E}[t_{\mathcal{E}}]$ is said to be *Schauder* if the coefficient functionals $\{c_n\}$ are $t_{\mathcal{E}}$-continuous.

Let now $\mathcal{D}[t] \subset \mathcal{H} \subset \mathcal{D}^\times[t^\times]$ be a rigged Hilbert space and $\{\xi_n\}$ a Schauder basis for $\mathcal{D}[t]$. Then, every $f \in \mathcal{D}$ can be written as $\sum_{n=0}^{\infty} c_n(f)\xi_n$, for suitable coefficients $c_n(f)$. Since every c_n is a *continuous* linear functional on $\mathcal{D}[t]$, there exists a sequence $\{\zeta_n\} \subset \mathcal{D}^\times$ such that

$$c_n(f) = \overline{\langle \zeta_n \mid f \rangle}, \quad \forall n \in \mathbb{N}, f \in \mathcal{D}.$$

For every $n \in \mathbb{N}$, the vector ζ_n is uniquely determined because of the continuity of the functional $c_n(f)$. If we take $f = \xi_k$, then it is clear that $c_n(\xi_k) = \delta_{n,k}$. Hence $\langle \zeta_n \mid \xi_k \rangle = \delta_{n,k}$; i.e., the sequences $\{\xi_n\}$ and $\{\zeta_n\}$ are *biorthogonal.*

We assume, from now on, that $\mathcal{D}[t]$ is complete and reflexive. As in [8] we give the following

Definition 3.3. *Let $\{\xi_n\}$ be a Schauder basis for $\mathcal{D}[t]$. Then $\{\xi_n\}$ is called*

- a generalized Riesz basis *if there exists a one-to-one operator $T \in C(\mathcal{D}, \mathcal{H})$ such that $\{T\xi_n\}$ is an orthonormal basis for $\mathcal{H}$.*
- *a* strict *generalized Riesz basis if the operator $T \in C(\mathcal{D}, \mathcal{H})$ which makes of $\{T\xi_n\}$ an orthonormal basis for $\mathcal{H}$ has a continuous inverse $T^{-1} : \mathcal{H}[\|\cdot\|] \to \mathcal{D}[t]$ (in particular, T^{-1} is a* bounded *operator in $\mathcal{H}$).*

Following [8, Proposition 3.14], if the rigged Hilbert space $\mathcal{D}[t] \subset \mathcal{H} \subset \mathcal{D}^\times[t^\times]$, with $\mathcal{D}[t]$ complete and reflexive, has a strict generalized Riesz basis $\{\xi_n\}$ then it is (equivalent to) a triplet of Hilbert spaces $\mathcal{H}_{+1}[\|\cdot\|_{+1}] \subset \mathcal{H} \subset \mathcal{H}_{-1}[\|\cdot\|_{-1}]$, with $\mathcal{H}_{+1}[\|\cdot\|_{+1}]$ the Hilbert space whose the norm is corresponding to the inner product

$$\langle \xi \mid \eta \rangle_{+1} := \langle T\xi \mid T\eta \rangle, \qquad \forall \xi, \eta \in \mathcal{D}$$

(where T is the continuous operator with continuous inverse such that $\{T\xi_n\}$ is an orthonormal basis of $\mathcal{H}$), and $\mathcal{H}_{-1}$ the Hilbert space conjugate dual of $\mathcal{H}_{+1}$ with norm $\|\Psi\|_{-1} := \|\Psi\|_{+1}^\times = \|(T^\dagger)^{-1}\Psi\|$, for every $\Psi \in \mathcal{D}^\times$. Moreover, $\{\xi_n\}$ is an orthonormal basis for $\mathcal{H}_{+1}$ and the dual sequence $\{\zeta_n\}$ is an orthonormal basis for $\mathcal{H}_{-1}$.

Example 3.4. *Let $\mathcal{H}$ be a separable Hilbert space and $\{e_n\}$ an orthonormal basis of $\mathcal{H}$. Then, for every $f \in \mathcal{H}$, $f = \sum_{k=0}^{\infty} \langle f \mid e_k\rangle e_k$ in $\mathcal{H}[\|\cdot\|]$. Let $\{\lambda_k\} \subset \mathbb{R}^+$ be an unbounded sequence and Λ a finite set (even empty) of indices such that $\lambda_k \in]0,1[$, when $k \in \Lambda$. Let us consider the subspace*

$$D(T) = \left\{ f \in \mathcal{H} : \sum_{k=0}^{\infty} \lambda_k^2 |\langle f \mid e_k\rangle|^2 < \infty \right\}$$

and let, for $f \in D(T)$,

$$Tf = T\left(\sum_{n=0}^{\infty} \langle f \mid e_n\rangle e_n\right) = \sum_{n=0}^{\infty} \lambda_n \langle f \mid e_n\rangle e_n.$$

Notice that $Te_n = \lambda_n e_n$, for every $n \in \mathbb{N}$. The operator T is self-adjoint on $D(T)$. Let $\mathcal{H}_{+1} = D(T)$ endowed with the topology induced by the norm $\|\cdot\|_{+1} = \|T\cdot\|$

Define $\xi_k = \frac{1}{\lambda_k} e_k$, $k \in \mathbb{N}$. Clearly, $T\xi_k = e_k$, for every $k \in \mathbb{N}$ and T maps continuously $\mathcal{H}_{+1}$ into $\mathcal{H}$. Moreover, $\{\xi_k\}$ is a basis for $\mathcal{H}_{+1}$ since

$$f = \sum_{n=0}^{\infty} \langle f \mid e_n\rangle e_n = \sum_{n=0}^{\infty} \lambda_n \langle f \mid e_n\rangle \xi_n,$$

for all $f \in \mathcal{H}$ and, therefore, for all $f \in \mathcal{H}_{+1}$.

T is one-to-one since, if $Tf = 0$, with 0 the zero vector, it is equivalent to

$$\|Tf\| = 0 \Leftrightarrow \sum_{k=0}^{\infty} \lambda_k^2 |\langle f \mid e_k\rangle|^2 = 0 \Leftrightarrow \langle f \mid e_k\rangle = 0,\ \forall k \in \mathbb{N} \Leftrightarrow f = 0.$$

Moreover, T is onto $\mathcal{H}$ since for every $g = \sum_{k=0}^{\infty} \langle g \mid e_k\rangle e_k \in \mathcal{H}$ there exists $f = \sum_{k=0}^{\infty} \langle g \mid e_k\rangle \xi_k \in \mathcal{H}_{+1}$ such that $Tf = g$. The vector f belongs to $\mathcal{H}_{+1}$, indeed

$$\|f\|_{+1} = \left\| \sum_{k=0}^{\infty} \langle g \mid e_k\rangle \xi_k \right\|_{+1} = \left\| T\left(\sum_{k=0}^{\infty} \langle g \mid e_k\rangle \xi_k \right) \right\| = \|g\| < \infty.$$

The inverse of T, $T^{-1} : \mathcal{H}[\|\cdot\|] \to \mathcal{H}_{+1}$ is defined as

$$T^{-1}f := \sum_{k=0}^{\infty} \langle f \mid e_k\rangle \xi_k = \sum_{k=0}^{\infty} \frac{1}{\lambda_k} \langle f \mid e_k\rangle e_k$$

and it is everywhere defined. In fact, $\sup_{f\in\mathcal{H},\|f\|=1}\{\|T^{-1}f\|_{+1}\} \le C < \infty$, with $C = \max\{1, \max_{k\in\Lambda} \frac{1}{\lambda_k}\}$, then T^{-1} is bounded. Moreover, $\{\xi_k\}$ is clearly a Schauder basis. Hence, $\{\xi_n\}$ is a strict generalized Riesz basis for $\mathcal{H}_{+1}$. We can find explicitly the biorthogonal sequence $\{\zeta_n\} \subset \mathcal{H}_{-1}$. Since T maps $\mathcal{D}[t]$ into $\mathcal{H}[\|\cdot\|]$ continuously, $T^\dagger$ has a continuous extension (which we denote by the same symbol) from $\mathcal{H}[\|\cdot\|]$ into $\mathcal{H}_{-1}[\|\cdot\|_{-1}]$. The continuity of T and $f = \sum_{n=0}^{\infty} \lambda_n \langle f \mid e_n\rangle \xi_n = \sum_{n=0}^{\infty} \overline{\langle \zeta_n \mid f\rangle} \xi_n$ imply

$$Tf = T\left(\sum_{n=0}^{\infty} \lambda_n \langle f \mid e_n\rangle \xi_n \right) = \sum_{n=0}^{\infty} \lambda_n \langle f \mid e_n\rangle T\xi_n = \sum_{n=0}^{\infty} \lambda_n \langle f \mid e_n\rangle e_n$$

and, on the other hand

$$Tf = \sum_{n=0}^{\infty} \overline{\langle \zeta_n \mid f \rangle} T\xi_n = \sum_{n=0}^{\infty} \overline{\langle \zeta_n \mid f \rangle} e_n, \quad \forall f \in \mathcal{D}.$$

This, in turn, implies that $\overline{\langle \zeta_n \mid f \rangle} = \lambda_n \langle f \mid e_n \rangle = \langle Tf \mid e_n \rangle$, *for every* $f \in \mathcal{H}_{+1}$. *Hence* $\zeta_n = T^\dagger e_n = \lambda_n e_n$, *for every* $n \in \mathbb{N}$. *Clearly, for every* $n, k \in \mathbb{N}$,

$$\langle \zeta_k \mid \xi_n \rangle = \left\langle T^\dagger e_k \mid \xi_n \right\rangle = \langle e_k \mid T\xi_n \rangle = \langle e_k \mid e_n \rangle = \delta_{k,n}.$$

By these two systems of biorthogonal vectors we could now construct the operators defined in Section 3.2. However, because of the particular form of these vectors, the resulting operators would be particularly simple, since they coincide with those constructed using the e_k*'s only. For this reason we will not give them here.*

3.2 Operators defined by a strict generalized Riesz basis and its dual basis

We gather here the definitions and results given in [8] about some operators defined via strict generalized Riesz bases and their dual bases.
Let $\{\xi_n\}$ be a Schauder basis for $\mathcal{D}$, $\{\zeta_n\}$ the family of vectors of $\mathcal{D}^\times$ such that $c_n(f) = \overline{\langle \zeta_n \mid f \rangle}$, for every $f \in \mathcal{D}$, $n \in \mathbb{N}$. Then every $f \in \mathcal{D}$ can be written as

$$f = \sum_{n=0}^{\infty} \overline{\langle \zeta_n \mid f \rangle} \xi_n,$$

in the sense of the topology t of $\mathcal{D}$. Moreover, we can also write also

$$f = \sum_{k=0}^{\infty} \langle f \mid \xi_k \rangle \zeta_k$$

in weak sense, i.e. $\langle \Psi \mid \sum_{k=0}^{n} \langle f \mid \xi_k \rangle \zeta_k \rangle \to \langle \Psi \mid f \rangle, \quad \forall \Psi \in \mathcal{D}^\times$. We now gather the operators defined in [8], analogous to those in Section 2, but in the more general setting of strict generalized Riesz basis. If $\boldsymbol{\alpha} = \{\alpha_n\}$ is a sequence of complex numbers we can formally define, for suitable f,

$$\hat{H}^{\boldsymbol{\alpha}}_{\xi,\zeta} f = \sum_{n=0}^{\infty} \alpha_n (\xi_n \otimes \overline{\zeta}_n) f = \sum_{n=0}^{\infty} \alpha_n \overline{\langle \zeta_n \mid f \rangle} \xi_n \tag{3}$$

$$\hat{H}^{\boldsymbol{\alpha}}_{\zeta,\xi} f = \sum_{n=0}^{\infty} \alpha_n (\zeta_n \otimes \overline{\xi}_n) f = \sum_{n=0}^{\infty} \alpha_n \langle f \mid \xi_n \rangle \zeta_n. \tag{4}$$

$$\hat{S_\xi}^{\boldsymbol{\alpha}} f = \sum_{n=0}^{\infty} \alpha_n (\xi_n \otimes \overline{\xi}_n) f = \sum_{n=0}^{\infty} \alpha_n \langle f \mid \xi_n \rangle \xi_n \tag{5}$$

$$\hat{S}_{\zeta}^{\boldsymbol{\alpha}} f = \sum_{n=0}^{\infty} \alpha_n (\zeta_n \otimes \overline{\zeta}_n) f = \sum_{n=0}^{\infty} \alpha_n \overline{\langle \zeta_n \mid f \rangle} \zeta_n \tag{6}$$

Of course, these are the simplest operators that can be defined via a Schauder basis $\{\xi_n\}$ and its dual basis $\{\zeta_n\}$, and they strongly reflect those introduced in Section 2.

Our goal in this section is the study of operators (3)-(6) in the case where $\{\xi_n\}$ is a strict generalized Riesz basis. Since, as showed in [8, Section 3.2], the existence of a strict generalized Riesz basis forces the rigged Hilbert space $\mathcal{D} \subset \mathcal{H} \subset \mathcal{D}^\times$, if $\mathcal{D}$ is complete and reflexive, to be in fact a triplet of Hilbert spaces, we confine our considerations to triplets of Hilbert spaces only.
Let $\mathcal{H}_{+1} \subset \mathcal{H} \subset \mathcal{H}_{-1}$ be a triplet of Hilbert spaces and $\{\xi_n\}$ a strict generalized Riesz basis for $\mathcal{H}_{+1}$.

Clearly, the operator formally defined by (3), can take values in $\mathcal{H}_{+1}$ or in $\mathcal{H}$ or even in $\mathcal{H}_{-1}$ and of course its domain may vary according to the different topologies that make the series on the right hand side convergent. Of course, analogous considerations can be made for the operators defined in (4), (5) and (6). It is worth remarking that for the operators $\hat{H}_{\zeta,\xi}^{\boldsymbol{\alpha}}$ and $\hat{S}_{\xi}^{\boldsymbol{\alpha}}$ the series on the right hand side of (4), (5) may converge also for some $f \in \mathcal{H}_{-1}$.

Now we examine more closely one of the possibilities mentioned above, and in particular we will restrict to the case in which $\hat{H}_{\xi,\zeta}^{\boldsymbol{\alpha}} f \in \mathcal{H}_{+1}$, for every $f \in \mathcal{H}_{+1}$, which is indeed a rather strong, but useful, assumption. Under this assumption, we will deduce the following diagram:

$$\mathcal{H}_{+1} \quad \overset{\hat{H}_{\xi,\zeta}^{\boldsymbol{\alpha}}}{\longrightarrow} \quad \mathcal{H}_{+1} \tag{7}$$

$$\hat{S}_{\xi}^{\boldsymbol{\alpha}} \uparrow \qquad\qquad \downarrow \hat{S}_{\zeta}^{\boldsymbol{\alpha}} \tag{8}$$

$$\mathcal{H}_{-1} \quad \overset{\hat{H}_{\zeta,\xi}^{\boldsymbol{\alpha}}}{\longleftarrow} \quad \mathcal{H}_{-1} \tag{9}$$

with obvious meaning. Let us define

$$\begin{cases} D(\hat{H}_{\xi,\zeta}^{\boldsymbol{\alpha}}) = \left\{ f \in \mathcal{H}_{+1};\, \sum_{n=0}^{\infty} \alpha_n \overline{\langle \zeta_n \mid f \rangle} \xi_n \text{ exists in } \mathcal{H}_{+1} \right\} \\ \hat{H}_{\xi,\zeta}^{\boldsymbol{\alpha}} f = \sum_{n=0}^{\infty} \alpha_n \overline{\langle \zeta_n \mid f \rangle} \xi_n,\ f \in D(\hat{H}_{\xi,\zeta}^{\boldsymbol{\alpha}}) \end{cases}$$

and

$$\begin{cases} D(\hat{H}_{\zeta,\xi}^{\boldsymbol{\alpha}}) = \{ \Psi \in \mathcal{H}_{-1};\, \sum_{n=0}^{\infty} \alpha_n \langle \Psi \mid \xi_n \rangle \zeta_n \text{ exists in } \mathcal{H}_{-1} \} \\ \hat{H}_{\zeta,\xi}^{\boldsymbol{\alpha}} \Psi = \sum_{n=0}^{\infty} \alpha_n \langle \Psi \mid \xi_n \rangle \zeta_n,\ \Psi \in D(\hat{H}_{\zeta,\xi}^{\boldsymbol{\alpha}}). \end{cases}$$

Then we have the following

$$\mathcal{D}_\xi := \text{span}\{\xi_n\} \subset D(\hat{H}^{\boldsymbol{\alpha}}_{\xi,\zeta}) \subset \mathcal{H}_{+1}; \tag{10}$$
$$\mathcal{D}_\zeta := \text{span}\{\zeta_n\} \subset D(\hat{H}^{\boldsymbol{\alpha}}_{\zeta,\xi}) \subset \mathcal{H}_{-1};$$
$$\hat{H}^{\boldsymbol{\alpha}}_{\xi,\zeta}\xi_k = \alpha_k\xi_k,\ \ k \in \mathbb{N}_0\ ; \tag{11}$$
$$\hat{H}^{\boldsymbol{\alpha}}_{\zeta,\xi}\zeta_k = \alpha_k\zeta_k,\ \ k \in \mathbb{N}_0\ .$$

Hence, $\hat{H}^{\boldsymbol{\alpha}}_{\xi,\zeta}$ and $\hat{H}^{\boldsymbol{\alpha}}_{\zeta,\xi}$ are densely defined, [8]. By Remark 3.2 in [7], the operators $(T^\dagger)^{-1}$ and $(T^{-1})^\dagger$ do coincide. Besides, by hypothesis, T is continuous and there exists T^{-1}, which is a bounded operator in $\mathcal{H}$, as well as its Hilbert adjoint $(T^{-1})^*$.

Before continuing, we recall that if $X : D(X) \subseteq \mathcal{H}_{+1} \to \mathcal{H}_{+1}$ is a closed map and $D(X)$ is dense in $\mathcal{H}_{+1}$, then there exists a closed densely defined map $X^\dagger : D(X^\dagger) \subseteq \mathcal{H}_{-1} \to \mathcal{H}_{-1}$ such that

$$\langle \Phi \mid X\xi\rangle = \left\langle X^\dagger \Phi \mid \xi\right\rangle, \quad \forall \xi \in \mathcal{H}_{+1},\ \Phi \in \mathcal{H}_{-1}.$$

If X is also closed as an operator in $\mathcal{H}$, then its Hilbert adjoint X^* exists and $X^* = X^\dagger \restriction D(X^*)$ where $D(X^*) = \{\phi \in \mathcal{H} :\ X^\dagger\phi \in \mathcal{H}\}$.

The following proposition has been proved in [8].

Proposition 3.5. *The following statements hold.*

(1) $D(\hat{H}^{\boldsymbol{\alpha}}_{\xi,\zeta}) = \{f \in \mathcal{H}_{+1}; \sum_{n=0}^{\infty} |\alpha_n|^2 |\langle \zeta_n \mid f\rangle|^2 < \infty\}$,
$D(\hat{H}^{\boldsymbol{\alpha}}_{\zeta,\xi}) = \{\Psi \in \mathcal{H}_{-1}; \sum_{n=0}^{\infty} |\alpha_n|^2 |\langle \Psi \mid \xi_n\rangle|^2 < \infty\}$.

(2) $\hat{H}^{\boldsymbol{\alpha}}_{\xi,\zeta}$ *and* $\hat{H}^{\boldsymbol{\alpha}}_{\zeta,\xi}$ *are closed operators respectively in* $\mathcal{H}_{+1}[\|\cdot\|_{+1}]$ *and in* $\mathcal{H}_{-1}[\|\cdot\|_{-1}]$.

(3) $\left(\hat{H}^{\boldsymbol{\alpha}}_{\xi,\zeta}\right)^\dagger = \hat{H}^{\overline{\boldsymbol{\alpha}}}_{\zeta,\xi}$, *where* $\overline{\boldsymbol{\alpha}} = \{\overline{\alpha}_n\}$.

(4) $\hat{H}^{\boldsymbol{\alpha}}_{\xi,\zeta}$ *is bounded in* $\mathcal{H}_{+1}$ *if, and only if,* $\hat{H}^{\boldsymbol{\alpha}}_{\zeta,\xi}$ *is bounded in* $\mathcal{H}_{-1}$ *and if, and only if,* $\boldsymbol{\alpha}$ *is a bounded sequence. In particular* $\hat{H}^{\boldsymbol{1}}_{\xi,\zeta} = frm[o]--_{\mathcal{H}_{+1}}$ *and* $\hat{H}^{\boldsymbol{1}}_{\zeta,\xi} = frm[o]--_{\mathcal{H}_{-1}}$.

Remark 3.6. *In [7] S. Di Bella, C. Trapani and one of the authors (GB) gave a definition of spectrum for continuous operators acting in a rigged Hilbert space* $\mathcal{D} \subset \mathcal{H} \subset \mathcal{D}^\times$. *We refer to that paper for precise definitions and results. So a natural question is: what is the spectrum (in that sense) of the operator* $\hat{H}^{\boldsymbol{\alpha}}_{\xi,\zeta}$ *defined above? Let us assume that the sequence* $\boldsymbol{\alpha}$ *is bounded, so that* $\hat{H}^{\boldsymbol{\alpha}}_{\xi,\zeta}$ *is a bounded operator in* $\mathcal{H}_{+1}$. *The analysis is, in this case, particularly simple since, as usual, the set of eigenvalues consists exactly of the* α_k*'s and, if* λ *does not belong to the closure* $\overline{\{\alpha_k; k \in \mathbb{N}_0\}}$ *of the set of eigenvalues, then the inverse of* $\hat{H}^{\boldsymbol{\alpha}}_{\xi,\zeta} - \lambda frm[o]--_{\mathcal{H}_{+1}}$ *exists as a bounded operator in* $\mathcal{H}_{+1}$. *Hence, as expected,* $\sigma(\hat{H}^{\boldsymbol{\alpha}}_{\xi,\zeta}) = \overline{\{\alpha_k; k \in \mathbb{N}_0\}}$. *The situation for* $\hat{H}^{\boldsymbol{\alpha}}_{\zeta,\xi}$ *is analogous.*

Let us now consider the operators formally given by (5) and (6). They are, in fact, defined as follows:

$$\begin{cases} D(\hat{S}_\xi^{\boldsymbol{\alpha}}) = \{\Psi \in \mathcal{H}_{-1}; \sum_{n=0}^{\infty} \alpha_n \langle \Psi \mid \xi_n \rangle \xi_n \text{ exists in } \mathcal{H}_{+1}\} \\ \hat{S}_\xi^{\boldsymbol{\alpha}} \Psi = \sum_{n=0}^{\infty} \alpha_n \langle \Psi \mid \xi_n \rangle \xi_n, \qquad \Psi \in D(\hat{S}_\xi^{\boldsymbol{\alpha}}) \end{cases}$$

$$\begin{cases} D(\hat{S}_\zeta^{\boldsymbol{\alpha}}) = \left\{f \in \mathcal{H}_{+1}; \sum_{n=0}^{\infty} \alpha_n \overline{\langle \zeta_n \mid f \rangle} \zeta_n \text{ exists in } \mathcal{H}_{-1}\right\} \\ \hat{S}_\zeta^{\boldsymbol{\alpha}} f = \sum_{n=0}^{\infty} \alpha_n \overline{\langle \zeta_n \mid f \rangle} \zeta_n, \qquad f \in D(\hat{S}_\zeta^{\boldsymbol{\alpha}}) \end{cases}.$$

It is clear that

$$\mathcal{D}_\zeta \subset D(\hat{S}_\xi^{\boldsymbol{\alpha}}) \text{ and } \hat{S}_\xi^{\boldsymbol{\alpha}} \zeta_k = \alpha_k \xi_k, \ k \in \mathbb{N}_0 \ ; \tag{12}$$

$$\mathcal{D}_\xi \subset D(\hat{S}_\zeta^{\boldsymbol{\alpha}}) \text{ and } \hat{S}_\zeta^{\boldsymbol{\alpha}} \xi_k = \alpha_k \zeta_k, \ k \in \mathbb{N}_0 \tag{13}$$

Hence, $\hat{S}_\xi^{\boldsymbol{\alpha}}$ and $\hat{S}_\zeta^{\boldsymbol{\alpha}}$ are densely defined (in different Hilbert spaces), and the following results can be established:

Proposition 3.7. *The following statements hold.*

(1) $D(\hat{S}_\xi^{\boldsymbol{\alpha}}) = \{\Psi \in \mathcal{H}_{-1}; \sum_{n=0}^{\infty} |\alpha_n|^2 | \langle \Psi \mid \xi_n \rangle |^2 < \infty\} = D(\hat{H}_{\zeta,\xi}^{\boldsymbol{\alpha}})$,
$D(\hat{S}_\zeta^{\boldsymbol{\alpha}}) = \{f \in \mathcal{H}_{+1}; \sum_{n=0}^{\infty} |\alpha_n|^2 | \langle \zeta_n \mid f \rangle |^2 < \infty\} = D(\hat{H}_{\xi,\zeta}^{\boldsymbol{\alpha}})$.

(2) $\hat{S}_\xi^{\boldsymbol{\alpha}}$ *and* $\hat{S}_\zeta^{\boldsymbol{\alpha}}$ *are closed.*

(3) $\left(\hat{S}_\xi^{\boldsymbol{\alpha}}\right)^\dagger = \hat{S}_\xi^{\overline{\boldsymbol{\alpha}}}$ *and* $\left(\hat{S}_\zeta^{\boldsymbol{\alpha}}\right)^\dagger = \hat{S}_\zeta^{\overline{\boldsymbol{\alpha}}}$*, where* $\overline{\boldsymbol{\alpha}} = \{\overline{\alpha_n}\}$.

(4) If $\{\alpha_n\} \subset \mathbb{R}$ *(respectively,* $\{\alpha_n\} \subset \mathbb{R}^+$*) then* $\hat{S}_\xi^{\boldsymbol{\alpha}}$ *and* $\hat{S}_\zeta^{\boldsymbol{\alpha}}$ *are self-adjoint (respectively, positive self-adjoint). Furthermore,* $\hat{S}_\xi^{\boldsymbol{\alpha}}$ *is bounded if and only if* $\hat{S}_\zeta^{\boldsymbol{\alpha}}$ *is bounded and if and only if* $\boldsymbol{\alpha}$ *is a bounded sequence.*

(5) If $\boldsymbol{\alpha} = \mathbf{1}$*, then* $\hat{S}_\xi := \hat{S}_\xi^{\mathbf{1}}$ *and* $\hat{S}_\zeta := \hat{S}_\zeta^{\mathbf{1}}$ *are bounded positive self-adjoint operators respectively on* $\mathcal{H}_{-1}$ *and* $\mathcal{H}_{+1}$ *and they are inverses of each other, that is* $\hat{S}_\xi = (\hat{S}_\zeta)^{-1}$*, and* $\hat{S}_\xi = T^{-1}(T^{-1})^\dagger$*,* $\hat{S}_\zeta = T^\dagger T$*, where* $T \in C(\mathcal{H}_{+1}, \mathcal{H})$ *is the operator such that* $T\xi_n = e_n, \forall n \in \mathbb{N}_0$ *and* $\{e_n\}$ *is an orthonormal basis for* $\mathcal{H}$.

Let $\boldsymbol{\alpha} = \{\alpha_n\}$ be a sequence of complex numbers. By easy computations (see also [8]) it results that $D(\hat{S}_\zeta \hat{H}_{\xi,\zeta}^{\boldsymbol{\alpha}}) = D(\hat{H}_{\zeta,\xi}^{\boldsymbol{\alpha}} \hat{S}_\zeta) = D(\hat{S}_\zeta^{\boldsymbol{\alpha}})$, $D(\hat{S}_\xi \hat{H}_{\zeta,\xi}^{\boldsymbol{\alpha}}) = D(\hat{H}_{\xi,\zeta}^{\boldsymbol{\alpha}} \hat{S}_\xi) = D(\hat{S}_\xi^{\boldsymbol{\alpha}})$, and

$$\begin{cases} \hat{S}_\zeta \hat{H}_{\xi,\zeta}^{\boldsymbol{\alpha}} = \hat{H}_{\zeta,\xi}^{\boldsymbol{\alpha}} \hat{S}_\zeta = \hat{S}_\zeta^{\boldsymbol{\alpha}}, \\ \hat{S}_\xi \hat{H}_{\zeta,\xi}^{\boldsymbol{\alpha}} = \hat{H}_{\xi,\zeta}^{\boldsymbol{\alpha}} \hat{S}_\xi = \hat{S}_\xi^{\boldsymbol{\alpha}}. \end{cases} \tag{14}$$

Remark 3.8. *Equations like* (14) *are, again, examples of intertwining relations similar to those already introduced in (5). The main difference, here, is that they are defined on Hilbert spaces different from* $\mathcal{H}$.

3.3 Lowering and raising operators defined on a RHS

With the aim of prosecuting what done in [8], in this section we further extend the notions in Section 2.1, introducing generalized lowering and raising operators which can be used to factorize the original Hamiltonian. Consider, for an opportune f

$$\hat{A}^{\boldsymbol{\gamma}}_{\xi,\zeta} f = \sum_{n=1}^{\infty} \gamma_n (\xi_{n-1} \otimes \overline{\zeta}_n) f = \sum_{n=1}^{\infty} \gamma_n \overline{\langle \zeta_n \mid f \rangle} \xi_{n-1},$$

$$\hat{A}^{\boldsymbol{\gamma}}_{\zeta,\xi} f = \sum_{n=1}^{\infty} \gamma_n (\zeta_{n-1} \otimes \overline{\xi}_n) f = \sum_{n=1}^{\infty} \gamma_n \langle f \mid \xi_n \rangle \zeta_{n-1},$$

$$\hat{B}^{\boldsymbol{\gamma}}_{\xi,\zeta} f = \sum_{n=0}^{\infty} \gamma_{n+1} (\xi_{n+1} \otimes \overline{\zeta}_n) f = \sum_{n=0}^{\infty} \gamma_{n+1} \overline{\langle \zeta_n \mid f \rangle} \xi_{n+1},$$

$$\hat{B}^{\boldsymbol{\gamma}}_{\zeta,\xi} f = \sum_{n=0}^{\infty} \gamma_{n+1} (\zeta_{n+1} \otimes \overline{\xi}_n) f = \sum_{n=0}^{\infty} \gamma_{n+1} \langle f \mid \xi_n \rangle \zeta_{n+1}.$$

More precisely,

$$\begin{cases} D(\hat{A}^{\boldsymbol{\gamma}}_{\xi,\zeta}) = \left\{ f \in \mathcal{H}_{+1};\, \sum_{n=1}^{\infty} \gamma_n \overline{\langle \zeta_n \mid f \rangle} \xi_{n-1} \text{ exists in } \mathcal{H}_{+1} \right\} \\ \hat{A}^{\boldsymbol{\gamma}}_{\xi,\zeta} f = \sum_{n=1}^{\infty} \gamma_n \overline{\langle \zeta_n \mid f \rangle} \xi_{n-1}, \qquad f \in D(\hat{A}^{\boldsymbol{\gamma}}_{\xi,\zeta}) \end{cases}$$

$$\begin{cases} D(\hat{A}^{\boldsymbol{\gamma}}_{\zeta,\xi}) = \{ \Psi \in \mathcal{H}_{-1};\, \sum_{n=1}^{\infty} \gamma_n \langle \Psi \mid \xi_n \rangle \zeta_{n-1} \text{ exists in } \mathcal{H}_{-1} \} \\ \hat{A}^{\boldsymbol{\gamma}}_{\zeta,\xi} \Psi = \sum_{n=1}^{\infty} \gamma_n \langle \Psi \mid \xi_n \rangle \zeta_{n-1}, \qquad \Psi \in D(\hat{A}^{\boldsymbol{\gamma}}_{\zeta,\xi}) \end{cases}$$

$$\begin{cases} D(\hat{B}^{\boldsymbol{\gamma}}_{\xi,\zeta}) = \left\{ f \in \mathcal{H}_{+1};\, \sum_{n=0}^{\infty} \gamma_{n+1} \overline{\langle \zeta_n \mid f \rangle} \xi_{n+1} \text{ exists in } \mathcal{H}_{+1} \right\} \\ \hat{B}^{\boldsymbol{\gamma}}_{\xi,\zeta} f = \sum_{n=0}^{\infty} \gamma_{n+1} \overline{\langle \zeta_n \mid f \rangle} \xi_{n+1}, \qquad f \in D(\hat{B}^{\boldsymbol{\gamma}}_{\xi,\zeta}) \end{cases}$$

$$\begin{cases} D(\hat{B}^{\boldsymbol{\gamma}}_{\zeta,\xi}) = \{ \Psi \in \mathcal{H}_{-1};\, \sum_{n=0}^{\infty} \gamma_{n+1} \langle \Psi \mid \xi_n \rangle \zeta_{n+1} \text{ exists in } \mathcal{H}_{-1} \} \\ \hat{B}^{\boldsymbol{\gamma}}_{\zeta,\xi} \Psi = \sum_{n=0}^{\infty} \gamma_{n+1} \langle \Psi \mid \xi_n \rangle \zeta_{n+1}, \qquad \Psi \in D(\hat{B}^{\boldsymbol{\gamma}}_{\zeta,\xi}). \end{cases}$$

Then

Proposition 3.9. *The following statements hold.*

(1)

$$D(\hat{A}^{\boldsymbol{\gamma}}_{\xi,\zeta}) = \left\{ f \in \mathcal{H}_{+1};\, \sum_{n=1}^{\infty} |\gamma_n|^2 |\langle \zeta_n \mid f \rangle|^2 < \infty \right\} = D(\hat{H}^{\boldsymbol{\gamma}}_{\xi,\zeta}),$$

$$D(\hat{A}^{\boldsymbol{\gamma}}_{\zeta,\xi}) = \left\{ \Psi \in \mathcal{H}_{-1};\, \sum_{n=1}^{\infty} |\gamma_n|^2 |\langle \Psi \mid \xi_n \rangle|^2 < \infty \right\} = D(\hat{H}^{\boldsymbol{\gamma}}_{\zeta,\xi}),$$

$$D(\hat{B}^{\boldsymbol{\gamma}}_{\xi,\zeta}) = \left\{ f \in \mathcal{H}_{+1};\ \sum_{n=0}^{\infty} |\gamma_{n+1}|^2 |\langle \zeta_n | f \rangle|^2 < \infty \right\} = D(\hat{H}^{\{\gamma_{n+1}\}}_{\xi,\zeta}),$$

$$D(\hat{B}^{\boldsymbol{\gamma}}_{\zeta,\xi}) = \left\{ \Psi \in \mathcal{H}_{-1};\ \sum_{n=0}^{\infty} |\gamma_{n+1}|^2 |\langle \Psi | \xi_n \rangle|^2 < \infty \right\} = D(\hat{H}^{\{\gamma_{n+1}\}}_{\zeta,\xi});$$

(2) *$\hat{A}^{\boldsymbol{\gamma}}_{\xi,\zeta}$, $\hat{B}^{\boldsymbol{\gamma}}_{\xi,\zeta}$ and $\hat{A}^{\boldsymbol{\gamma}}_{\zeta,\xi}$, $\hat{B}^{\boldsymbol{\gamma}}_{\zeta,\xi}$ are densely defined closed operators respectively in $\mathcal{H}_{+1}[\|\cdot\|_{+1}]$ and $\mathcal{H}_{-1}[\|\cdot\|_{-1}]$;*

(3) *$\left(\hat{A}^{\boldsymbol{\gamma}}_{\xi,\zeta}\right)^\dagger = \hat{B}^{\overline{\boldsymbol{\gamma}}}_{\zeta,\xi}$ and $\left(\hat{A}^{\boldsymbol{\gamma}}_{\zeta,\xi}\right)^\dagger = \hat{B}^{\overline{\boldsymbol{\gamma}}}_{\xi,\zeta}$;*

(4) *$\hat{A}^{\boldsymbol{\gamma}}_{\xi,\zeta}$ and $\hat{B}^{\boldsymbol{\gamma}}_{\xi,\zeta}$ are bounded in $\mathcal{H}_{+1}$ if, and only if, $\hat{A}^{\boldsymbol{\gamma}}_{\zeta,\xi}$ and $\hat{B}^{\boldsymbol{\gamma}}_{\zeta,\xi}$ are bounded in $\mathcal{H}_{-1}$ and if, and only if, $\boldsymbol{\gamma}$ is a bounded sequence. Also, for any sequence $\boldsymbol{\gamma}$, bounded or not,*

$$\hat{A}^{\boldsymbol{\gamma}}_{\xi,\zeta}\xi_{n+1} = \gamma_{n+1}\xi_n, \qquad \hat{A}^{\boldsymbol{\gamma}}_{\zeta,\xi}\zeta_{n+1} = \gamma_{n+1}\zeta_n,\ n = 0, 1, \dots;$$
$$\hat{B}^{\boldsymbol{\gamma}}_{\xi,\zeta}\xi_n = \gamma_{n+1}\xi_{n+1}, \qquad \hat{B}^{\boldsymbol{\gamma}}_{\zeta,\xi}\zeta_n = \gamma_{n+1}\zeta_{n+1},\ n = 0, 1, \dots .$$

Proof. (1): Let $f \in \mathcal{H}_{+1}$. Since $\{\xi_n\}$ is an orthonormal basis for $\mathcal{H}_{+1}$, we have

$$\left\| \sum_{k=n}^{m} \gamma_k \overline{\langle \zeta_k | f \rangle} \xi_{k-1} \right\|^2_{+1} = \sum_{k=n}^{m} |\gamma_k|^2 |\langle \zeta_k | f \rangle|^2 \tag{15}$$

which shows that $f \in D(\hat{A}^{\boldsymbol{\gamma}}_{\xi,\zeta})$ if and only if $\sum_{n=1}^{\infty} |\gamma_n|^2 |\langle \zeta_n | f \rangle|^2 < \infty$, if and only if $f \in D(\hat{H}^{\boldsymbol{\gamma}}_{\xi,\zeta})$.

The proofs for $\hat{A}^{\boldsymbol{\gamma}}_{\zeta,\xi}$, $\hat{B}^{\boldsymbol{\gamma}}_{\xi,\zeta}$ and $\hat{B}^{\boldsymbol{\gamma}}_{\zeta,\xi}$ are similar.

(2): The operator are densely defined, indeed

$$\begin{aligned}
&\mathcal{D}_\xi := \operatorname{span}\{\xi_n\} \subset D(\hat{A}^{\boldsymbol{\gamma}}_{\xi,\zeta})\ ;\\
&\mathcal{D}_\xi := \operatorname{span}\{\xi_n\} \subset D(\hat{B}^{\boldsymbol{\gamma}}_{\xi,\zeta})\ ;\\
&\mathcal{D}_\zeta := \operatorname{span}\{\zeta_n\} \subset D(\hat{A}^{\boldsymbol{\gamma}}_{\zeta,\xi})\ ;\\
&\mathcal{D}_\zeta := \operatorname{span}\{\zeta_n\} \subset D(\hat{B}^{\boldsymbol{\gamma}}_{\zeta,\xi})\ ;\\
&\hat{A}^{\boldsymbol{\gamma}}_{\xi,\zeta}\xi_{k+1} = \gamma_{k+1}\xi_k, \quad \hat{A}^{\boldsymbol{\gamma}}_{\zeta,\xi}\zeta_{k+1} = \gamma_{k+1}\zeta_k,\ k = 0, 1, \dots ;\\
&\hat{B}^{\boldsymbol{\gamma}}_{\xi,\zeta}\xi_k = \gamma_{k+1}\xi_{k+1},\ \hat{B}^{\boldsymbol{\gamma}}_{\zeta,\xi}\zeta_k = \gamma_{k+1}\zeta_{k+1},\ k = 0, 1, \dots .
\end{aligned} \tag{16}$$

Hence, all the previous four operators are densely defined [8].

In order to prove the closedness of $\hat{A}^{\boldsymbol{\gamma}}_{\xi,\zeta}$, let $\{f_n\}$ be an arbitrary sequence in $D(\hat{A}^{\boldsymbol{\gamma}}_{\xi,\zeta})$ such that $f_n \overset{\|\cdot\|_{+1}}{\to} f$ and $\hat{A}^{\boldsymbol{\gamma}}_{\xi,\zeta} f_n \overset{\|\cdot\|_{+1}}{\to} g$. Then, for every $\epsilon > 0$, there exists $N \in \mathbb{N}$ such that

$$\left\| \hat{A}^{\boldsymbol{\gamma}}_{\xi,\zeta} f_n - \hat{A}^{\boldsymbol{\gamma}}_{\xi,\zeta} f_m \right\|_{+1} = \left\| \sum_{k=1}^{\infty} \gamma_k \overline{\langle \zeta_k | f_n - f_m \rangle} \xi_{k-1} \right\|_{+1} < \epsilon, \quad \forall n, m \geq N$$

and, for every $n, m \geq N$,

$$\sum_{k=1}^{\infty} |\gamma_k|^2 |\langle \zeta_k | f_n - f_m \rangle|^2 = \left\| \sum_{k=1}^{\infty} \gamma_k \overline{\langle \zeta_k | f_n - f_m \rangle} e_{k-1} \right\|^2$$

$$= \left\| \sum_{k=1}^{\infty} \gamma_k \overline{\langle \zeta_k | f_n - f_m \rangle} T\xi_{k-1} \right\|^2$$

$$= \left\| \sum_{k=1}^{\infty} \gamma_k \overline{\langle \zeta_k | f_n - f_m \rangle} \xi_{k-1} \right\|_{+1}^2 < \epsilon^2.$$

Hence, for every $M \in \mathbb{N}$ and $n, m \geq N$,

$$\sum_{k=1}^{M} |\gamma_k|^2 |\langle \zeta_k | f_n - f_m \rangle|^2 < \epsilon^2.$$

Since the Hilbert norm $\|\cdot\|_{+1}$ is stronger than $\|\cdot\|$ in $\mathcal{H}$, letting $m \to \infty$, we get

$$\sum_{k=1}^{M} |\gamma_k|^2 |\langle \zeta_k | f_n - f \rangle|^2 \leq \epsilon^2, \quad \forall n \geq N. \tag{17}$$

Therefore

$$\sum_{k=1}^{M} |\gamma_k|^2 |\langle \zeta_k | f \rangle|^2 \leq 2 \sum_{k=1}^{M} |\gamma_k|^2 |\langle \zeta_k | f - f_N \rangle|^2 + 2 \sum_{k=1}^{M} |\gamma_k|^2 |\langle \zeta_k | f_N \rangle|^2$$

$$\leq 2\epsilon^2 + 2 \sum_{k=1}^{M} |\gamma_k|^2 |\langle \zeta_k | f_N \rangle|^2$$

and, letting $M \to \infty$,

$$\sum_{k=1}^{\infty} |\gamma_k|^2 |\langle \zeta_k | f \rangle|^2 \leq 2\epsilon^2 + 2 \sum_{k=1}^{\infty} |\gamma_k|^2 |\langle \zeta_k | f_N \rangle|^2 < \infty.$$

This implies that $f \in D(\hat{A}^{\boldsymbol{\gamma}}_{\xi,\zeta})$.

Moreover, by (17), we get

$$\|\hat{A}^{\boldsymbol{\gamma}}_{\xi,\zeta} f_n - \hat{A}^{\boldsymbol{\gamma}}_{\xi,\zeta} f\|_{+1} = \sum_{k=1}^{\infty} |\gamma_k|^2 |\langle \zeta_k | f_n - f \rangle|^2 \leq \epsilon^2, \quad \forall n \geq N.$$

Hence

$$\hat{A}^{\boldsymbol{\gamma}}_{\xi,\zeta} f_n \overset{\|\cdot\|_{+1}}{\to} \hat{A}^{\boldsymbol{\gamma}}_{\xi,\zeta} f.$$

Thus, $\hat{A}^{\boldsymbol{\gamma}}_{\xi,\zeta}$ is closed.

(3): Let first show that $\hat{B}^{\overline{\boldsymbol{\gamma}}}_{\zeta,\xi} \subset (\hat{A}^{\boldsymbol{\gamma}}_{\xi,\zeta})^\dagger$ and $\hat{B}^{\overline{\boldsymbol{\gamma}}}_{\xi,\zeta} \subset (\hat{A}^{\boldsymbol{\gamma}}_{\zeta,\xi})^\dagger$. In order to prove the former inclusion we have to show that $D(\hat{B}^{\overline{\boldsymbol{\gamma}}}_{\zeta,\xi}) \subset D((\hat{A}^{\boldsymbol{\gamma}}_{\xi,\zeta})^\dagger)$ with

$$D((\hat{A}^{\boldsymbol{\gamma}}_{\xi,\zeta})^\dagger) = \{\Psi \in \mathcal{H}_{-1} | \, \exists \Phi \in \mathcal{H}_{-1} \text{ such that } \langle \Phi | f\rangle = \left\langle \Psi | \, \hat{A}^{\boldsymbol{\gamma}}_{\xi,\zeta} f \right\rangle, \quad \forall f \in D(\hat{A}^{\boldsymbol{\gamma}}_{\xi,\zeta})\Big\},$$

(and in this case $(\hat{A}^{\boldsymbol{\gamma}}_{\xi,\zeta})^\dagger \Psi = \Phi$) and $\hat{B}^{\overline{\boldsymbol{\gamma}}}_{\zeta,\xi}\Psi = (\hat{A}^{\boldsymbol{\gamma}}_{\xi,\zeta})^\dagger\Psi$. Indeed, let $\Psi \in D(\hat{B}^{\overline{\boldsymbol{\gamma}}}_{\xi,\zeta})$

$$\begin{aligned}
\left\langle \hat{B}^{\overline{\boldsymbol{\gamma}}}_{\zeta,\xi}\Psi \,|\, f \right\rangle &= \left\langle \sum_{k=0}^{\infty} \overline{\gamma_{k+1}} \langle \Psi | \, \xi_k\rangle \, \zeta_{k+1} \,|\, f \right\rangle \\
&= \sum_{k=0}^{\infty} \langle \Psi | \, \xi_k\rangle \, \overline{\gamma_{k+1}} \, \langle \zeta_{k+1} \,|\, f\rangle \\
&= \sum_{k=0}^{\infty} \langle \Psi | \, \xi_k\rangle \left(\sum_{n=1}^{\infty} \overline{\gamma_n} \, \langle \zeta_n \,|\, f\rangle \, \delta_{k,n-1} \right) \\
&= \sum_{k=0}^{\infty} \langle \Psi | \, \xi_k\rangle \left(\sum_{n=1}^{\infty} \overline{\gamma_n} \, \langle \zeta_n \,|\, f\rangle \, \langle \zeta_k \,|\, \xi_{n-1}\rangle \right) \\
&= \left\langle \sum_{k=0}^{\infty} \langle \Psi | \, \xi_k\rangle \, \zeta_k \,|\, \sum_{n=1}^{\infty} \gamma_n \overline{\langle \zeta_n \,|\, f\rangle} \xi_{n-1} \right\rangle = \left\langle \Psi | \, \hat{A}^{\boldsymbol{\gamma}}_{\xi,\zeta} f \right\rangle,
\end{aligned}$$

hence, $\Psi \in D((\hat{A}^{\boldsymbol{\gamma}}_{\xi,\zeta})^\dagger)$, and $\hat{B}^{\overline{\boldsymbol{\gamma}}}_{\zeta,\xi}\Psi = (\hat{A}^{\boldsymbol{\gamma}}_{\xi,\zeta})^\dagger\Psi$, too. In an analogous way one can prove that $\hat{B}^{\overline{\boldsymbol{\gamma}}}_{\xi,\zeta} \subset (\hat{A}^{\boldsymbol{\gamma}}_{\zeta,\xi})^\dagger$.

Now it remains to show that $(\hat{A}^{\boldsymbol{\gamma}}_{\xi,\zeta})^\dagger \subset \hat{B}^{\overline{\boldsymbol{\gamma}}}_{\zeta,\xi}$ and $(\hat{A}^{\boldsymbol{\gamma}}_{\zeta,\xi})^\dagger \subset \hat{B}^{\overline{\boldsymbol{\gamma}}}_{\xi,\zeta}$. Let $\Psi \in D((\hat{A}^{\boldsymbol{\gamma}}_{\xi,\zeta})^\dagger)$; then there exists $\Phi \in \mathcal{H}_{-1}$ such that

$$\left\langle \Psi | \, \sum_{n=1}^{\infty} \gamma_n \overline{\langle \zeta_n \,|\, f\rangle} \xi_{n-1} \right\rangle = \langle \Phi \,|\, f\rangle, \quad \forall f \in D(\hat{A}^{\boldsymbol{\gamma}}_{\xi,\zeta}).$$

By (4) and (3), $\mathcal{D}_\xi \subseteq D(\hat{A}^{\boldsymbol{\gamma}}_{\xi,\zeta})$ and $\hat{A}^{\boldsymbol{\gamma}}_{\xi,\zeta}\xi_k = \gamma_k \xi_{k-1}$, $k = 1, 2, \dots$. Thus, $\langle \Psi | \, \gamma_k \xi_{k-1}\rangle = \langle \Phi | \, \xi_k\rangle$, $k = 1, 2, \dots$. Hence

$$\sum_{j=0}^{\infty} |\gamma_{j+1}|^2 |\langle \Psi | \, \xi_j\rangle|^2 = \sum_{k=1}^{\infty} |\gamma_k|^2 |\langle \Psi | \, \xi_{k-1}\rangle|^2 = \sum_{k=1}^{\infty} |\langle \Phi | \, \xi_k\rangle|^2 =$$

$$= \sum_{k=1}^{\infty} \left|\left\langle (T^{-1})^\dagger \Phi | \, e_k \right\rangle\right|^2 = \|(T^{-1})^\dagger \Phi\|^2.$$

This implies that $\Psi \in D(\hat{B}^{\overline{\boldsymbol{\gamma}}}_{\zeta,\xi})$. The remaining inclusion is proved in a similar way.

(4): Let $\boldsymbol{\gamma}$ a bounded sequence, then there exist $M > 0$ such that

$$\|\hat{A}^{\boldsymbol{\gamma}}_{\xi,\zeta} f\|_{+1} = \left\| \sum_{k=1}^{\infty} \gamma_k \overline{\langle \zeta_k \,|\, f\rangle} \xi_{k-1} \right\|_{+1} \leq M \|f\|_{+1};$$

indeed,

$$\begin{aligned}\left\|\sum_{k=1}^{\infty}\overline{\langle\zeta_k\,|\,f\rangle}\xi_{k-1}\right\|_{+1} &= \sum_{k=1}^{\infty}|y_k|^2|\langle\zeta_k\,|\,f\rangle|^2\\ &\le |y_0|^2|\langle\zeta_0\,|\,f\rangle|^2+\sum_{k=1}^{\infty}|y_k|^2|\langle\zeta_k\,|\,f\rangle|^2\\ &= \left\|\sum_{k=0}^{\infty}\overline{\langle\zeta_k\,|\,f\rangle}\xi_k\right\|_{+1} = \|f\|_{+1}.\end{aligned}$$

then $\hat{A}^{\mathbf{y}}_{\xi,\zeta}$ is bounded in $\mathcal{H}_{+1}$.

By an analogous argument one can prove our remaining claim. This completes the proof. □

As in Section 2.1, we now focus our attention on some products of these operators. Again, to simplify the notation, and without any loss of generality, we will assume here that $y_0 = 0$.

Proposition 3.10. *The following statements hold.*

(1)

$$D(\hat{A}^{\mathbf{y}}_{\xi,\zeta}\hat{B}^{\mathbf{y}}_{\xi,\zeta}) = \left\{f\in\mathcal{H}_{+1};\ \sum_{n=0}^{\infty}|y_{n+1}|^2|\langle\zeta_n\,|\,f\rangle|^2<\infty \text{ and } \sum_{n=0}^{\infty}|y_{n+1}|^4|\langle\zeta_n\,|\,f\rangle|^2<\infty\right\}$$

and

$\hat{A}^{\mathbf{y}}_{\xi,\zeta}\hat{B}^{\mathbf{y}}_{\xi,\zeta}\subset\sum_{n=0}^{\infty}y_{n+1}^2\xi_n\otimes\overline{\zeta}_n=\hat{H}^{\{y_{n+1}^2\}}_{\xi,\zeta}$.

(2)

$$D(\hat{B}^{\mathbf{y}}_{\xi,\zeta}\hat{A}^{\mathbf{y}}_{\xi,\zeta}) = \left\{f\in\mathcal{H}_{+1};\ \sum_{n=1}^{\infty}|y_{n}|^2|\langle\zeta_n\,|\,f\rangle|^2<\infty \text{ and } \sum_{n=0}^{\infty}|y_{n}|^4|\langle\zeta_n\,|\,f\rangle|^2<\infty\right\}$$

and

$\hat{B}^{\mathbf{y}}_{\xi,\zeta}\hat{A}^{\mathbf{y}}_{\xi,\zeta}\subset\sum_{n=0}^{\infty}y_{n}^2\xi_n\otimes\overline{\zeta}_n=\hat{H}^{\{y_{n}^2\}}_{\xi,\zeta}$.

Suppose that $0\equiv|y_0|\le|y_1|\le|y_2|\le\cdots$. *Then*

(3) $D(\hat{B}^{\mathbf{y}}_{\xi,\zeta}\hat{A}^{\mathbf{y}}_{\xi,\zeta})=\{f\in\mathcal{H}_{+1};\sum_{n=0}^{\infty}|y_n|^4|\langle\zeta_n\,|\,f\rangle|^2<\infty\}=D(\hat{H}^{\{y_n^2\}}_{\xi,\zeta})$;
$D(\hat{A}^{\mathbf{y}}_{\xi,\zeta}\hat{B}^{\mathbf{y}}_{\xi,\zeta})=\{f\in\mathcal{H}_{+1};\ \sum_{n=0}^{\infty}|y_{n+1}|^4|\langle\zeta_n\,|\,f\rangle|^2<\infty\}=D(\hat{H}^{\{y_{n+1}^2\}}_{\xi,\zeta})$;
$\hat{B}^{\mathbf{y}}_{\xi,\zeta}\hat{A}^{\mathbf{y}}_{\xi,\zeta}\subset\hat{A}^{\mathbf{y}}_{\xi,\zeta}\hat{B}^{\mathbf{y}}_{\xi,\zeta}$;

$\hat{B}^{\boldsymbol{\gamma}}_{\xi,\zeta}\hat{A}^{\boldsymbol{\gamma}}_{\xi,\zeta} = \sum_{n=1}^{\infty}\gamma_n^2\xi_n\otimes\overline{\zeta}_n = \hat{H}^{\{\gamma_n^2\}}_{\xi,\zeta}$;
$\hat{A}^{\boldsymbol{\gamma}}_{\xi,\zeta}\hat{B}^{\boldsymbol{\gamma}}_{\xi,\zeta} = \sum_{n=0}^{\infty}\gamma_{n+1}^2\xi_n\otimes\overline{\zeta}_n = \hat{H}^{\{\gamma_{n+1}^2\}}_{\xi,\zeta}$.

(4) $\hat{A}^{\boldsymbol{\gamma}}_{\xi,\zeta}\hat{B}^{\boldsymbol{\gamma}}_{\xi,\zeta}f - \hat{B}^{\boldsymbol{\gamma}}_{\xi,\zeta}\hat{A}^{\boldsymbol{\gamma}}_{\xi,\zeta}f = \hat{H}^{\{\gamma_{n+1}^2\}}_{\xi,\zeta}f - \hat{H}^{\{\gamma_n^2\}}_{\xi,\zeta}f = \left(\sum_{n=0}^{\infty}(\gamma_{n+1}^2-\gamma_n^2)\xi_n\otimes\overline{\zeta}_n\right)f$, *for every* $f \in D(\hat{H}^{\{\gamma_{n+1}^2\}}_{\xi,\zeta})$.

Proof. We put for shortness $A := \hat{A}^{\boldsymbol{\gamma}}_{\xi,\zeta}$ and $B := \hat{B}^{\boldsymbol{\gamma}}_{\xi,\zeta}$.
(1): By (1) and (2) of Proposition 3.9 we have

$$\begin{aligned} D(AB) &= \left\{f \in \mathcal{H}_{+1};\ \sum_{n=0}^{\infty}|\gamma_{n+1}|^2|\langle\zeta_n \mid f\rangle|^2 < \infty \text{ and } \sum_{n=1}^{\infty}|\gamma_n|^2|\langle\zeta_n \mid Bf\rangle|^2 < \infty\right\} \\ &= \left\{f \in \mathcal{H}_{+1};\ \sum_{n=0}^{\infty}|\gamma_{n+1}|^2|\langle\zeta_n \mid f\rangle|^2 < \infty \text{ and } \sum_{n=0}^{\infty}|\gamma_{n+1}|^4|\langle\zeta_n \mid f\rangle|^2 < \infty\right\} \\ &\subset \left\{f \in \mathcal{H}_{+1};\ \sum_{n=0}^{\infty}|\gamma_{n+1}|^4|\langle\zeta_n \mid f\rangle|^2 < \infty\right\} = D\left(\sum_{n=0}^{\infty}\gamma_{n+1}^2\xi_n\otimes\overline{\zeta}_n\right) \\ &= D(\hat{H}^{\{\gamma_{n+1}^2\}}_{\xi,\zeta}) \end{aligned}$$

and

$$ABf = \left(\sum_{n=0}^{\infty}\gamma_{n+1}^2\xi_n\otimes\overline{\zeta}_n\right)f, \quad \forall f \in D(AB).$$

Hence,

$$AB \subset \sum_{n=0}^{\infty}\gamma_{n+1}^2\xi_n\otimes\overline{\zeta}_n.$$

(2): The proof is similar to that of (1) and we omit it.
(3): If $\boldsymbol{\gamma} = \{\gamma_n\}$ is bounded then both A and B are bounded, and so (3) (and (4) too) holds. As for the general case, we put $N_1 = \max\{n \in \mathbb{N}_0; |\gamma_n| \le 1\}$.

Let us suppose that $f \in D(\hat{H}^{\{\gamma_n^2\}}_{\xi,\zeta})$, i.e. $\sum_{n=0}^{\infty}|\gamma_n|^4|\langle\zeta_n \mid f\rangle|^2 < \infty$. Then we have

$$\sum_{n=1}^{\infty}|\gamma_n|^2|\langle\zeta_n \mid f\rangle|^2 \le \sum_{n=1}^{N_1}|\gamma_n|^2|\langle\zeta_n \mid f\rangle|^2 + \sum_{n=N_1+1}^{\infty}|\gamma_n|^4|\langle\zeta_n \mid f\rangle|^2 < \infty.$$

This implies that

$$D(BA) = \left\{f \in \mathcal{H}_{+1};\ \sum_{n=0}^{\infty}|\gamma_n|^4|\langle\zeta_n \mid f\rangle|^2 < \infty\right\} = D(\hat{H}^{\{\gamma_n^2\}}_{\xi,\zeta})$$

and

$$BA = \sum_{n=1}^{\infty}\gamma_n^2\xi_n\otimes\overline{\zeta}_n.$$

In similar way, we have

$$D(AB) = \left\{f \in \mathcal{H}_{+1};\ \sum_{n=0}^{\infty}|\gamma_{n+1}|^4|\langle\zeta_n \mid f\rangle|^2 < \infty\right\}$$

and

$$AB = \sum_{n=0}^{\infty} \gamma_{n+1}^2 \xi_n \otimes \overline{\zeta}_n.$$

Clearly, $D(AB) \subset D(BA)$.

(4): This follows easily from (3). □

It is clear that the lowering and raising operators considered above constitute a further generalization of those considered in Section 2.1 and, as before, they can be used to factorize the original Hamiltonian. In particular, if the sequence $\boldsymbol{\alpha} = \{\alpha_n\}$ is such that $0 = \alpha_0 < \alpha_1 < \alpha_2 < \ldots$, and if we take $\gamma_n = \sqrt{\alpha_n}$ here, the sequence $\boldsymbol{\gamma}$ satisfies the assumptions above, and we find, for instance that $\hat{B}^{\boldsymbol{\gamma}}_{\xi,\zeta}\hat{A}^{\boldsymbol{\gamma}}_{\xi,\zeta} = \sum_{n=0}^{\infty} \alpha_n \xi_n \otimes \overline{\zeta}_n = \hat{H}^{\boldsymbol{\alpha}}_{\xi,\zeta}$. Then $\hat{H}^{\boldsymbol{\alpha}}_{\xi,\zeta}$ can be factorized.

We end this section with the following example.

Example 3.11. *Let us consider the triplet of Sobolev spaces $W^{1,2}(\mathbb{R}) \subset \mathcal{L}^2(\mathbb{R}) \subset W^{-1,2}(\mathbb{R})$ and the set $\{e_n(x)\}$ of Hermite functions, which is, as it is well known, an orthonormal basis for $\mathcal{L}^2(\mathbb{R})$. Let us denote by T the multiplicative operator $T := 1 + x^2$, with $T : W^{1,2}(\mathbb{R}) \to \mathcal{L}^2(\mathbb{R})$, then it is continuous w.r. to the graph topology. Furthermore, T is invertible with bounded inverse in $\mathcal{H} = \mathcal{L}^2(\mathbb{R})$, $T^{-1} := \frac{1}{1+x^2}$. Consider the sequences $\{\xi_n\}$, with $\xi_n = T^{-1}e_n$ and $\{\zeta_n\}$, with $\zeta_n = T^{\dagger}e_n$, for every $n \in \mathbb{N}_0$. Then $\{\xi_n\}$ is a basis for $W^{1,2}(\mathbb{R})$. Indeed, for every $f \in D(T)$, we want to prove that there exists, uniquely determined, $\{c_n\} \subset \mathbb{C}$ such that $f = \sum_{n=0}^{\infty} c_n\xi_n$ in $\mathcal{H}_{+1}[\|\cdot\|_{+1}] = W^{1,2}(\mathbb{R})[\|T\cdot\|_2]$, i.e. $\forall\varepsilon > 0$ there exists $\overline{n} \in \mathbb{N}$, such that $\forall n \in \mathbb{N}$ with $n \geq \overline{n}$, it results*

$$\left\| f - \sum_{k=0}^{n} c_k\xi_k \right\|_{+1} < \varepsilon.$$

Now,

$$\begin{aligned} \left\| f - \sum_{k=0}^{n} c_k\xi_k \right\|_{+1} &= \left\| T\left(f - \sum_{k=0}^{n} c_k\xi_k \right) \right\|_2 \\ &= \left\| (1+x^2)\left(f - \sum_{k=0}^{n} c_k(1+x^2)^{-1}e_k \right) \right\|_2 \\ &= \left\| (1+x^2)f - \sum_{k=0}^{n} c_k e_k \right\|_2 < \varepsilon, \end{aligned}$$

from which we deduce that $c_n(f) = \langle (1+x^2)f \mid e_n\rangle = \langle f \mid (1+x^2)^{\dagger}e_n\rangle = \overline{\langle \zeta_n \mid f\rangle}$, for every $n \in \mathbb{N}_0$ and they are obviously continuous. Hence $\{\xi_n\}$ is a strict generalized Riesz basis for $W^{1,2}(\mathbb{R})$, and $\{\zeta_n\}$ is its dual basis.

Now, following what done in Section 3.2, we can define $\hat{H}^{\boldsymbol{\alpha}}_{\xi,\zeta} = \sum_{n=0}^{\infty} \alpha_n\xi_n \otimes \overline{\zeta_n}$ and we can see, using the above relations between ξ_n, ζ_n and e_n, that

$$\hat{H}^{\boldsymbol{\alpha}}_{\xi,\zeta} = T^{-1}\hat{H}^{\boldsymbol{\alpha}}_{e,e}T,$$

with $\hat{H}^{\boldsymbol{\alpha}}_{e,e} = \sum_{n=0}^{\infty} \alpha_n e_n \otimes \overline{e_n}$.

Remark 3.12. *What happens in the Example 3.11, formally, holds always true, by the same definitions of the operators involved. In other words, we always can formally write*

$$\hat{H}^{\boldsymbol{\alpha}}_{\xi,\zeta} = T^{-1}\hat{H}^{\boldsymbol{\alpha}}_{e,e}T,$$

with $\mathcal{E} = \{e_n\}$ *an orthonormal basis of the Hilbert space* $\mathcal{H}$*, and* T *is the operator used in the definition of the generalized Riesz bases.*

4 Conclusions

We have discussed several recent and new results on non self-adjoint Hamiltonians defined, in the way it is discussed all along the article, by means of Riesz bases or by strict generalized Riesz bases. We have also discussed the possibility of factorizing these Hamiltonians in terms of (generalized) lowering and raising operators.

There are many questions that can be posed in this framework. The most interesting (and difficult) is very likely the following: under what conditions can a closed operator H be constructed from a Riesz basis in the way we did in Section 2? We guess there is no general answer to this problem as well as there is no general answer for the corresponding question in the case of self-adjoint operators, where only certain classes of them are known to have a discrete simple spectrum.

A further interesting open question concerns the possibility of repeating a similar procedure as that described, for instance, by formulas (1), (3) etc, when the sets $\mathcal{F}_\varphi$ and $\mathcal{F}_\Psi$ are not Riesz bases but just bases or, even less, $\mathcal{D}$-quasi bases in the sense of [2]; in other words, is it possible to reconstruct an operator H from these bases in the same spirit of what we did here for Riesz bases? Work on this matter is in progress and we hope to discuss this case in detail in a future paper.

Acknowledgment: This work has been supported by the Università degli Studi di Palermo and by the Gruppo Nazionale per la Fisica Matematica (GNFM) and the Gruppo Nazionale per l'Analisi Matematica, la Probabilità e le loro Applicazioni (GNAMPA) of the Istituto Nazionale di Alta Matematica (INdAM). Both authors are grateful to Prof. Camillo Trapani for so many useful suggestions.

Bibliography

[1] Antoine, J.-P., and Trapani, C. *Metric operators, generalized hermiticity and lattices of hilbert spaces*, in *Non-selfadjoint operators in quantum physics*, F. Bagarello, J.P. Gazeau, F.H. Szafraniek and M. Znoijl Eds., John Wiley and Sons, to appear in 2015.

[2] Bagarello, F. *From self to non self-adjoint harmonic oscillators: physical consequences and mathematical pitfalls*, Phys. Rev. A, 88, 032120, (2013).

[3] Bagarello, F. *Non-selfadjoint operators in quantum physics: Mathematical aspects*, in *Non-selfadjoint operators in quantum physics*, F. Bagarello, J.P. Gazeau, F.H. Szafraniek and M. Znoijl Eds., John Wiley and Sons, to appear in 2015.

[4] F. Bagarello, J.P. Gazeau, F.H. Szafraniek and M. Znoijl Eds. *Non-selfadjoint operators in quantum physics*, John Wiley and Sons, to appear in 2015.

[5] Bagarello, F., Inoue, A., Trapani, C. *Non-self-adjoint hamiltonians defined by Riesz bases*, J. Math. Phys., 55, 033501, (2014).

[6] Bagarello, F., Inoue, A., Trapani, C. *Weak commutation relations of unbounded operators: nonlinear extensions*, J. Math. Phys., 53, 123510, (2012).

[7] Bellomonte, G., Di Bella, S., Trapani, C. *Operators in rigged Hilbert spaces: Some spectral properties*, J. Math. Anal. Appl., **411** (2014) 931–946.

[8] Bellomonte, G., Trapani, C. *Generalized Riesz bases in Rigged Hilbert Spaces*, submitted.

[9] Bender, C.M., Boettcher, S. *Real Spectra in Non-Hermitian Hamiltonians Having PT-Symmetry*, Phys. Rev. Lett. 80, 5243, (1998).

[10] Christensen, O. *An Introduction to Frames and Riesz Bases*, Birkhäuser, Boston, (2003).

[11] Cooper, F., Khare, A., and Sukhatme, U. *Supersimmetry and quantum mechanics*, World Scientific, Singapore (2001).

[12] Davies, E.B., Kuijlaars, B.J. *Spectral asymptotics of the non self-adjoint harmonic oscillator*, J. London Math. Soc., 70 (2004) 420–426.

[13] Dorey, P., Dunning, C., Tateo, R. *Spectral equivalences, Bethe ansatz equations, and reality properties in PT-symmetric quantum mechanics*, J. Phys. A, 34, 5679, (2001).

[14] Heil, C. *A basis theory primer: expanded edition*, Springer, New York, (2010)

[15] Kuru, S., Demircioglu, B., Onder, M., Vercin, A. *Two families of superintegrable and isospectral potentials in two dimensions*, J. Math. Phys, 43, (2002) 2133–2150.

[16] Kuru, S., Tegmen, A., Vercin, A. *Intertwined isospectral potentials in an arbitrary dimension*, J. Math. Phys, 42 (2001) 3344–3360.

[17] Messiah, A., *Quantum mechanics*, vol. 2, North Holland Publishing Company, Amsterdam, 1962.

[18] Samani, K.A., Zarei, M. *Intertwined hamiltonians in two-dimensional curved spaces*, Ann. of Phys., 316 (2005) 466–482.

[19] Siegl, P., Krejčiřík, D. *On the metric operator for the imaginary cubic oscillator*, Phys. Rev. D, 86,121702(R), (2012).

Vladimir I. Chilin and Sofya S. Masharipova

Functional calculus on algebras of operators generated by a self-adjoint operator in Pontryagin space Π_1

Abstract: We construct a functional calculus on the algebras generated by self-adjoint operator in Pontryagin space Π_1. Different classes of such algebras by Shulman's classification are considered.

Keywords: Operator algebras on Pontryagin space Π_1, operator algebras on spaces with indefinite metrics, functional calculus, linear operators on spaces with indefinite metrics

MSC: 47B50

Introduction

The work is devoted to construction of a functional calculus for the algebras generated by self-adjoint operator in Pontryagin space Π_1.

The infinity-dimensional spaces with indefinite metrics as new object were considered in works of P. Dirac [4] and W. Pauli [15] in the quantum field theory, but first description of such spaces was given by L. Pontryagin [16]. Further, the theory of spaces with indefinite metrics and operators on it was developed by M. Krein and I. Iokhvidov [5],[6] and P. Nevanlinna [14]. Reader can refer the development of geometry of the spaces with indefinite metrics and of the spectral theory of operators in such spaces in [8], [1] and [1].

Operator algebras in such spaces were considered in works of M. Naimark [13] and P. Ismagilov [7]. Later, in [3] and [19], [20], the commutative symmetrical algebras of operators on Pontryagin space Π_1 were studied.

This works is continuing the study of the commutative symmetrical algebras of operators on Pontryagin space Π_1. We are using the technics of Shulman's models and a classification, given by V. Shulman in [18], [18] (see also a fundamental monograph [9]). We are constructing a functional calculus for commutative symmetrical algebras of operators on Π_1 of classes 0, $2a$, and $3a$ (in classification of Shulman).

Vladimir I. Chilin, National University of Uzbekistan, Department of Mathematics, Tashkent 700095, Uzbekistan; Email: chilin@ucd.uz
Sofya S. Masharipova, Department of Mathematics, University of Pittsburgh - Johnstown, 450 Schoolhouse Road, Johnstown, PA 15904, U.S.A.; Email: ssm57@pitt.edu

https://doi.org/10.1515/9783110413557-004

The author is planning to devote a different work for construction of a functional calculus for commutative symmetrical algebras of operators on Π_1 in the complicated case of the algebra of the type 1.

1 Preliminaries

Let $(\mathfrak{H}, [\cdot\,\cdot])$ be a Hilbert space, P_+ - an orthogonal projector on $\mathfrak{H}$ and $P_- = P_+^\perp$. Let us consider an operator $J = P_+ - P_-$. Then $J^* = J = J^{-1} \in B(\mathfrak{H})$, where $B(\mathfrak{H})$ is an algebra of all bounded operators on H. Let us define a new bilinear form $(\cdot, \cdot)$ on H:

$$(\xi, \eta) = [J\xi, \eta]$$

for $\xi, \eta \in \mathfrak{H}$.

Definition 1.1. A triple $(\mathfrak{H}, [\cdot, \cdot], (\cdot, \cdot))$ is called a Krein space.

Let $(H, [\cdot, \cdot], (\cdot, \cdot))$ be a Krein space. It is easy to see that $(\xi, \eta) = \overline{(\eta, \xi)}$ for any $\xi, \eta \in \mathfrak{H}$, but $(\cdot, \cdot)$ is not an inner product on H, because there exists such a vector (actually, many of its) $\eta \in H$ that $(\eta, \eta) \leq 0$.

Definition 1.2. A vector $\xi \in H$ is called *positive* (corresp. *negative* or *neutral*), if $(\xi, \xi) > 0$ (corresp. $(\xi, \xi) < 0$ or $(\xi, \xi) = 0$).

Any subspace $L \subset H$ is called positive (corresp. non-negative, neutral, non-positive, negative) if for any non-zero vector $\xi \in L$ we have: $(\xi, \xi) > 0$ (corresp. $(\xi, \xi) \geq 0$, $(\xi, \xi) = 0$, $(\xi, \xi) \leq 0$ or $(\xi, \xi) < 0$).

Definition 1.3. The vectors $\xi, \xi \in \mathfrak{H}$ we will call *J-orthogonal*, if $(\xi, \xi) = 0$.

Using this definition, we can define now J-orthogonality of subsets of $\mathfrak{H}$ and J-orthogonal complement of $L^\perp$ of subset $L \subset \mathfrak{H}$.

Let us define $H_+ = P_+\mathfrak{H}$ and $N_- = P_-\mathfrak{H}$. Then for any non-zero vectors $\xi_+ \in H_+$ and $\xi_- \in H_-$ we have: $(\xi_+, \xi_+) > 0$ and $(\xi_-, \xi_-) < 0$. So, then H_+ is a positive subspace in $\mathfrak{H}$ and H_- is a negative subspace in $\mathfrak{H}$, by Definition 1.2. Then H is a direct J-orthogonal sum of H_+ and H_-: $H = H_+ \oplus H_-$.

For simplicity, we will denote now a space $(\mathfrak{H}, [\cdot, \cdot], (\cdot, \cdot))$ as $(H, (\cdot, \cdot))$ or simply $\mathfrak{H}$.

Definition 1.4.

(1) A subspace $L \subset \mathfrak{H}$ is called *a non-degenerated subspace*, if $L \cap L^+ = \{0\}$.

(2) If $L \cap L^+ \neq \{0\}$, then L is called *a degenerated subspace*.

For any linear bounded operator $A \in B(\mathfrak{H})$ there exists only one linear bounded operator $B \in B(\mathfrak{H})$ such that $(A\xi, \eta) = (\xi, B\eta)$ for any $\xi, \eta \in \mathfrak{H}$ (see [1]). The operator B is called a J-adjoint operator for A, and we will use a symbol A^* for B. It is easy to check that the triple $(B(\mathfrak{H}), ||\cdot||, *)$ is a B^*-algebra, and then the operation $* : B(\mathfrak{H}) \to B(\mathfrak{H})$ is an involution on $B(\mathfrak{H})$, and $||A|| = ||A^*||$ for any operator $A \in B(\mathfrak{H})$. But, generally speaking, the equality $||AA^*|| = ||A||^2$ is true not for all $A \in B(\mathfrak{H})$, and therefore the B^*-algebra $(B(\mathfrak{H}), ||\cdot||, *)$ is not C^*-algebra (see [12]).

An algebra $R \subset B(\mathfrak{H})$ is called a symmetric algebra, or J-algebra, if for any $A \in R$ we have: $A^* \in R$.

Definition 1.5. The algebras $R_1 \subset B(\mathfrak{H}_1)$ and $R_2 \subset B(\mathfrak{H}_2)$ are called *the equivalent algebras*, if there exists an isometric isomorphism $V : \mathfrak{H}_1 \to \mathfrak{H}_2$ such that $V(R_1) = R_2$.

If $H = H_+ \oplus H_-$, where $dimH_+ = k$, then H is called a *Pontryagin* space of type Π_k.

In this article we study the symmetric algebras of bounded operators in space Π_1 (which means $dimH_+ = 1$).

Two neutral vectors $\xi, \eta in \Pi_k$ are called *skew-connected*, if $(\xi, \eta) \neq 0$. If $(\xi, \eta) = 1$, then the vectors ξ and η are called the *bi-orthogonal* vectors. It is well known that for any two skew-connected vectors ξ, η the subspace $span(\xi, \eta) \subset H$, generated by ξ and η is a Pontryagin space Π_1.

Theorem 1.6. (see [5]) For any closed subspace $L \subset H$, where H is a Pontryagin space Π_k, the following are equivalent:
(i) L is non-degenerated;
(ii) $H = L \oplus L^+$.

Let us consider an algebra $U \subset B(H)$. A linear mapping $p : U \to H$ is called a *quasi-vector of algebra U*, if $p(AB) = Ap(B)$ for any $A, B \in U$. The set of all quasi-vectors of the algebra U is a linear space. We denote it as $k(U)$. A simple example of quasi-vector is $p(A) = A\chi$, where χ is a fixed vector in H. This type of quasi-vectors is called *space quasi-vector*.For any quasi-vector $p \in k(U)$ we can define a mapping $\hat{p} : U \to H \oplus H$ acting as $\hat{p}(A) = p(A) \oplus p(A^*)$, where $A \in U$.We consider now the symmetric algebras of bounded operators in Pontryagin space Π_1. We can divide the set of all such algebras into two subsets. The first subset will contain so called non-degenerated algebras, which means that those algebras have only non-degenerated invariant subspaces. Really, if S is non-degenerated algebra then there exists such a decomposition $\mathfrak{H} = \mathfrak{H}_+ \oplus \mathfrak{H}_-$, generating an inner product $[.,.]$ in $\mathfrak{H}$, that S is a C^*-algebra with respect to the involution defined by this inner product $[.,.]$.

The second subset will contain only not non-degenerated algebras. We can outline now a classification of such algebras, given by V. Shulman in [18].

For any neutral subspace $L \subset \mathfrak{H}$, let us define:

$$I_L = \{\tilde{A} \in B(\mathfrak{H}) : \ \tilde{A}L \subset L, \tilde{A}^*L \subset L\},$$

$$\Phi_L = \{\tilde{A} \in B(\mathfrak{H}) : \ \tilde{A}^\perp \subset L\}.$$

Since $dimL = 1$, then it is an eigen-subspace for any $\tilde{A} \in I_L$. We will use for the corresponding eigen-value a notation $\lambda_{\tilde{A}}$. Let us consider a subspace $K \subset \mathfrak{H}$, such that K is skew-connected with L. Then $H = L^\perp \cap K^\perp$ is a Hilbert space with induced inner product $(.,.)$. For any $\tilde{A} \in I_L$ let us define an operator $A \in B(\mathfrak{H})$ by following way: $A\chi = \pi\tilde{A}\chi$, $\chi \in \mathfrak{H}$, where π is an orthogonal projection from $\mathfrak{H}$ on H. For ant subset $R \subset I_L$, let us define $R_1 = \{A : \ \tilde{A} \in R\}$. Now we consider two vectors $\xi \ln L$ and $\eta \in K$, such that $(\xi, \eta) = 1$. We are fixing the vectors and considering the mappings $x_{\tilde{A}} = \pi\tilde{A}\eta$, $y_{\tilde{A}} = (\tilde{A}\eta, \eta)$, $y_{\tilde{A}} = x_{\tilde{A}^*}$ and $\mu_{\tilde{A}} = \lambda_{\tilde{A}^*}$.

It is well known that the norm

$$|||\tilde{A}||| = ||A||_{B(\mathfrak{H})} + ||x_{\tilde{A}}||_{\mathfrak{H}} + ||y_{\tilde{A}}||_{\mathfrak{H}} + |\lambda_{\tilde{A}}| + |\mu_{\tilde{A}}| + |y_{\tilde{A}}|$$

on I_L is equivalent to norm $||\tilde{A}||_{B(\mathfrak{H})}$ (see [18]).

Let us consider now $N = span\{\xi_0, \eta_0\}$, where ξ_0 and η_0 are some neutral vectors, such that $(\xi_0, \eta_0) = 1$. Then N is a Pontryagin space of type Π_1 and with $dimN = 2$. Let H_1 be any Hilbert space. Consider an orthogonal sum $H \oplus N$. Then a set, including an operator $A \in H_1$, vectors $x, y \in H$ and complex numbers λ, μ, y is defining the operator

$$\tilde{A} =< A, x, y, \lambda, \mu, y >\in B(H) \oplus N),$$

such that:
(*i*) $A\xi_0 = \lambda\xi_0$;
(*ii*) $\tilde{A} = \mu\eta_0 + x + y\xi_0$;
(*iii*) $\tilde{A}h = Ah + (h, y)\xi_0$ for any $h \in H$.
We can represent any operator from I_L in such a form.

Definition 1.7. A closed symmetric subalgebra of $B(\mathfrak{H})$, containing a unit **1**, is called a *general operator algebra*, if it is non-degenerated (which means it has a neutral invariant subspace.

Let us note for a general operator algebra A and for a neutral S-invariant subspace L, we always have: $S \subset I_L$.

Let us consider $S_0 = \{\tilde{A} \in S : \lambda_{\tilde{A}} = \mu_{\tilde{A}} = 0\}$, and $U = (S_0)_h$. A symmetric ideal $chS = S \cap \Phi_L$ of the algebra S is called its L-*characteristic* .

All general operator algebras on H we can classify by six types - so-called Shulman models (see [18]): the types 0, 1, $2a$, $2b$, $3a$ and $3b$.

Definition 1.8. If L-characteristic of algebra S is equal to zero for some neutral subspace $L \subset \mathfrak{H}$, then S is called an algebra of type 0.

Proposition 1.9. For any general operator algebra S of type 0 there exist a Hilbert space H and a C^*-algebra U in H (which does not contain an identity) that the algebra $M^0(H, U) = \{< A + \lambda\mathbf{1}, 0, 0, \lambda, \lambda, 0 >: A \in U, \lambda \in \mathbb{C}\}$ of operators in $H \oplus N$ is unitary equivalent to S.

Definition 1.10. If a general operator algebra S is not of type 0 and its L-characteristic is contained in S_0 for some neutral subspace $L \subset \mathfrak{H}$, then S is called an algebra of type 1.

Proposition 1.11. For any algebra S of type 1 there exist a Hilbert space H, a $*$-algebra U of bounded linear operators on H, a quasi-vector p from $K_U(U)$, a closed subspace H_F which is invariant for U and orthogonal to $p(U)$, a closed subspace $H_S \subset KerU$, which is orthogonal to H_F, a closed anti-linear involutive operator operator V on H_0, with a domain D_V, such that the algebra

$$M^1(H, H_F, H_C, V, U, p) = \{< A + \lambda\mathbf{1}, x_1 + y + p(A), x_2 + Vy + p(A^*),$$

$$\lambda, \lambda, \gamma >, A \in U, x_1, x_2 \in H_F, y \in D_V, \lambda, \gamma \in \mathbb{C}\}$$

is unitary equivalent to S.

Let us consider $KerU = \{x \in H, Ax = 0, \forall A \in U\}$. Then $KerU$ is a closed subspace in H.

Definition 1.12.
(1) If a general operator algebra S is not of type 0 or 1, and for the some neutral subspace L a functional $\tilde{A} \to \lambda_{\tilde{A}}$ is real-valued and $ch(S) \cap S = \{0\}$, then an algebra S is said to be of type $2a$.
(2) If a general operator algebra S is not of type 0 or 1, and for the same neutral subspace L a functional $\tilde{A} \to \lambda_{\tilde{A}}$ is real-valued, and $ch(S) \cap S \neq \{0\}$, then an algebra S is said to be of type $2b$.
(3) If a general operator algebra S is not of type 0, 1, $2a$ or $2b$, and for the same neutral subspace L we have: $ch(S) \cap S = \{0\}$, then an algebra S is said to be of type $3a$.
(4) If a general operator algebra S is not of type 0, 1, $2a$ or $2b$, and for the same neutral subspace L we have: $ch(S) \cap S \neq \{0\}$, then an algebra S is said to be of type $3b$.

The following propositions will describe the structure of the general operator algebras of types $2a$, $2b$, $3a$ and $3b$ (see [18]).

Proposition 1.13. Let S be a general operator algebra of type $2a$. Then there exist a Hilbert space H and a C^*-algebra U with an identity $\mathbf{1} \in U$ on H such that the algebra

$$M^{2a}(H,U) = \{< A, 0, 0, \lambda, \lambda, 0 >, A \in U, \lambda \in \mathbb{C}\}$$

of operators from $B(N \oplus H)$ is unitary equivalent to S.

Proposition 1.14. Let S be a general operator algebra of type $2a$. Then there exist a Hilbert space H, a C^*-algebra U with an identity $\mathbf{1} \in U$ on H and a U-invariant closed subspace $H_F \subset H$ such that the algebra

$$M^{2b}(H,U) = \{< A, x_1, x_2, \lambda, \lambda, y >, A \in U, x_1, x_2 \in U, \lambda, y \in \mathbb{C}\}$$

of operators from $B(N \oplus H)$ is unitary equivalent to S.

Proposition 1.15. Let S be a general operator algebra of type $3a$. Then there exist a Hilbert space H and a C^*-algebra U with an identity $\mathbf{1} \in U$ on H such that the algebra

$$M^{3a}(H,U) = \{< A, 0, 0, \lambda, \mu, 0 >, A \in U, \lambda, \mu \in \mathbb{C}\}$$

of operators from $B(N \oplus H)$ is unitary equivalent to S.

Proposition 1.16. Let S be a general operator algebra of type $2a$. Then there exist a Hilbert space H, a C^*-algebra U with an identity $\mathbf{1} \in U$ on H and a U-invariant closed subspace $H_F \subset H$ such that the algebra

$$M^{3b}(H,U) = \{< A, x_1, x_2, \lambda, \mu, y >, A \in U, x_1, x_2 \in U, \lambda, \mu, y \in \mathbb{C}\}$$

of operators from $B(N \oplus H)$ is unitary equivalent to S.
References in [1], [18], [9].

2 Commutative symmetric algebras of operators in $B(H)$, generated by a self-adjoint operator

Let $\mathfrak{H}$ be a Pontryagin space, of type Π_1, L be a neutral subspace in H such that $L = span(\xi_0)$, K be a subspace of H, which is skew-connected with L, $\eta_0 \in K$ - a vector, such that $(\xi_0, \eta_0) = 1$. Consider $N = L + K$ and $H = N^{\perp}$. Then $\mathfrak{H}$ is a Hilbert space with an inner product $(,)$. Consider $I_L = \{\tilde{A} \in B(\mathfrak{H}) : \tilde{A}L \subset L, \tilde{A}^*L \subset L\}$.

Lemma 2.1. An operator $\tilde{A} \in I_L$ is a self-adjoint operator if and only if it can be represented as

$$\tilde{A} =< A, x, x, \lambda, \bar{\lambda}, y >,$$

where $A \in B(H)$, $A = A^*$, $x \in H$, $\lambda \in \mathbb{C}$, $\gamma \in \mathbb{R}$.

Proof. $\Rightarrow$) We know that $A \in I_L$ can be written in the form

$$\tilde{A} =< A, x, y, \lambda, \mu, \gamma >,$$

where $A \in B(H)$, $x, y \in H$, $\lambda, \mu, \gamma \in \mathbb{C}$. Let us remind how the operator $\tilde{A}$ is acting on $\mathfrak{H}$:

(*i*) $A\xi_0 = \lambda\xi_0$;

(*ii*) $\tilde{A}\eta_0 = \mu\eta_0 + x + \gamma\xi_0$;

(*iii*) $\tilde{A}h = Ah + (h, y)\xi_0$ for any $h \in H$.

Let us assume that a operator $\tilde{A}$ is self-adjoint. Then $(\tilde{A}\eta_0, h) = (\eta_0, \tilde{A}h)$ for all $h \in H$. Then $(\tilde{A}\eta_0, h) = (\mu\eta_0 + x + \gamma\xi_0, h) = (x, h)$ and $(\eta_0, \tilde{A}h) = (\eta_0, Ah + (h + y)\xi_0) = \overline{(h, y)} = (y, h)$, and then $(x, h) = (y, h)$ for all $h \in \mathfrak{H}$. Therefore, $x = y$.

Since

$$(\tilde{A}\xi_0, \eta_0) = (\xi_0, \tilde{A}\eta_0),$$

$$(\tilde{A}\xi_0, \eta_0) = (\lambda\xi_0, \eta_0) = \lambda$$

and

$$(\xi_o, \tilde{A}\eta_0) = (\xi_0, \mu\eta_0 + x + \gamma\xi_0) = \bar{\mu},$$

then $\lambda = \bar{\mu}$.

Since $\tilde{A}\eta_0, \eta_0) = (\eta_0, \tilde{A}\eta_0)$ and $(\tilde{A}\eta_0, \eta_0) = \gamma$, $(\eta_0, \tilde{A}\eta_0) = \bar{\gamma}$, then $\gamma = \bar{\gamma}$, which means $\gamma \in \mathbb{R}$.

Let h, q be some vectors from $\mathfrak{H}$. Since $(\tilde{A}h, q) = (h, \tilde{A}q)$, $(\tilde{A}h, q) = (h, Aq + (h, y)\xi_0, q) = (Ah, q)$ and $(h, \tilde{A}q) = (h, Aq + (y, q)\xi_0) = (h, Aq)$, then $(Ah, q) = (h, Aq)$ for all $h, q \in \mathfrak{H}$. Then A is a self-adjoint operator from $B(H)$. Therefore, a self-adjoint operator $\tilde{A} \in I_L$ has a representation:

$$\tilde{A} =< A, x, x, \lambda, \bar{\lambda}, \gamma >,$$

where $A \in B(H)$, $A = A^*$, $x \in H$, $\lambda \in \mathbb{C}$, $\gamma \in \mathbb{R}$.

$\Leftarrow$) Let $\tilde{A}$ be an operator from I_L which can be written as

$$\tilde{A} =< A, x, x, \lambda, \bar{\lambda}, \gamma >,$$

where $A \in B(H)$, $A = A^*$, $x \in H$, $\lambda \in \mathbb{C}$, $\gamma \in \mathbb{R}$. Let us consider some vectors a, b from $\mathfrak{H}$. Then a and b have representations: $a = \alpha_1\xi_0 + \alpha_2\eta_0 + h$, $b = \beta_1\xi_0 + \beta_2\eta_0 + q$ for some $\alpha_1, \alpha_2, \beta_1, \beta_2 \in \mathbb{C}$, $h, q \in H$. Consider $(\tilde{A}a, b)$:

$$(\tilde{A}a, b) = (\tilde{A}(\alpha_1\xi_0 + \alpha_2\eta_0 + h), \beta_1\xi_0 + \beta_2\eta_0 + q) =$$

$$= (\lambda\alpha_1\xi_0 + \bar{\lambda}\alpha_2\eta_0 + \alpha_2 x + \alpha_2\gamma\xi_0 + Ah + (h,x)\xi_0, \beta_1\xi_0 + \beta_2\eta_0 + q) =$$
$$= \lambda\alpha_1\bar{\beta_2} + \bar{\lambda}\alpha_2\beta_1 + \alpha_2(x,q) + \alpha_2\gamma\bar{\beta_2} + (Ah,q) + \bar{\beta_2}(h,x).$$

From other side,

$$(a, \tilde{(}A)b) = (\alpha_1\xi_0 + \alpha_2\eta_0 + h, \tilde{(}A)(\beta_1\xi_0 + \beta_2\eta_0 + q)) =$$

$$= (\alpha_1\xi_0 + \alpha_2\eta_0 + h, \lambda\beta_1\xi_0 + \bar{\lambda}\beta_2\eta_0 + \beta_2 x + \beta_2\gamma xi_0 + Aq + (q,x)\xi_0) =$$
$$= \alpha_1\lambda\bar{\beta_2} + \alpha_2\bar{\lambda}\bar{\beta_1} + (h, Aq) + \alpha_2\overline{(q,x)} + \alpha_2\beta_2\gamma + (h, \beta_2 x) =$$
$$= \alpha_1\lambda\bar{\beta_2} + \alpha_2\bar{\lambda}\bar{\beta_1} + (Ah, q) + \alpha_2(x,q) + \alpha_2\bar{\beta_2}\gamma + \bar{\beta_2}(h,x).$$

Then $(\tilde{A}a, b) = (a, \tilde{A}b)$ for any arbitrary vectors from H. Therefore, $\tilde{A}$ is a self-adjoint operator from I_L. □

Let $\tilde{T}$ be a self-adjoint operator from I_L. Consider

$$\mathcal{P}(\tilde{T} = \{\sum_{i=1}^{n} \alpha_i \tilde{T}^i, \alpha_i \in \mathbb{C}, i = 1, 2, ...n, \}$$

where n is a natural number. Then $\mathcal{P}(\tilde{T})$ is a commutative symmetrical algebra of operators in $B(H)$, generated by $\tilde{T}$, and $\mathcal{P}(\tilde{T}, \mathbf{1}) = \mathcal{P}(\tilde{T}) + \mathbb{C}\mathbf{1}$ is a commutative symmetric algebra of operators in $B(\mathfrak{H})$, generated by $\mathcal{P}(\tilde{T})$ and $\mathbf{1}$.

Let $S(\tilde{T})$ be a closure of $\mathcal{P}(\tilde{T})$ and $S(\tilde{T}, \mathbf{1})$ be a closure of $\mathcal{P}(\tilde{T}, \mathbf{1})$ in the uniform topology. We can easy see that $S(\tilde{T})$ and $S(\tilde{T}, \mathbf{1})$ are contained in I_L.

For given polynomial $P(t) = \sum_{i=0}^{n} \alpha_i t^i$, $\alpha_i \in \mathbb{C}$, $t \in \mathbb{R}$, let us define $P(T) = \sum_{i=0}^{n} \alpha_i T^i$, $T \in B(H)$ and $P(\tilde{T}) = \sum_{i=0}^{n} \alpha_i \tilde{T}^i$, $T \in I_L$. Since for arbitrary operators $\tilde{A}, \tilde{B} \in I_L$, say, $\tilde{A} =< A, x_1, y_1, \lambda_1, \mu_1, \gamma_1 >$ and $\tilde{B} =< B, x_2, y_2, \lambda_2, \mu_2, \gamma_2 >$ we have

$$\tilde{A} + \tilde{B} =< A + B, x_1 + x_2, y_1 + y_2, \lambda_1 + \lambda_2, \mu_1 + \mu_2, \gamma_1 + \gamma_2 >,$$

$$\tilde{A} \cdot \tilde{B} =< AB, \mu_2 x_1 + Ax_2, B^* y_1 + \lambda_1 y_2, \lambda_1\lambda_2, \mu_1\mu_2, \mu_2\gamma_1 + \lambda_1\gamma_2 + (x_2, y_1 >,$$

$$\alpha\tilde{A} =< \alpha A, \alpha x_1, \bar{\alpha} y_1, \alpha\lambda_1, \alpha\mu_1, \alpha\gamma_1 >,$$

then the operator $P(\tilde{T})$ has a form: $P(\tilde{T}) =< P(T), x', y', P(\lambda), P(\bar{\lambda}), \gamma' >$, where x' and y' are some vectors from $\mathfrak{H}$ and $\gamma' \in \mathbb{C}$.

Proposition 2.2. Let $\tilde{T}$ be a self-adjoint operator from I_L, such that $\tilde{T}\xi_0 = \lambda\xi_0$, $\lambda \in \mathbb{C}$, and let A be an arbitrary operator from $S(\tilde{T}, \mathbf{1})$ (or from $S(\tilde{T})$). Then $\tilde{A}$ has a representation $\tilde{A} =< A, x, y, \alpha, \alpha, \gamma >$, where $x, y \in \mathfrak{H}$ and $\alpha, \gamma \in \mathbb{C}$.

Proof. Since $\tilde{A} \in S(\tilde{T}, \mathbf{1})$ (or $\tilde{A} \in S(\tilde{T})$), then there exists a sequence of polynomials $\{P_n(\tilde{T})\}$ such that $P_n(\tilde{T}) \to \tilde{A}$ in topology of norm in $B(H)$. Since the operator $\tilde{A}$ is contained in I_L, then $\tilde{A}$ has a form: $\tilde{A} =< A, x, y, \alpha, \beta, \gamma >$. From other side, any polynomial $P_n(\tilde{T})$ we can represent in a form: $P_n(\tilde{T}) =< P_n(T), x_n, y_n, P_n(\lambda), P_n(\lambda), \gamma_n >$. Since a convergence in I_L is equivalent to convergence coordinate-wise, then $\alpha = \lim_n P_n(\lambda) = \beta$, and then $\tilde{A} =< A, x, y, \alpha, \alpha, \gamma >$. □

Theorem 2.3. Let $\tilde{T}$ be a self-adjoint operator from $B(\mathfrak{H})$. Then the algebra $S(\tilde{T}, \mathbf{1})$ is only a non-degenerated algebra of types 0,1, $2a$ or $3a$.

Proof. Let us consider a positive vector $h_0 \in \mathfrak{H}$, such that $\tilde{T}h_0 = \lambda h_0$. Then $\tilde{A}h_0 = \lambda(\tilde{A})h_0$ for all operators $\tilde{A} \in S(\tilde{T}, \mathbf{1})$. Let us define a set $M = \{\xi \in H, \tilde{A}\xi = \lambda(\tilde{A})\xi$ for any $\tilde{A}$ from $\tilde{A} \in S(\tilde{T}, \mathbf{1})$. Then there exists a neutral vector $\eta_{)}$, such that $\tilde{T}\eta_0 = \beta\eta_0$ and $\beta \neq \lambda$. Since a dimension of any non-negative subspace in $\mathfrak{H}$ is no more then 1, then $(h_0, \eta_0) \neq 0$. It is easy to see that we can choose h_0 and η_0 such that $(h_0, h_0) = 2$ and $(h_0, \eta_0) = 2$. Let us consider $\xi_0 = h_0 - \eta_0$. Then $\bar{\beta} = (\xi_0, \tilde{T}\eta_0) = (\tilde{T}\xi_0, \eta_0) = (\lambda h_0 - \beta\eta_{)}, \eta_0) = \lambda$ and $\lambda = 1/2(\tilde{T}h_0, h_0) = 1/2(h_0, \tilde{T}h_0) = \bar{\lambda}$, and then $\lambda = \beta$. Therefore, $\eta_0 \in M$, and then we can see that $dimM = 1$, an operator $\tilde{T}$ does not possess any neutral invariant vectors and the algebra $S(\tilde{T}, \mathbf{1})$ is a non-degenerated algebra.

Assume now that $dimM > 1$. Then M is a Pontryagin space of type Π_1, and then there exist some neutral vectors $\xi_0, \eta_0 \in M$. Now we will construct an algebra $S(\tilde{T}, \mathbf{1})$, using Shulman's model for vectors ξ_0 and η_0.

Let us define a Hilbert space $H = M^{\perp}$, and for any operator $\tilde{A}$ from $\tilde{A} \in S(\tilde{T}, \mathbf{1})$ let us define an operator $A_1 \in B(H)$: $A_1 h = \pi \circ \tilde{A}h$, where $h \in \mathfrak{H}$ and pi is an orthogonal projection from H to $\mathfrak{H}$. Then any operator $\tilde{A}$ has a representation $\tilde{A} =< A, 0, 0, \alpha, \alpha, 0 >$ and then $\tilde{T} =< T, 0, 0, \lambda, \lambda, 0 >$.

Assume now that $\lambda \in \sigma(T)$, where $\sigma(T)$ is a spectrum of the operator T. Consider an arbitrary operator $\tilde{B}$ from algebra $S(\tilde{T}, \mathbf{1})$. Then there exists a sequence of polynomials $\{P_n(\tilde{T})\}$ such that $P_n(\tilde{T}) \to \tilde{B}$ in topology of norm in $B(\mathfrak{H})$. If $B = 0$, then $P_n(\tilde{T}) \to 0$, and since $\lambda \in \sigma(T)$, then $P_n(\lambda) \to 0$. Then $\tilde{B} \equiv 0$, and then $chS(\tilde{T}, \mathbf{1}) = \{0\}$. That means that the algebra $S(\tilde{T}, \mathbf{1})$ is of type 0.

Assume now that $\lambda \notin \sigma(T)$. Then we can show that the algebra $S(\tilde{T}, \mathbf{1})$ is not the algebra of type 0. Really, if $chS(\tilde{T}, \mathbf{1}) = \{0\}$, then on the $*$-algebra $U = \{A \in B(H) : A = A_1$, where A_1 is some operator, $A_1 \in S(\tilde{T}, \mathbf{1})$, an Hermitian character $f(A) = \alpha$ is well-defined (where $\tilde{A} =< A, 0, 0, \alpha, \alpha, 0 >\in S(\tilde{T}, \mathbf{1})$ and $f(A) = \alpha = (\tilde{A}\xi_0, \eta_0)$ is weakly continuous. Then (see [11]) there exists a non-zero vector $h \in H$, such that $Ah = f(A)h = \alpha$ and then for the operator T we have: $Th = \lambda h$. Then $\lambda \in \sigma(T)$, which makes a contradiction. So, we had proven that $chS(\tilde{T}, \mathbf{1}) \neq \{0\}$. Then the algebra $S(\tilde{T}, \mathbf{1})$ is not of type 0.

In other side, the algebra $S(\tilde{T}, \mathbf{1})$ can be represented as

$$S(\tilde{T}, \mathbf{1}) = \{< a, 0, 0, \alpha, \alpha, 0 >: A \in U, \alpha \in \mathbb{C}\},$$

where U is a $*$-algebra with $\mathbf{1}$, generated by an operator T. Since the algebra $S(\tilde{T}, \mathbf{1})$ is closed in the norm $|||\tilde{A}||| = ||A||_{B(H)} + |\alpha|$ and by [] the norms $|||.|||$ and $||.||_{B(\mathfrak{H})}$ are equivalent, then U is a $*$-algebra in $B(\mathfrak{H})$. Therefore, the algebra $S(\tilde{T}, \mathbf{1})$ is of type $2a$.

In the case when the operator $\tilde{T}$ has not any positive invariant vectors. That means that all non-negative invariant vectors are neutral. Assume that all eigen-values of are real-valued. Since the operator $\tilde{A}$ is self-adjoint, then $\tilde{T}$ has a representation $\tilde{A} =< A, x, y, \alpha, \alpha, y >$, where $A \in B(H)$, $x, y \in H$, $\alpha, y \in \mathbb{C}$.

Assume that $chS(\tilde{T}, \mathbf{1}) = \{0\}$. Then the algebra $S(\tilde{T}, \mathbf{1})$ is of type 0 and its has two neutral skew-connected eigen-vectors, and then the algebra $S(\tilde{T}, \mathbf{1})$ and the operator $\tilde{T}$ have a positive invariant vector, which is equal to the sum of neutral vectors. This is impossible, so, $chS(\tilde{T}, \mathbf{1}) \neq \{0\}$.

Now we will show that if $\lambda \in \sigma(T)$ then $chS(\tilde{T}, \mathbf{1}) \subset S_0(\tilde{T}, \mathbf{1}) = \{A \in S(\tilde{T}, \mathbf{1}) : \lambda_{\tilde{A}} = 0$. For that we need to show that $A = 0$ implies $\alpha = 0$ (here $\tilde{A} =< A, x, y, \alpha, \alpha, y >\in S(\tilde{T}, \mathbf{1})$. For any $\tilde{A}$ there exists a sequence of polynomials $\{P_n(\tilde{T})\}$ such that $P_n(\tilde{T}) \to \tilde{A}$ in topology of norm in $B(\mathfrak{H})$. Here we can consider $\{P_n(\tilde{T})\}$ as a $\{P_n(\tilde{T})\} =< P_n(T), x_n, y_n, P_n(\lambda), P_n(\lambda), y_n >$. Since convergence in I_L is equivalent to convergence coordinate-wise, then $||P_n(T)||_{B(H)} \to 0$ implies $P_n(\lambda) \to 0$ because $\lambda \in \sigma(T)$, and then $\alpha = 0$. Therefore, $chS(\tilde{T}, \mathbf{1}) \subset S_0(\tilde{T}, \mathbf{1})$, but $chS(\tilde{T}, \mathbf{1}) \neq \{0\}$, and the algebra $S(\tilde{T}, \mathbf{1})$ is of type 1.

Assume now that $\lambda \notin \sigma(T)$. We will show now that then the algebra $S(\tilde{T}, \mathbf{1})$ is not of type 1.

We know that if $S(\tilde{T}, \mathbf{1})$ is an algebra of type 1, then we can correctly define a Hermitian character $f(A) = \alpha$ on the algebra $S_1 = A \in B(H)$, where $< A, x, y, \alpha, \alpha, y >\in S(\tilde{T}, \mathbf{1})$. This character is weakly continuous. By Lemma 2.12 from [11], there exists such a vector $h_0 \in H$ that $Ah_0 = f(A)h_0 = \alpha h_0$ and $Th_0 = \lambda h_0$, where $\lambda \in \sigma(T)$. This is impossible, so, $\lambda \notin \sigma(T)$ and then the algebra $S(\tilde{T}, \mathbf{1})$ is not of type 1.

Then here we have two possibilities:
(i) $chS(\tilde{T}, \mathbf{1}) \cap S_0(\tilde{T}, \mathbf{1}) = \{0\}$, or
(ii) $chS(\tilde{T}, \mathbf{1}) \cap S_0(\tilde{T}, \mathbf{1}) \neq \{0\}$.

In first case the algebra $S(\tilde{T}, \mathbf{1})$ is of type $2a$, and it has eigen-vectors which are neutral and skew-connected with the same eigen-value. Therefore, the algebra $S(\tilde{T}, \mathbf{1})z$ and the operator $\tilde{T}$ have a positive invariant vector, which is impossible.

In second case the algebra $S(\tilde{T}, \mathbf{1})$ is of type $2b$. We will show now that this is impossible. let us construct the algebra $S(\tilde{T}, \mathbf{1})$, using the Shulman's model with a pair of neutral skew-connected vectors ξ_0, η_0. So, we are now in assumption that the algebra $S(\tilde{T}, \mathbf{1})$ is of type $2b$. Consider a non-zero operator $\tilde{A} =< 0, y, z, 0, 0, \alpha >\in chS(\tilde{T}, \mathbf{1}) \cap S_0(\tilde{T}, \mathbf{1})$. Since the algebra $S(\tilde{T}, \mathbf{1})$ is commutative, then $\tilde{A}\tilde{T}\eta_0 = \tilde{T}\tilde{A}\eta_0$, which means that $\tilde{A}(\lambda\eta_0 + x + y\eta_0) = \lambda y + \lambda\alpha\xi_0 + (y, x)\xi_0 = T(y + \alpha\xi_0) = Ty + \lambda\alpha\xi_) + (y, x)\xi_0)$, and then $Ty = \lambda y$. Using similar way, we can conclude from the equation

$\tilde{A}\tilde{T}h = \tilde{T}\tilde{A}h$ that $A(Th + (h,x)\xi_0) = (Th,z)\xi_0 = \tilde{T}((h,z)\xi_0) = \lambda(h,z)\xi_0$, and then $(h, Tz) = (h, \lambda z)$ for any $h \in H$ and $Tz = \lambda z$. Since $\lambda \notin \sigma(T)$, then $y = z = 0$. Therefore,

$$chS(\tilde{T}, \mathbf{1}) \cap S_0(\tilde{T}, \mathbf{1}) = \{< 0, 0, 0, 0, 0, y >, y \in \mathbb{C}\}.$$

Let us consider the pair of arbitrary operators $\tilde{B} =< 0, z_1, z_2, \alpha, \alpha, y_1 >$ and $\tilde{C} =< 0, y_1, y_2, \beta, \beta, y_2 >$ from $chS(\tilde{T}, \mathbf{1})$. Then the operators $\tilde{B}_0 =< 0, z_1, z_2, \alpha, \alpha, 0 >$ and $\tilde{C}_0 =< 0, y_1, y_2, \beta, \beta, 0 >$ are also contained in $chS(\tilde{T}, \mathbf{1})$. Consider an operator $\tilde{C}_1 = \mathbf{1} - \frac{1}{\beta}\tilde{C}_0 =< 0, y_1', y_2', 0, 0, 0 >$, where $y_1' = -\frac{1}{\beta}y_1, y_2' = -\frac{1}{\beta}y_2$. Then $\tilde{C}_1 \in S(\tilde{T}, \mathbf{1})$. Since the algebra $S(\tilde{T}, \mathbf{1})$ is commutative, then $\tilde{B}_0\tilde{C}_1\eta_0 = \tilde{C}_1\tilde{B}_0\eta_0$. Then we can see that $\tilde{B}_0\tilde{C}_1\eta_0 = \tilde{B}_0 y_1' = (y_1', z_1)\xi_0 = \tilde{C}_1\tilde{B}_0\eta_0 = \tilde{C}_1(\alpha\eta_0 + z_1)\alpha y_1' + z_1 + (z_1, y_2')\xi_0)$, and then $\alpha y_1' + z_1 = 0$. Similarly, since $\tilde{B}_0\tilde{C}_1 h = \tilde{C}_1\tilde{B}_0 h$ we can conclude that that $\tilde{B}_0\tilde{C}_1 h = \tilde{B}_0(h + (h, y_2')\xi_0) = (h, z_2)\xi_0 + \alpha(h, y_2')\xi_0 = \tilde{C}_1\tilde{B}_0 h = \tilde{C}_1(h, z_2)\xi_0 = 0$. Then $(h, z_2 + \alpha y_2') = 0$ for any $h \in H$, and then $z_2 + \alpha y_2' = 0$. Therefore, according [41], $H_F = \{x_{\tilde{A}} : \tilde{A} \in chS(\tilde{T}, \mathbf{1}), \tilde{A} =< 0, x_{\tilde{A}}, y_{\tilde{A}}, \alpha, \alpha, y >, y_{\tilde{A}} = 0\}$ has a dimension equal 1. Then $\tilde{T}x = \omega x$, for some number $\omega \in \mathbb{C}, \omega \neq \lambda$.

Let us define a vector

$$\eta_1 = \eta_0 - \frac{(x,x)}{2(\lambda-\omega)^2}\xi_0 + \frac{1}{\lambda-\omega}x.$$

Then

$$(\eta_1, \eta_1) = -\frac{(x,x)}{2(\lambda-\omega)^2} - \frac{(x,x)}{2(\lambda-\omega)^2} + \frac{(x,x)}{(\lambda-\omega)^2} = 0,$$

and $(\eta_1, \xi_0) = 1$.

Since the operator $< 0, 0, 0, 0, 0, y >\in S(\tilde{T}, \mathbf{1})$, then we can re-write a operator $\tilde{T}$ in the form $\tilde{T} =< T, x, x, \lambda, \lambda, y' >$, where $y' = \frac{(x,x)}{\omega-\lambda}$. Then

$$\tilde{T}'\eta_1 = \tilde{T}'(\eta_0 - \frac{(x,x)}{2(\lambda-\omega)^2}\xi_0 + \frac{1}{\lambda-\omega}x) =$$

$$= \lambda\eta_0 + x + \frac{(x,x)}{\omega-\lambda}\xi_0 - \frac{\lambda(x,x)}{2(\lambda-\omega)^2}\xi_0 + \frac{\omega}{\lambda-\omega}x + \frac{(x,x)}{\lambda-\omega}\xi_0 =$$

$$= \lambda\eta_0 + \frac{\lambda}{\lambda-\omega}x - \frac{\lambda(x,x)}{2(\lambda-\omega)^2}\xi_0 = \lambda(\eta_0 + \frac{1}{\lambda-\omega}x - \frac{(x,x)}{2(\lambda-\omega)^2}\xi_0) = \lambda\eta_1.$$

Since $S(\tilde{T}, \mathbf{1}) = S(\tilde{T}', \mathbf{1})$, we see that the algebra $S(\tilde{T}, \mathbf{1})$ has two neutral invariant skew-connected vectors with the same eigen-value. Then the algebra $S(\tilde{T}, \mathbf{1})$ is an algebra of type $2a$, and we have a contradiction with assumed type $2b$.

Assume now that the operator $\tilde{T}$ has non-positive invariant (eigen)-vectors and in the set of neutral eigen-vectors we can choose a vector ξ_0 such that its eigen-value is not real: $T\xi_0 = \lambda\xi_0, \lambda \neq \bar{\lambda}$. Since the algebra $S(\tilde{T}, \mathbf{1})$ is commutative, then there exists a neutral skew-connected with ξ_0 vector η_0. Then for any $\tilde{A} \in S(\tilde{T}, \mathbf{1})$ we have: $\tilde{A}\xi_0 = \alpha\xi_0, \tilde{A}\eta_0 = \bar{\alpha}\eta_0$. let us consider a Hilbert space $H = \{\xi_0, \eta_0\}^\perp$ and an operator

$A \in B(H)$, $A = \tilde{A}|_H$, where $\tilde{A} \in S(\tilde{T}, \mathbf{1})$. Then any operator $\tilde{A}$ from $S(\tilde{T}, \mathbf{1})$ has a form $\tilde{A} =< A, 0, 0, \alpha, \beta, 0 >$. Consider an algebra U of all $= A \in B(H)$ such that the operator $< A, 0, 0, \alpha, \beta, 0 >$ is contained in the algebra $S(\tilde{T}, \mathbf{1}$. Since the algebra $S(\tilde{T}, \mathbf{1}$ is closed in topology of norm $|||\tilde{A}||| = ||A||_{B(H)} + |\alpha| + |\beta|$ and the norms $|||.|||$ and $||.||_{B(H)}$ are equivalent on U, then U is a C^*-algebra, in $B(H)$. Therefore, the algebra $S(\tilde{T}, \mathbf{1})$ is an algebra of type $3a$. □

3 Functional representation of commutative symmetric algebras with a positive eigen-vector

Let S be a commutative non-degenerated uniformly closed symmetrical subalgebra in $B(\mathfrak{H})$. Then there exists such a decomposition $\mathfrak{H} = H_1 \oplus H_2$, generating an inner product $[,]$ in $\mathfrak{H}$, that S is a C^*-algebra with respect to the decomposition $\mathfrak{H} = H_1 \oplus H_2$ in a Hilbert space $(\mathfrak{H}, [,])$. Consider $J = P_+ - P_-$, where P_+ is a projection from $\mathfrak{H}$ to H_+ and P_- is a projection from $\mathfrak{H}$ to H_-. Then $[\xi, \eta] = (J\xi, \eta)$ and $T^* = JT^{\bar{*}}J$ for any $T \in B(\mathfrak{H})$, where $*$ is an operation of conjugation for T with respect to an inner product $(.,.)$ and $\bar{*}$ is an operation of conjugation for T with respect to an inner product $[.,.]$. Then $T^{\bar{*}} \in S$ for any $T\, inS$, and since S is a commutative C^*-algebra with an identity with respect to an involution $T \to T^{\bar{*}}$, then there exists a compact set K such that S is $\star$-isomorphic to the C^*-algebra of all continuous complex-valued functions $C(K)$ on K.

Let Φ be a $\star$-isomorphism from $C(K)$ on $(S, \bar{*})$. Let us define on $C(K)$ another $\star$-operation: $f^{\#} = \Phi^{-1}((\Phi(f)^*))$. It is easy to see that this operation is an involution. Since

$$||f^{\#}||_{C(K)} = ||\Phi^{-1}(\Phi(f)^*)||_{C(K)} = ||\Phi(f)^*||_{B(\mathfrak{H})} = ||\Phi(f)||_{B(\mathfrak{H})} = ||f||_{C(K)},$$

then the algebra $(C(K), \#)$ is a Banach $\star$-algebra.

Since Φ is a linear bijective mapping from $(C(K), \#)$ into S and since $\Phi(f)\Phi(g) = \Phi(fg)$ and $\Phi(f^{\#} = \Phi(f)^*$, then Φ is a $\star$-isomorphism from a Banach $\star$-algebra $(C(K), \#)$ on S and $||f||_{C(K)} = ||\Phi(f)||_{B(\mathfrak{H})}$. This is a proof of a

Proposition 3.1. Φ is an isometric $\star$-isomorphism from Banach $\star$-algebra $(C(K), \#)$ on $(S, *)$.

Now let S be a commutative symmetric operator algebra of type 0. Then we know that each operator $\tilde{A} \in S$ has a form

$$\tilde{A} =< A + \alpha\mathbf{1}, 0, 0, \alpha, \alpha, 0 >, A \in U, \alpha \in \mathbb{C},$$

where U is a commutative C^*-algebra without an identity, acting on the space $H = \{\xi_0, \eta_0\}^\perp$. Consider a C^*-algebra $\hat{U} = U + \mathbb{C}\mathbf{1}$. Is is easy to see that any character ω on U has a unique continuation to the character $\hat{\omega}$ on $\hat{U}$, and, moreover, $\hat{\omega}(A + \alpha\mathbf{1}) = \omega(A) + \alpha$. From other side, if $\hat{\omega}$ is a character on $\hat{U}$, then $\omega = \hat{\omega}|_U$ is also a character on U, except the case when $\hat{\omega}|_U = 0$, and then $\hat{\omega}$ will be a unique character (denote it as $\hat{\omega}_0$), for which $\hat{\omega}_0(\mathbf{1}) = 1$ and $\hat{\omega}_0|_U = 0$. Therefore, $\sigma(\hat{U}) = \sigma(U) \cup \{\hat{\omega}_0\}$ or a spectrum $\sigma(\hat{U})$ of the algebra $\hat{U}$. For any operator $\hat{A} \in \hat{U}$ let us define a mapping $\phi_{\hat{A}} : \sigma(\hat{U}) \to \mathbb{C}$ as $\phi_{\hat{A}}(\hat{\omega}) = \hat{\omega}(\hat{A})$. It is known that $\phi_{\hat{A}} \in C(\sigma(\hat{U}))$ and that a mapping $\Phi : \hat{U} \to C(\sigma(\hat{U}))$, acting as $\Phi(\hat{A}) = \phi_{\hat{A}}$, is a Gelfand $\star$-isomorphism from $\hat{U}$ to $C(\sigma(\hat{U}))$. For any function $f \in C(\sigma(\hat{U}))$ we define a mapping $\psi(f) =< \Phi^{-1}(f), 0, 0, f(\hat{\omega}_0), f(\hat{\omega}_0), 0 >$.

Proposition 3.2. A mapping ψ is a topological $\star$-isomorphism from $C(\sigma(\hat{U}))$ on S.

Proof. We know that ψ is an injective $\star$-homomorphism from $C(\sigma(\hat{U}))$ on S. We have to show that ψ is a surjective mapping. Consider any $\tilde{A} \in S$. Then $\tilde{A} =< A + \alpha\mathbf{1}, 0, 0, \alpha, \alpha, 0 >$, where $A \in U, \alpha \in \mathbb{C}$. For the operator $\hat{A} = A + \alpha\mathbf{1}$ we have then: $\phi_{\hat{A}}(\hat{\omega}) = \hat{\omega}(\hat{A}) = \hat{\omega}(A + \alpha\mathbf{1})$ for all $\hat{\omega} \in \sigma(\hat{U})$. In particular, $\phi_{\hat{A}}(\hat{\omega}_0) = \hat{\omega}_0(A + \alpha\mathbf{1}) = \alpha$. Therefore, $\psi(\phi_{\hat{A}}) = \tilde{A}$, and then ψ is a $\star$-isomorphism from $C(\sigma(\hat{U}))$ on S. Since the norm $||.||_{B(\mathfrak{H})}$ is equivalent to the norm $|||.|||$ on I_L, and $|||A||| = ||A + \alpha\mathbf{1}||_{B(\mathfrak{H})} + 2|\alpha| = ||\Phi(A + \alpha\mathbf{1})||_{C(\sigma(\hat{U}))} + 2|\phi_{\hat{A}}(\hat{\omega}_0)|$, then ψ is a topological $\star$-isomorphism. □

Let us consider now a commutative algebra S of type $2a$. Then there exist two skew-connected neutral vectors $\xi_0, \eta_0 \in \mathfrak{H}$ such that one can represent S as $S = \{< A, 0, 0, \alpha, \alpha, 0 >: A \in U, \alpha \in \mathbb{C}\}$, where U is a commutative C^*-algebra with an identity, acting on a Hilbert space $H = \{\xi_0, \eta_0\}^\perp$. Consider a linear functional $\lambda : S \to \mathbb{C}$, acting as $\lambda(\tilde{A}) = \lambda(< A, 0, 0, \alpha, \alpha, 0 >) = \alpha$. Easy to see that $\lambda(\tilde{A})\lambda(\tilde{B}) = \lambda(\tilde{A}\tilde{B})$ and that $\lambda|_U = 0$. Therefore, λ is a continuous character on S.

Let us show that any character h on U we can extend to a character $\tilde{h}$ on S. Assume that $\tilde{h}(< A, 0, 0, \alpha, \alpha, 0 >) = h(A)$ for any $A \in U, \alpha \in \mathbb{C}$. Easy to see that $\tilde{h} : S \to \mathbb{C}$ is non-zero linear function, and that $\tilde{h}(\tilde{A})\tilde{h}(\tilde{B}) = \tilde{h}(\tilde{A}\tilde{B})$, and then $\tilde{h}$ is a character on S, wherein $\tilde{h}(\tilde{A}^*) = h(A^*) = \overline{h(A)} = \overline{\tilde{h}(\tilde{A})}$.

Proposition 3.3. Let $\tilde{h} : S \to \mathbb{C}$ be an arbitrary continuous character on S. Then only two opportunities are possible:
(a) $h = \tilde{h}|_U$ is a non-zero character on U;
(b) If $h = \tilde{h}|_U = 0$, then $\tilde{h} = \lambda$.

Proof. If h is a non-zero character on U, then, obviously, $\tilde{h} \neq \lambda$. Assume that $h(A) = 0$ for all $A \in U$, and there exists a number $\alpha \in \mathbb{C}$ such that $\tilde{h}(< 0, 0, 0, \alpha, \alpha, 0 >) \neq 0$.

Assume that $\tilde{h}(< 0, 0, 0, 1, 1, 0 >) = k$ for some number k. Then $\tilde{h}(< 0, 0, 0, \alpha, \alpha, 0 >) = \alpha k \neq 0$ and $k^2 = \tilde{h}(< 0, 0, 0, 1, 1, 0 >) \cdot \tilde{h}(< 0, 0, 0, 1, 1, 0 >) = k$, and then $k^2 = k$, so, $k = 1$. Therefore, $\tilde{h}(\tilde{A}) = \alpha = \lambda(\tilde{A})$. □

Define $\sigma = \{\tilde{h} : h \in \sigma(U)\}$ and consider a set $K = \sigma \cup \{\lambda\}$. Let us consider on K a topology τ, given by following way: $G \in \tau \Longleftrightarrow G \in \tau_0$ or $G = \{\lambda\} \cup G'$, $G' \in \tau_0$, where τ_0 is a topology of Gelfand in $\sigma(U) = \sigma$ for which $\sigma(U)$ is a compact. The proof of the following Proposition is obvious.

Proposition 3.4. (K, τ) is a compact.

Consider now a commutative C^*-algebra $C(K, \tau)$ of all continuous complex-valued functions on (K, τ). For any function $\hat{f} \in C(K, \tau)$ the function $f = \hat{f}_{\sigma(U)}$ is contained in $C(\sigma(U))$. Let Φ be a Gelfand $\star$-isomorphism from U on $C(\sigma(U))$. For any $\hat{f} \in C(K, \tau)$ consider $\Psi(\hat{f}) =< \Phi^{-1}(f), 0, 0, \hat{f}(\lambda), \hat{f}(\lambda), 0 >$.

Proposition 3.5. Ψ is a topological $\star$-isomorphism from $C(K, \tau)$ on S.

Proof. It is easy to see that Ψ is a topological $\star$-homomorphism from $C(K, \tau)$ to S. We need to show only that Ψ is a surjective mapping. Let $\tilde{A} \in S$, then $\tilde{A} =< A, 0, 0, \alpha, \alpha, 0 >$, where $A \in U$, $\alpha \in \mathbb{C}$. Then, there exist a function $f \in C(\sigma(U))$, such that $f = \Phi(A)$. Let us define on (K, τ) the function $\hat{f}$ as $\hat{f}(t) = f(t)$ for $t \in \sigma(U)$ and $\hat{f}(\lambda) = \alpha$. Then easy to see that $\hat{f} \in C(K, \tau)$ and $\Psi(\hat{f}) = \tilde{A}$. Then Ψ is a surjective mapping. Repeating the end of the proof of Proposition 4.2, we obtain that Ψ is a topological $\star$-isomorphism. □

Combining Propositions 3.1, 3.2 and 3.5, we obtain the following theorem.

Theorem 3.6. Let S be a commutative symmetric operator algebra with a positive eigen-vector in Pontryagin space $\mathfrak{H}$. Then there exists a compact K, such that S is topologically $\star$-isomorphic to the C^*-algebra $C(K)$.

Corollary 3.7. let S be an algebra with an identity, generated by self-adjoint operator $\tilde{T} \in B(\mathfrak{H}$ which has a positive eigen-vector. Then there exist a compact K and a topological $\star$-isomorphism $\Phi : C(K) \to S$ such that $\Phi(f_0) = 1$, $\Phi(f_1) = \tilde{T}$, where $f_0(t) \equiv 1$, $f_1(t) = t$.

4 Functional representation of commutative algebras of type 3a

In this section we will study the structure of commutative algebras of type $3a$.
Let S be a commutative algebra of type $3a$. Then there exist a pair of skew-connected neutral vectors $\xi_0.\eta_0 \in \mathfrak{H}$, for which $\tilde{A}\xi_0 = \lambda(\tilde{A})\xi_0, \tilde{A}\eta_0 = \mu(\tilde{A})\eta_0, \lambda(\tilde{A}) = \overline{\mu(\tilde{A}^*)}$ for all $\tilde{A} \in S$ and the algebra S in Shulman's model can be realized as

$$S = \{\tilde{A} =< A, 0, 0, \lambda(\tilde{A}), \mu(\tilde{A}), 0 >: A \in U, \lambda(\tilde{A}), \mu(\tilde{A}) \in \mathbb{C}\},$$

where U is a commutative C^*-algebra with an identity, acting on $H = \{\xi_0, \eta_0\}^\perp$. We will show that any character h on U can be extended to a character $\tilde{h}$ on S. Assume that $\tilde{h}(< A, 0, 0, \alpha, \beta, 0 >) = h(A)$ for all $A \in U, \alpha, \beta \in \mathbb{C}$, It is clear that $\tilde{h} : S \to \mathbb{C}$ is a non-zero linear function and that $\tilde{h}(\tilde{A})\tilde{h}(\tilde{B}) = h(A)h(B) = h(AB) = \tilde{h}(\tilde{A}\tilde{B})$, which means that $\tilde{h}$ is a character on S, therein $\tilde{h}(\tilde{A}^*) = h(A^*) = \overline{h(A)} = \overline{\tilde{h}(\tilde{A})}$. Further, it is easy to see that $\lambda(\tilde{A})$ and $\mu(\tilde{A})$ are continuous characters on S.

Proposition 4.1. Let $\tilde{h} : S \to \mathbb{C}$ be an arbitrary continuous character on S. Then only two cases are possible:
(a) $h = \tilde{h}|_U$ is a non-zero character on U;
(b) If $h = \tilde{h}|_U \equiv 0$, then $\tilde{h} = \lambda$ or $\tilde{h} = \mu$.

The proof is similar to proof of Proposition 3.3. □

Let (K_0, τ_0) be a spectrum of the C^*-algebra U. Assume that $K = K_0 \cup \{\lambda\} \cup \{\mu\}$, where $\lambda \neq \mu$, λ and μ are characters from Proposition 5.1. Easy to see that $\lambda =\notin K_0$ and $\mu \notin K_0$. Let us consider a topology τ on K, such that $G \in \tau$ if and only if one of following conditions are carrying out: (i) $G \in \tau_0$; (ii) $G = \{\lambda\} \cup G'$; (iii) $G = \{\mu\} \cup G'$, or (iv) $G = \{\lambda, \mu\} \cup G'$, where $G' \in \tau_0$. Again, it is easy to see that (K, τ) is a compact.
Let us consider now a commutative algebra $C(K, \tau)$ of all continuous complex-valued functions on (K, τ) and define on $C(K, \tau)$ an operation "#":
(i) $f^\#(t) = \bar{f}(t)$, if $t \in K_0$;
(ii) $f^\#(t) = \bar{f}(\mu)$, if $t = \lambda$;
(iii) $f^\#(t) = \bar{f}(\lambda)$, if $t = \mu$.

Proposition 4.2. $(C(K, \tau), \#)$ is a Banach $\star$-algebra.

Proof. Directly using a definition of the operation "#", we can check that $(f(t)g(t))^\# = f^\#(t)g^\#(t)$, $(f(t) + g(t))^\# = f^\#(t) + g^\#(t)$ and $(\alpha f(t))^\# = \bar{\alpha} f^\#(t)$, $f, g \in C(K, \tau)$. Then

$$||f^{\#}||_{C(K)} = \max_{t\in K} |f^{\#}(t)| = \max\{\max_{t\in K_0} |\bar{f}(t)|, \max\{|\bar{f}(\mu|), |\bar{f}(\lambda)|\} =$$

$$= \max_{t\in K} |f(t)| = ||f||_{C(K)}.$$

Then $||f_0||_{C(K)} = 1$ for $f_0(t) = 1$ for all $t \in K$. Therefore, $(C(K,\tau), \#)$ is a Banach ⋆-algebra. □

Note 4.3. The algebra $(C(K,\tau), \#)$ is not a C^*-algebra, because, generally speaking, $||f||^2_{C(K,\tau)} \neq ||f^{\#}f||_{C(K,\tau)}$.

Now for any operator $\tilde{A} \in S$ define $\phi_{\tilde{A}}(\tilde{h}) = \tilde{h}(\tilde{A})$, where $\tilde{h} \in K$. then $\phi_{\tilde{A}} \in C(K,\tau)$ for all $\tilde{A} \in S$. For any function $\hat{f}\ inC(K,\tau)$ the function $f = \hat{f}|_{K_0}$ is such that $f \in C(K_0, \tau_0) \equiv C(\sigma(U))$. Let Φ be a Gelfand ⋆-isomorphism from U on $C(\sigma(U))$. For any function $\hat{f} \in C(K,\tau)$ we define $\Psi(\hat{f}) =< \Phi^{-1}(f), 0, 0, \hat{f}(\lambda), \hat{f}(\mu), 0 >$.

Proposition 4.4. Ψ is a topological ⋆-isomorphism from $(C(K,\tau), \#)$ on S.

Proof. It is easy to see that Ψ is a linear injective mapping from $(C(K,\tau), \#)$ into S, wherein $\Psi(\hat{f}\hat{g}) = \Psi(\hat{f})\Psi(\hat{g})$, $\hat{f}, \hat{g} \in (C(K,\tau), \#)$. We need to show that $\Psi(\hat{f}^{\#}) = \Psi(\hat{f})^{\#}$. First, $\Psi(\hat{f}^{\#} =< \Phi^{-1}(\hat{f}), 0, 0, \bar{\hat{f}}(\mu), \bar{\hat{f}}(\lambda), 0\ >$. From other side, $\Psi(\hat{f})^{\#} =< \Phi^{-1}(f), 0, 0, \hat{f}(\lambda), \hat{f}(\mu), 0 >^* =< \Phi^{-1}(\bar{f}), 0, 0, \bar{\hat{f}}(\mu), \bar{\hat{f}}(\lambda), 0 >$. Therefore, Ψ is an injective ⋆-homomorphism from $(C(K,\tau), \#)$ into S. Let us show that Ψ is surjective. Let $\tilde{A} \in S$; then $\tilde{A} =< A, 0, 0, \alpha, \beta, 0 >$, where $A \in U$, $\alpha, \beta \in \mathbb{C}$. Then there exists a function f such that $f = \Phi(A) \in C(\sigma(U))$ Let us define a function $\hat{f}$, assuming that $\hat{f}(t) = f(t)$ for $t \in \sigma(U)$, $\hat{f}(\lambda) = \alpha$ and $\hat{f}(\mu) = \beta$. Then $\hat{f}\ inC(K,\tau)$ and $\Psi(\hat{f}) = \tilde{A}$. Then Ψ is a surjective mapping. Since the norms $||.||_{B(\mathfrak{H})}$ and $|||.|||$ are equivalent on I_L and the norm $|||A||| = ||A||_{B(\mathfrak{H})} + |\alpha| + |\beta| = ||\Phi^{-1}(f)||_{C(\sigma(A)} + |\hat{f}(\lambda)| + |\hat{f}(\mu)|$, then Ψ is a topological ⋆-isomorphism. □

Let S be an algebra of type $3a$, generated by a self-adjoint operator $\tilde{T} \in B(\mathfrak{H})$. Then the operator $\tilde{T}$ does not possess any positive eigen-vectors, and $\tilde{T}\xi_0 = \lambda\xi_0$ for some neutral vector $\xi_0 \in \mathfrak{H}$, $\lambda \neq \bar{\lambda}$.

Corollary 4.5. Let S be an algebra of type $3a$, generated by a self-adjoint operator $\tilde{T}$ such that $\tilde{T}\xi_0 = \lambda\xi_0$, $(\xi_0, \xi_0) = 0$, $\lambda \neq \bar{\lambda}$. Then there exist a compact K_0 and topological ⋆-isomorphism from $(C(K), \#)$ on S (where $K = K_0 \cup \{\lambda, \bar{\lambda}\}$), such that $\Psi(f_0) = 1$ and $\Psi(f_1) = T$ for $f_0(t) = 1, f_1(t) = t$.

Bibliography

[1] Azizov, T., Iokhvidov, I. *Linear Operators in Spaces with an Indefinite Metric*; Pure and Applied Mathematics: A Wiley Series of Texts, Monographs and Tracts (Book 7), Wiley, New York, 1989.

[2] Bognar J. *Indefinite inner product spaces*; Springer, Berlin, 1974.

[3] Bendersky, O., Chilin, V., Litvinov, S. *A description of commutative symmetric operator algebras in Pontryagin space Π_1*; J. Operator Theory, 37 (1997), 201–222.

[4] Dirac, P. A. M. *The physical interpretation of quantum mechanics*; Proc. Royal Soc. London, Ser. A, 180 (1942), 1–40.

[5] Iokhvidov, I., Krein, M. *Spectral theory of operators in spaces with an indefinite metrics, I*; Proceedings of Moscow Math. Soc., 5 (1956), 367–432.

[6] Iokhvidov, I., Krein, M. *Spectral theory of operators in spaces with an indefinite metrics, II*; Proceedings of Moscow Math. Soc., 8 (1959), 413–496.

[7] Ismagilov, P. *On rings of operators in a space with an indefinite metrics*; Docl. Acad. Sci. USSR 158 no 2 (1964), 268–270.

[8] Iokhvidov, I., Krein, M., Langer, H. *Introduction to the Spectral Theory of Operators in Spaces with an Indefinite Metrics*; Academic - Verlag, Berlin, 1982.

[9] Kissin, E., Shulman, V. *Representations on Krein Spaces and Derivations of C^*-Algebras*; Monographs and Surveys in Pure and Applied Mathematics (Book 89), Chapman and Hall/CRC, London, 1997.

[10] Litvinov, S. Bicyclic WJ^*-algebras in Pontryagin space of type Π_1; Func. Analysis and Appl., 26 No 3 (1992), 46–54.

[11] Litvinov, S. *Commutative symmetric operators in Pontryagin space Π_1*; Ph.D. Thesis, Tashkent State University, Tashkent, 1987.

[12] Loginov, A., Shulman, V. *Irreditible J-symmetric algebras of operators in spaces with an indefinite metrics*; Docl. Acad. Sci. USSR 240 No 1 (1978), 21–23.

[13] Naimark, M. *Commutative algebras of operators in space Π_1*; Rev. roum. math. pures et appl., 9:6 (1964), 499–529.

[14] Nevanlinna, R. *Uber metrische linear raume, IV: Zur Theoric der Unterraume*; Ann. Acad. Sci. Fenn., Ser. A1, 163 (1954).

[15] Pauli, W. *On Dirac's new method of field quantization*; Rev. Modern Phys., 15 (1943), 175–207.

[16] Pontryagin, L. *Hermitian operators in space with an indefinite metrics I.*; USSR Acad. Izvestia, Ser. Math., 8 (1944), 243–248.

[17] Shulman, V. *On algebras of operators in spaces with an indefinite metrics of type Π_1*; Docl. Acad. Sci. USSR 201 (1971), 44–47.

[18] Shulman, V. *Banach symmetric algebras in space of type Π_1*; Math. Sbornik, 89, 131 No. 2 (1972), 264–279.

[19] Shtraus, V. *A functional representation of an algebra, generated by a self-adjoint operator, in Pontryagin space*; Func. Analysis and Appl., 20 No. 1 (1986), 91–92.

[20] Shtraus, V. *On stucture of family of commuting J-self-adjoint operators*; Ukrainian Math J., 41 No.10 (1989), 1431–1433.

Oleg Friedman and Alexander A. Katz

On Gelfand-Naimark type Theorems for unital abelian complex and real locally C*-, and locally JB-algebras

Abstract: A type Λ for the unital locally C^*-algebra B is introduced, and the notions of a local *-homomorphisms between two algebras of the same type Λ is studied. The structure of the spectrum $\mathfrak{M}(B)$ is discussed and it is shown how to topologize $\mathfrak{M}(B)$ to turn it to a certain compactly-generated topological space with an extra filtration property, so that B is locally *-isometric to the functional algebra $C(\mathfrak{M}(B))$. After that the structure of $\mathfrak{M}(B)$ is used to obtain versions of Arens-Kaplansky type theorem for real unital abelian locally C^*-algebras and Gelfand-Naimark type theorem for unital abelian locally JB-algebras.

Keywords: Locally C^*-algebras, real locally C^*-algebras, locally JB-algebras, Gelfand-Naimark type theorems

MSC: Primary 46H30, 46H70; Secondary 46H15, 46H05

1 Introduction

It is believed that C^*-algebras were first considered primarily for their use in Quantum Mechanics to model algebras of Physical Observables. This line of research started with Heisenberg's Matrix Mechanics and, in a more mathematically developed form, with Jordan in the early 1930's. Subsequently von Neumann attempted to establish a general framework for these algebras which culminated in a series of papers on Rings of Operators. These papers considered a special class of C^*-algebras which are now known as von Neumann algebras (see [12]).

In early 1940s, C^*-algebras were baptized in the work [15] of Gelfand and Naimark. They yielded an abstract characterization of C^*-algebras making no reference to the functional algebras or the operator algebras acting on a Hilbert space.

One of the most fundamental results in the theory of C^*-algebras was discovered by Gelfand and Naimark in the same seminal paper [15]. It says that any unital abelian C^*-algebra B is isometrically *-isomorphic to the algebra $C(X)$ of all continu-

Oleg Friedman, Department of Mathematical Sciences, University of South Africa, P.O. Box 392, UNISA, Theo van Wijk Building, Muckleneuk Campus, Pretoria 0003, Republic of South Africa; Current address: Lander College for Men/Touro College, 75-31 150th Street, Kew Gardens Hills, NY 11367, USA; Email: friedman001@yahoo.com
Alexander A. Katz, Department of Mathematics and Computer Science, St. John's College of Liberal Arts and Sciences, St. John's University, SJH-334-G, 8000 Utopia Parkway, Queens, NY 11439, USA; Email: katza@stjohns.edu
https://doi.org/10.1515/9783110413557-005

ous complex-valued functions on a compact Hausdorff topological space X (the spectrum of the algebra). Moreover, any *-homomorphism of unital abelian C^*-algebras is related to a continuous map of the underlying Hausdorff compacts. Briefly speaking, the category of unital abelian C^*-algebras with their *-homomorphisms is dual to the category of compact Hausdorff spaces and their continuous mappings.

A real counterpart of the theory of C^*-algebras was born a few years later following their complex brethren (see [17] and [27] for details and bibliography). Using the fact that the complexification

$$B = A \dot{+} iA$$

of each real C^*-algebra A can be endowed with a structure of a C^*-algebra, and the process of complexification naturally generates an order 2 involutory linear antiautomorphism

$$\Phi : B \to B$$

on B as

$$\Phi(x + iy) = x^* + iy^*,$$

where $x, y \in A$, and

$$A \cong \left\{a \in B : \Phi(a) = a^*\right\}.$$

Similarly, the aforementioned process of complexification naturally generates an order 2 involutory conjugate-linear automorphism

$$\Psi : B \to B$$

on B as

$$\Psi(x + iy) = x - iy,$$

where $x, y \in A$, and

$$A \cong \{a \in B : \Psi(a) = a\},$$

and

$$\Phi(a) = \Psi(a^*) = (\Psi(a))^*,$$

for all $a \in B$.

Arens and Kaplansky were able in late 1940's to extend Gelfand-Naimark duality to a duality between pairs (B, Ψ) and (X, h), where B is an abelian complex C^*-algebra, Ψ, X are as above, and

$$h : X \to X$$

is called a topological involution on X, and is an order 2 homeomorphism on X (see [4] for details).

The real Jordan analogues of C^*-algebras, so called ***JB*-algebras,** were introduced in 1970's by Alfsen, Shultz and Størmer in [1]. In the same paper they produced a Gelfand-Naimark type theorem on the functional representation of the abelian unital

JB-algebras. The development of the subject is reflected in the monograph of Hanche-Olsen and Størmer [18], which we will use for further references on the general theory of JB-algebras.

The Hausdorff projective limits of projective families of Banach algebras as natural locally-convex generalizations of Banach algebras have been studied sporadically by many authors since early 1950's, when they were first introduced by Arens [5] and Michael [30]. The Hausdorff projective limits of projective families of C^*-algebras were first mentioned by Arens [5]. They have since been studied under various names (i.e. locally C^*-algebras, pro-C^*-algebras, b^*-algebras, LMC^*-algebras) by many authors. Development of the subject is reflected in the monograph of Fragoulopoulou [13]. We will follow Inoue [20] in the usage of the name **locally C^*-algebras** for these algebras. The Hausdorff projective limits of projective families of real C^*-algebras and resp. JB-algebras were first introduced under the name of **real locally C^*-algebras** and resp. **locally JB-algebras** by Katz and Friedman in [22].

In the study of functional topological algebras $C(X)$, where X is a topological space, the Shilov program devoted to the interaction of Functional Analysis and General Topology, following the ideas of the aforementioned Gelfand-Naimark duality, naturally asks: "To what extend does the spectrum determine the algebra and conversely, is the topological space recoverable from the structure of the algebra of functions associated with it?" So, it is very natural to try to generalize Gelfand-Naimark and Arens-Kaplansky type results to respectively locally C^*-algebras, real locally C^*-algebras and locally JB-algebras.

There were numerous attempts to extend aforementioned Gelfand-Naimark duality to locally C^*-algebras. The main difficulty was to find a proper category to which the spectrum normally belongs, such that:

1). within this category the spectrum admits a dual decomposition to the Arens-Michael decomposition of the algebra into a projective limit of a projective family of unital abelian C^*-algebras;

2). the algebra of functions associated with each of its objects should be complete;

3). the intrinsic structure of the spectrum is convenient in the sense of Steenrod (see [39]).

We introduce a type Λ for the unital locally C^*-algebra B, and the notions of a local *-homomorphisms between two algebras of the same type Λ. When a local *-homomorphism is a *-isomorphism, we come to a notion of a locally isometric *-isomorphism between two unital locally C^*-algebras. Then we describe the structure of the spectrum $\mathfrak{M}(B)$ and show how to topologize $\mathfrak{M}(B)$ to turn it into a certain compatly-generated topological space with an extra filtration property, so that B is locally isometrically *-isometric to the functional algebra $C(\mathfrak{M}(B))$. After that we use the structure of $\mathfrak{M}(B)$ to obtain a version of a Arens-Kaplansky type theorem for real unital abelian locally C^*-algebras and a version of a Gelfand-Naimark type theorem for unital abelian locally JB-algebras.

2 Preliminaries

First, we recall some basic notions on topological *-algebras. A *-algebra (or involutive algebra) is an algebra B over $\mathbb{C}$ with an involution

$$^* : B \to B,$$

such that

$$(a + \lambda b)^* = a^* + \overline{\lambda} b^*,$$

and

$$(ab)^* = b^* a^*,$$

for every $a, b \in B$ and $\lambda \in \mathbb{C}$.

A seminorm $\|.\|$ on a *-algebra B is a C^*-seminorm if it is submultiplicative, i.e.

$$\|ab\| \leq \|a\| \, \|b\| ,$$

and satisfies the C^*-condition, i.e.

$$\left\|a^* a\right\| = \|a\|^2 ,$$

for every $a, b \in B$. Note that the C^*-condition alone implies that $\|.\|$ is submultiplicative, and in particular

$$\left\|a^*\right\| = \|a\| ,$$

for every $a \in B$ (cf. for example [13]).

When a seminorm $\|.\|$ on a *-algebra B is a C^*-norm, and B is complete in the topology generated by this norm, B is called a **C^*-algebra.**

Let a be a self-adjoint element in a unital C^*-algebra B. By $Sp(a)$ we denote the Hausdorff compact set $\lambda \in \mathbb{C}$, such that the elements $a - \lambda \mathbf{1}_B$ are not invertible in B. The Spectral Theorem states that the abelian C^*-subalgebra $C^*(a)$ of B generated by a and $\mathbf{1}_B$ is isometrically *-isomorphic to the functional algebra $C(Sp(a))$ of all continuous $\mathbb{C}$-valued functions on $Sp(a)$ with supremum norm, in which case $Sp(a)$ is homeomorphic to the spectrum $\mathfrak{M}(C^*(a))$ (see [32] for details).

A real Banach *-algebra $(A, \|.\|_A)$, such that

$$A \cap iA = \{\mathbf{0}_A\}$$

is called a **real C^*-algebra**, if there exists an extension $\|.\|_B$ of $\|.\|_A$ from A on

$$B = B \dotplus iB,$$

such that $(B, \|.\|_B)$ is a C^*-algebra (see [17] and [27] for details).

A real Jordan Banach algebra A with identity and a norm $\|.\|$ is called **JB-algebra** if:

1).

$$\left\|a^2\right\| = \|a\|^2 \,;$$

2).

$$\|a\|^2 \leq \left\|a^2 + b^2\right\| ,$$

for all $a, b \in A$.

Let B be a real associate algebra. Then $(B, \circ)$, where " $\circ$ " be the symmetric multiplication

$$a \circ b = \frac{1}{2}(ab \mid ba),$$

$a, b \in B$, is a real Jordan algebra. A Jordan algebra which can be obtained in such a way is called *special*. A Jordan algebra which cannot be obtained in such a way is called **exceptional**. An example of an exceptional Jordan algebra is the algebra M_3^8 of all 3×3 symmetric matrices over Calley numbers (see [18]).

Let now B be a real or complex associative *-algebra. Then $(B_{sa}, \circ)$, where " $\circ$ " be the symmetric multiplication

$$a \circ b = \frac{1}{2}(ab + ba),$$

$a, b \in B_{sa}$, is real Jordan algebra.

Example 2.1. *Let B be a complex or real C^*-algebra. Then its self-adjoint part B_{sa} with the symmetric multiplication is a JB-algebra. Each norm closed Jordan subalgebra of B_{sa} is as well a JB-algebra.*

Example 2.2. *Let H be a complex Hilbert space, $B(H)$ be the C^*-algebra of all linear bounded operators on H. A JB-algebra which is an operator norm closed Jordan subalgebra of $B(H)_{sa}$ is called a JC-**algebra**.*

Each special JB-algebra is isometrically Jordan isomorphic to a JC-algebra (see [1]).

A homomorphic image of a special Jordan algebra does not have to be special. However, if you have a surjection from a special JB-algebra A onto a JB-algebra B, then B has to be special (see [1] for details).

Let A be a unital abelian (associative) JB-algebra. Then there exists a compact Hausdorff space X such that A is isometrically isomorphic to the Jordan algebra $(C^{\mathbb{R}}(X), \circ)$ of all real valued continuous functions on X, with symmetric multiplication (see [18]).

2.1 Complex locally C*-algebras

A topological *-algebra is a *-algebra B equipped with a topology making the operations (addition, multiplication, additive inverse, involution) jointly continuous. For a topological *-algebra B, one puts $\Gamma = P(B)$ for the set of continuous C^*-seminorms on A. One can see that $P(B)$ is a directed set with respect to pointwise ordering, because

$$\max\{\|.\|_\alpha, \|.\|_\beta\} \in P(B)$$

for every $\|.\|_\alpha, \|.\|_\beta \in P(B)$, where $\alpha, \beta \in \Lambda$, with Λ being a certain directed set.

For a topological *-algebra B, and $\|.\|_\alpha \in P(B)$, $\alpha \in \Lambda$,

$$\ker \|.\|_\alpha = \{a \in B : \|a\|_\alpha = 0\}$$

is a *-ideal in A, and $\|.\|_\alpha$ induces a C^*-norm (we as well denote it by $\|.\|_\alpha$) on the quotient

$$B_\alpha = B/\ker \|.\|_\alpha,$$

and B_α is automatically complete in the topology generated by the norm $\|.\|_\alpha$, thus is a C^*-algebra (see [13] for details). Each pair $\|.\|_\alpha, \|.\|_\beta \in P(B)$, such that

$$\beta \succeq \alpha,$$

$\alpha, \beta \in \Lambda$, induces a natural (continuous) surjective *-homomorphism

$$g_\alpha^\beta : B_\beta \to B_\alpha.$$

Let, again, Λ be a set of indices, directed by a relation (reflexive, transitive, antisymmetric) " $\preceq$ ". Let

$$\{B_\alpha, \alpha \in \Lambda\}$$

be a family of C^*-algebras, and g_α^β be, for

$$\alpha \preceq \beta,$$

the continuous linear surjective *-mappings

$$g_\alpha^\beta : B_\beta \longrightarrow B_\alpha,$$

so that

$$g_\alpha^\alpha(x_\alpha) = x_\alpha,$$

for all $\alpha \in \Lambda$, and

$$g_\alpha^\beta \circ g_\beta^\gamma = g_\alpha^\gamma,$$

whenever

$$\alpha \preceq \beta \preceq \gamma.$$

Let Ξ be the collections $\{g_\alpha^\beta\}$ of all such transformations. Let A be a *-subalgebra of the direct product algebra

$$\prod_{\alpha \in \Lambda} B_\alpha,$$

so that for its elements

$$x_\alpha = g_\alpha^\beta(x_\beta),$$

for all

$$\alpha \preceq \beta,$$

where

$$x_\alpha \in B_\alpha,$$

and

$$x_\beta \in B_\beta.$$

Definition 2.3. *The *-algebra B constructed above is called a* ***Hausdorff projective limit*** *of the projective family*

$$\{B_\alpha, \alpha \in \Lambda\},$$

relatively to the collection

$$\Xi = \{g_\alpha^\beta : \alpha, \beta \in \Lambda : \alpha \preceq \beta\},$$

and is denoted by

$$\underleftarrow{\lim} B_\alpha,$$

$\alpha \in \Lambda$, and

$$B \cong \underleftarrow{\lim} B_\alpha,$$

$\alpha \in \Lambda$, is called the Arens-Michael decomposition of A.

It is well known (see, for example [40]) that for each $x \in B$, and each pair $\alpha, \beta \in \Lambda$, such that $\alpha \preceq \beta$, there is a natural surjective projection

$$\pi_\beta : B \longrightarrow B_\beta,$$

defined by

$$\pi_\alpha(x) = g_\alpha^\beta(\pi_\beta(x)),$$

and each projection π_α for all $\alpha \in \Lambda$ is continuous.

Definition 2.4. *A topological *-algebra (B, τ) over $\mathbb{C}$ is called a* ***locally C^*-algebra*** *(**of type** Λ), where Λ is a directed set, if there exists a projective family of C^*-algebras*

$$\{B_\alpha; g_\alpha^\beta; \alpha, \beta \in \Lambda\},$$

so that

$$B \cong \underleftarrow{\lim} B_\alpha,$$

$\alpha \in \Lambda$, i.e. B is topologically *-iseomorphic (*-isomorphic and homeomorphic) to a projective limit of a projective family B_α, $\alpha \in \Lambda$, of C^*-algebras, i.e. there exists Arens-Michael decomposition of B of type (weight) Λ, composed entirely of C^*-algebras.

A topological *-algebra (B, τ_Γ) over $\mathbb{C}$ is a locally C^*-algebra of type Λ iff A is a complete Hausdorff topological *-algebra in which the topology τ_Γ is generated by a saturated separating family $\Gamma = P(B) = \{\|.\|_\alpha, \alpha \in \Lambda\}$ of C^*-seminorms (see [13] for details).

Each locally C^*-algebra (B, τ_Γ) has a type, determined by the saturated separating family Γ of C^*-seminorms, but a given *-algebra B can be generally speaking endowed with different topologies turning it into a locally C^*-algebra, perhaps of a different type (see [13] for details).

Example 2.5. *Every C^*-algebra is a locally C^*-algebra of type Λ for Λ being a one-point set.*

Example 2.6. *A closed *-subalgebra of a locally C^*-algebra of type Λ is a locally C^*-algebra of type Λ. Because a closed *-subalgebra of a locally C^*-algebra of type Λ can be a C^*-algebra, one can see that a C^*-algebra can have different types.*

Example 2.7. *The product $\prod\limits_{\alpha\in\Lambda} B_\alpha$ of C^*-algebras A_α, with the product topology, is a locally C^*-algebra of type Λ.*

Example 2.8. *Let X be a k-space- compactly generated Hausdorff space (this means that a subset $Y \subset X$ is closed iff $Y \cap K$ is closed for every compact subset $K_\alpha \subset X$, $\alpha \in \Lambda$). Then the algebra $C(X)$ of all continuous, not necessarily bounded complex-valued functions on X, with the topology of uniform convergence on compact subsets, is a locally C^*-algebra of type Λ. It is well known that all metrizable spaces and all locally compact Hausdorff spaces are compactly generated (see [13] for details).*

Example 2.9. *Let $Haus$ denotes the category of Hausdorff topological spaces and their continuous maps, and $Tych$ denotes its full subcategory of Tychonoff or completely regular spaces.*

*Let now X be a Tychnoff space. A **filtration** of X is a directed family*

$$F = \{X_\alpha, \alpha \in \Lambda\},$$

of compact subsets of X, such that:

(i) Ordered by inclusion, the family F is filtered, meaning:

a) Every one-point subset of X is in F,

b) A compact subset of an element of F is again in F,

c) Every finite union of X_α's belongs to some X_β in F (the saturation property);

(ii) The inductive limit of the family F being taken in the subcategory $Tych$ is homeomorphic to the space X, i.e. we can think that

$$X = \bigcup_{\alpha\in\Lambda} X_\alpha,$$

and the topology on X is a defined as follows: a subset

$$Y \subset X$$

is closed iff $Y \cap X_\alpha$ is closed for every compact element

$$X_\alpha \subset F,$$

$\alpha \in \Lambda$.

By a ***filtered space (of type** Λ**)*** *we understand a pair*

$$(X, F) = (X, X_\alpha, \alpha \in \Lambda),$$

consisting of a Tychonoff space X and a fixed filtration F on it.

By a morphism from a filtered space $(X, X_\alpha, \alpha \in \Lambda_1)$ of type Λ_1 to a filtered space $(Y, Y_t, t \in \Lambda_2)$ of type Λ_2 we mean a continuous map

$$u : X \to Y,$$

subject to the following extra condition: For every $\alpha \in \Lambda_1$ there exists $t \in \Lambda_2$, such that

$$u(X_\alpha) \subset Y_t.$$

With these morphisms the filtered spaces form a category denoted by Filt.

Two filtered spaces $(X, X_\alpha, \alpha \in \Lambda)$ and $(Y, Y_\alpha, \alpha \in \Lambda)$ of type Λ are called ***locally homeomorphic****, iff there exists a homeomorphism*

$$u : X \to Y,$$

such that for each $\alpha \in \Lambda$, $u(X_\alpha)$ is homeomorphic to Y_α.

Let now

$$(X, X_\alpha, \alpha \in \Lambda)$$

be a filtered space of type Λ. Let us consider the algebra $C(X)$ of all continuous complex-valued functions on X equipped with the supremum seminorms $\|.\|_\alpha$, $\alpha \in \Lambda$, corresponding to X_α. This family of seminorms

$$\Gamma = \{\|.\|_\alpha \, , \alpha \in \Lambda\},$$

generate on $C(X)$ a locally convex topology τ_Γ, such that $(C(X), \tau_\Gamma)$ becomes an unital abelian functional locally C^*-algebra of type Λ, which has the Arens-Michael decomposition

$$(C(X), \tau_\Gamma) = \varprojlim (C(X_\alpha), \|.\|_\alpha),$$

$\alpha \in \Lambda$, into a projective limit of a projective family of functional abelian unital C^*-algebras

$$(C(X_\alpha), \|.\|_\alpha),$$

$\alpha \in \Lambda$. (see [34] for details.)

Let B be a locally C^*-algebra. Then an element $a \in B$ is called **bounded**, if

$$\|a\|_\infty = \{\sup \|a\|_\alpha , \alpha \in \Lambda : \|.\|_\alpha \in P(B)\} < \infty.$$

The set of all bounded elements of B is denoted by $b(B)$.

It is well-known that for each locally C^*-algebra B, its set $b(B)$ of bounded elements of B is a locally C^*-subalgebra, which is a C^*-algebra in the norm $\|.\|_\infty$, such that it is dense in B in its topology (see for example [13]).

2.2 Real locally C*-algebras

Let us now recall some notions and results about real locally C^*-algebras.

Definition 2.10. *A topological* **-algebra* A *over* $\mathbb{R}$, *such that*

$$A \cap iA = \{\mathbf{0}_A\},$$

is called a ***hermitian real topological*** ****-algebra*** *if*

$$\sigma(a) \subset \mathbb{R},$$

for each

$$a \in A_{sa},$$

i.e.

$$a = a^*,$$

where $\sigma(a)$ *is a* ***local spectrum*** *of* A, *computed in its* ***complexification***

$$B = A \dotplus iA.$$

Definition 2.11. *A topological* **-algebra* A *over* $\mathbb{R}$ *is called a* ***real locally*** C^****-algebra*** *(of* ***type*** Λ*) if there exists a projective family of real* C^**-algebras*

$$\{A_\alpha; \overleftarrow{g}_\alpha^\beta; \alpha, \beta \in \Lambda\},$$

so that

$$A \cong \varprojlim A_\alpha,$$

i.e. A *is real topologically* **-isomorphic to a projective limit of a projective family of real* C^**-algebras, i.e. there exits an Arens-Michael decomposition of* A *composed entirely of real* C^**-algebras.*

Theorem 2.12 (Katz and Friedman [22]). *A topological* **-algebra* A *over* $\mathbb{R}$ *is a real locally* C^**-algebra iff* A *is a complete real hermitian Hausdorff topological* **-algebra in which topology is generated by a saturated separating family of* C^**-seminorms.*

Example 2.13. *Every real C^*-algebra is a locally C^*-algebra of type Λ, for Λ being a one-point set.*

Example 2.14. *A closed real *-subalgebra of a real locally C^*-algebra of type Λ is a real locally C^*-algebra of type Λ. Due to the fact that a closed *-subalgebra of a real locally C^*-algebra of type Λ can be a real C^*-algebra, one can see that a real C^*-algebra can have different types.*

Example 2.15. *The product $\prod\limits_{\alpha\in\Lambda} A_\alpha$ of real C^*-algebras A_α, with the product topology, is a real locally C^*-algebra of type Λ.*

Example 2.16. *Let X be a compactly generated Hausdorff space (this means that a subset $Y \subset X$ is closed iff $Y \cap K$ is closed for every compact subset $K \subset X$). Then the algebra $C(X)$ of all continuous, not necessarily bounded real-valued functions on X, with the topology of uniform convergence on compact subsets, is a real locally C^*-algebra.*

Katz and Friedman have presented in [22] that for each real locally C^*-algebra A, its complexification

$$B = A \dotplus iA,$$

can be endowed with topology turning it to a locally C^*-algebra. More specifically, the following theorem is valid:

Theorem 2.17 (Katz and Friedman [23]). *Let A be a real locally C^*-algebra of type Λ, and $P(A)$ be the saturated separating family of continuous C^*-seminorms $\widetilde{\|.\|}_\alpha$, $\alpha \in \Lambda$, generating the topology of A. Then there exists a saturated separating family $P(B)$ of continuous C^*-seminorms $\|.\|_\alpha$, $\alpha \in \Lambda$, on its complexification*

$$B = A \dotplus iA,$$

such that $P(B)$ generates a topology an B turning it into a locally C^*-algebra of type Λ, and such that $\|.\|_\alpha$ extends $\widetilde{\|.\|}_\alpha$ for each $\alpha \in \Lambda$.

Remark 2.18. *A detailed proof of the aforementioned result, which is using [24] and [21], will appear in a separate publication.*

Let A be a locally real C^*-algebra. Then an element $a \in A$ is called **bounded**, if

$$\widetilde{\|a\|}_\infty = \{\sup \widetilde{\|a\|}_\alpha, \alpha \in \Lambda : \widetilde{\|.\|}_\alpha \in P(A)\} < \infty.$$

The set of all bounded elements of A is denoted by $b(A)$.

It is known that for each real locally C^*-algebra A, its set $b(A)$ of bounded elements of A is a real locally C^*-subalgebra, which is a real C^*-algebra in the norm $\widetilde{\|.\|}_\infty$, such that it is dense in A in its topology, and

$$b(B) = b(A) \dotplus ib(A),$$

(cf. [22]), where $b(B)$ is the set of all bounded elements of

$$B = A \dotplus iA.$$

2.3 Locally JB-algebras

Definition 2.19. *A topological Jordan algebra A over $\mathbb{R}$ is called a* **locally** JB**-algebra** *if there exists a projective family of JB-algebras*

$$\{A_\alpha; g_\alpha^\beta; \alpha, \beta \in \Lambda\},$$

so that

$$A \cong \varprojlim A_\alpha,$$

$\alpha \in \Lambda$.

Otherwise speaking, A is a locally JB-algebra iff A is topologically Jordan-isomorphic to a projective limit of a projective family of JB-algebras, i.e. there exits its Arens-Michael decomposition

$$A \cong \varprojlim A_\alpha,$$

$\alpha \in \Lambda$, composed entirely of JB-algebras A_α. (see [22] for details).

Example 2.20. *Every JB-algebra is a locally JB-algebra.*

Example 2.21. *A closed complex Jordan subalgebra of a locally JB-algebra is a locally JB-algebra.*

Example 2.22. *The product $\prod_{\alpha \in \Lambda} A_\alpha$ of JB-algebras A_α, with the product topology, is a locally JB-algebra.*

Example 2.23. *Let X be a compactly generated Hausdorff space (this means that a subset $Y \subset X$ is closed iff $Y \cap K$ is closed for every compact subset $K \subset X$). Then the algebra $C(X)$ of all continuous, not necessarily bounded real-valued functions on X, with the topology of uniform convergence on compact subsets, is a locally JB-algebra. It is well known that all metrizable spaces and all locally compact Hausdorff spaces are compactly generated.*

Example 2.24. *Let B be a complex or real locally C^*-algebra. Then its self-adjoint part B_{sa} with the symmetric multiplication is a locally JB-algebra. Each projective topology closed Jordan subalgebra of B_{sa} is as well a locally JB-algebra.*

Example 2.25. *Let H be a complex locally Hilbert space (inductive limit of an inductive family of Hilbert spaces). Let $L(H)$ be the locally C^*-algebra of all continuous linear op-*

*erators on H with projective topology. A locally JB-algebra which is a projective topology closed Jordan subalgebra of $L(H)_{sa}$ is called a locally JC-**algebra**.*

Let B be a locally JB-algebra. Then an element $a \in B$ is called **bounded**, if

$$\|a\|_\infty = \{\sup \|a\|_\alpha \, , \alpha \in \Lambda : \|.\|_\alpha \in N(B)\} < \infty.$$

The set of all bounded elements of A is denoted by $b(B)$. It is known that $b(B)$ is a JB-algebra (see [22] for details).

3 Gelfand-Naimark type Theorem for unital abelian complex locally C*-algebras

Let (B, τ_Γ) be a locally C^*-algebra, where the Hausdorff locally-convex topology τ_Γ is generated by

$$\Gamma = P(B) = \{p_\alpha\}_{\alpha \in \Lambda},$$

(where Λ is a directed set)- a saturated separating family of C^*-seminorms on B. One has an Arens-Michael decomposition

$$B \cong \varprojlim B_\alpha,$$

where

$$B_\alpha = B/\ker(p_\alpha),$$

is a projective family of C^*-algebras.

Because a *-isomorphism between C^*-algebras is automatically an isometry (see for example [32]), the following definition makes sense.

Definition 3.1. *Let*

$$\psi : B \to C,$$

be a *-isomorphism from a locally C^*-algebra B of type Λ to a locally C^*-algebra C of type Λ. Morphism ψ is called a locally *-isometry iff for each Arens-Michael decomposition of B,

$$B \cong \varprojlim B_\alpha,$$

there exists some Arens-Michael decomposition of C,

$$C \cong \varprojlim C_\alpha,$$

$\alpha \in \Lambda$, such that $\psi(B_\alpha)$ is *-isomorphic to C_α, for each $\alpha \in \Lambda$.

At first, let us recall the following result due to Nassopoulos:

Proposition 3.2. *For a* **-homomorphism*

$$\psi : B \to C,$$

from a locally C^**-algebra* B *of type* Λ_1 *onto a locally* C^**-algebra* C *of type* Λ_2*, the following two statements are equivalent:*

(i) ψ *is continuous;*

(ii) ψ *is decomposable, in the sense that for each* $\beta \in \Lambda_2$ *there exists* $\alpha \in \Lambda_1$*, and a unique morphism*

$$\psi_\beta^\alpha : B_\alpha \to C_\beta,$$

so that

$$\psi_\beta^\alpha \circ {}_B\pi_\alpha = {}_C\pi_\beta \circ \psi,$$

where

$${}_B\pi_\alpha : B \to B_\alpha,$$

and

$${}_C\pi_\beta : C \to C_\beta,$$

are natural projections.

Proof. See [34] for details. □

The next proposition explains the true meaning of the notion of the locally *-isometric mapping.

Proposition 3.3. *Let*

$$\psi : B \to C$$

be an algebraic **-isomorphism from a locally* C^**-algebra* B *of type* Λ *to a locally* C^**-algebra* C *of type* Λ*. Then* ψ *is a locally* **-isometric mapping iff* ψ *is a homeomorphism.*

Proof. Let ψ be locally *-isometric mapping. Thus, due to Proposition 1 ψ and ψ^{-1} are both decomposable, and therefore both are continuous.

Inversly, if ψ and ψ^{-1} are both continuous, then due to Proposition 1 they are both decomposable, and as a result ψ is *-isometric. □

Corollary 3.4. *Let*

$$\psi : B \to C$$

be an algebraic **-isomorphism from a locally* C^**-algebra* B *of type* Λ_1 *to a locally* C^**-algebra* C *of type* Λ_2*. If* ψ *is a homeomorphism, then* Λ_1 *can be identified with* Λ_2 *(means* B *and* C *have the same type), and* ψ *is a locally* **-isometric mapping.*

Proof. From Proposition 2 it follows that ψ being a homeomorphic mapping is equivalent to ψ being locally *-isometric mapping. It implies that there exists a bijection between the projective families B_α, $\alpha \in \Lambda_1$ and C_β, $\beta \in \Lambda_2$, thus, there exists a bijection between Λ_1 and Λ_2. □

Now, let us proceed to the notion of a global spectrum of a locally C^*-algebra. By a character on a topological algebra we understand a non-zero complex-valued morphism on it. Let (B, τ_Γ) be a unital locally C^*-algebra, and $\mathfrak{M}(B)$ be the set of all continuous characters on B, that is

$$\mathfrak{M}(B) = \{\varphi \in A' : \varphi \neq 0 \text{ and } \varphi(xy) = \varphi(x)\varphi(y), \ \ \forall x, y \in B\}.$$

Let us endow $\mathfrak{M}(B)$ with the relative topology $s|_{\mathfrak{M}(B)}$ from B'_s, where by s we mean the weak *-topology $\sigma(B', B)$ on B'. That topology is the topology of simple or point-wise convergence on B'. The resulting Tychonoff topological space $(\mathfrak{M}(B), s)$ is called the **global** or **topological spectrum** of (B, τ_Γ). In what follows we will refer to $(\mathfrak{M}(B), s)$ by writing simply $\mathfrak{M}(B)$.

Let (B, τ_Γ) be a complex unital abelian locally C^*-algebra, where the Hausdorff locally-convex topology τ_Γ is generated by

$$\Gamma = P(B) = \{\|.\|_\alpha\}_{\alpha \in \Lambda},$$

(where Λ is a directed set) - a saturated separating family of C^*-seminorms on B. One has the Arens-Michael decomposition

$$B \cong \varprojlim B_\alpha,$$

where

$$B_\alpha = B/\ker(\|.\|_\alpha),$$

$\alpha \in \Lambda$ is a projective family of C^*-algebras. Let $\mathfrak{M}(B_\alpha)$ denote the topological spectrum of C^*-algebra B_α, $\alpha \in \Lambda$. It is well known that each $\mathfrak{M}(B_\alpha)$ is a Hausdorff compact (see for example [32]).

From [20] it follows that $\mathfrak{M}(B)$ can be identified with the inductive limit of the family $\mathfrak{M}(B_\alpha)$ equipped with direct limit topology, i.e.

$$\mathfrak{M}(B) = \bigcup_{\alpha \in \Lambda} \mathfrak{M}(B_\alpha), \text{ and } \mathfrak{M}(B) \cong \varinjlim \mathfrak{M}(B_\alpha).$$

Recall the following:

Definition 3.5. *A topological space X is called **functionally Hausdorff** if for any two points $x, y \in X$ there exists a continuous function*

$$f : X \to [0, 1],$$

such that

$$f(x) = 0 \textit{ and } f(y) = 1.$$

Remark 3.6. *This condition is stronger then being Hausdorff and weaker then complete regularity even for compactly generated spaces (see [36] for details).*

Now we can formulate the main result of the present section.

Theorem 3.7 (Gelfand-Naimark type Theorem). *Each unital abelian locally C^*-algebra (B, τ_Γ) of type Λ is locally *-isometric to the algebra $C(\mathfrak{M}(B))$ of all continuous complex-valued functions on the functionally Hausdorff space $\mathfrak{M}(B)$, such that their restrictions on each compact $\mathfrak{M}(B_\alpha)$, $\alpha \in \Lambda$, are continuous, equipped with the corresponding to $C(\mathfrak{M}(B_\alpha))$'s supremum seminorms $\widehat{\|.\|}_\alpha$, $\alpha \in \Lambda$. This family of seminorms generates on $C(\mathfrak{M}(B))$ a locally convex topology $\tau_{\widehat{\Gamma}}$,*

$$\widehat{\Gamma} = \{\widehat{\|.\|}_\alpha, \alpha \in \Lambda\},$$

such that $(C(\mathfrak{M}(B)), \tau_{\widehat{\Gamma}})$ becomes an unital abelian functional locally C^*-algebra of type Λ, for which there exists an Arens-Michael decomposition

$$(C(\mathfrak{M}(B)), \tau_{\widehat{\Gamma}}) \cong \varprojlim (C(\mathfrak{M}(B_\alpha)), \widehat{\|.\|}_\alpha),$$

$\alpha \in \Lambda$, into a projective limit of a projective family of functional abelian unital C^*-algebras

$$(C(\mathfrak{M}(B_\alpha)), \widehat{\|.\|}_\alpha),$$

$\alpha \in \Lambda$.

Proof. We will start with showing that if (B, τ_Γ) is unital abelian locally C^*-algebra of type Λ, and $\mathfrak{M}(B)$ is its global spectrum, then $\mathfrak{M}(B)$ with the inductive limit topology is a compactly generated, functionally Hausdorff space, and $\mathfrak{M}(B_\alpha)$, $\alpha \in \Lambda$ is a distinguished family of generating compacts in it, and B is locally $*$-isometric to the algebra $C(\mathfrak{M}(B))$ of all continuous complex-valued functions on the space $\mathfrak{M}(B)$, such that their restrictions on each compact $\mathfrak{M}(B_\alpha)$, $\alpha \in \Lambda$, are continuous.

Indeed, let (B, τ_Γ) be an unital abelian locally C^*-algebra of type Λ, and

$$B \cong \varprojlim B_\alpha,$$

be its Arens-Michael decomposition as a projective limit of the projective family of unital abelian C^*-algebras B_α, $\alpha \in \Lambda$.

From the fact that each B_α, $\alpha \in \Lambda$, is unital and abelian it follows (see for example [32]) that each space $\mathfrak{M}(B_\alpha)$, $\alpha \in \Lambda$, is a Hausdorff compact, thus, $\mathfrak{M}(B)$ is compactly generated. Due to the fact that

$$\mathfrak{M}(B) \cong \varinjlim \mathfrak{M}(B_\alpha),$$

$\alpha \in \Lambda$, when $\mathfrak{M}(B)$ is equipped with the inductive limit topology, one can easily see that $\mathfrak{M}(B_\alpha)$, $\alpha \in \Lambda$ is a distinguished family of generating compacts in $\mathfrak{M}(B)$.

Again, as each B_α, $\alpha \in \Lambda$, is unital and abelian it follows (see for example [32]) that each algebra B_α is isometrically *-isomorphic to $C(\mathfrak{M}(B_\alpha))$. Let now

$$\varphi_\alpha : B_\alpha \to C(\mathfrak{M}(B_\alpha)),$$

$$a_\alpha \mapsto f_{a_\alpha},$$

where $a_\alpha \in B_\alpha, f_{a_\alpha} \in C(\mathfrak{M}(B_\alpha))$ be that isomorphism, where for each $h_\alpha \in \mathfrak{M}(B_\alpha)$,

$$h_\alpha(a_\alpha) = f_{a_\alpha}(h_\alpha).$$

As B_α, $\alpha \in \Lambda$, is a projective family of C^*-algebras, for each pair $\alpha, \beta \in \Lambda$, such that

$$\alpha \preceq \beta,$$

there exists a surjective *-homomorphism

$$g_\alpha^\beta : B_\beta \to B_\alpha,$$

such that

1).

$$g_\alpha^\alpha = id;$$

2).

$$g_\alpha^\gamma = g_\alpha^\beta \circ g_\beta^\gamma,$$

for any $\alpha, \beta, \gamma \in \Lambda$, $\alpha \preceq \beta \preceq \gamma$,

and a surjective *-homomorphism

$$\pi_\alpha : B \to B_\alpha,$$

such that

$$\pi_\alpha = g_\alpha^\beta \circ \pi_\beta,$$

for each pair $\alpha, \beta \in \Lambda$, such that $\alpha \preceq \beta$.

Let f_{a_β} be an arbitrary function from $C(\mathfrak{M}(B_\beta))$, such that

$$\Phi_\beta^{-1}(f_{a_\beta}) = a_\beta,$$

$\beta \in \Lambda$. Let us define for $\alpha \preceq \beta$, $\alpha, \beta \in \Lambda$,

$$\tilde{g}_\alpha^\beta : C(\mathfrak{M}(B_\beta)) \to C(\mathfrak{M}(B_\alpha)),$$

$$\tilde{g}_\alpha^\beta(\Phi_\beta(a_\beta)) \mapsto \Phi_\alpha(g_\alpha^\beta(a_\beta)),$$

where $a_\beta \in B_\beta$. One can easily see that with these morphisms the family $C(\mathfrak{M}(B_\alpha))$ is a projective family, and let $\varprojlim C(\mathfrak{M}(B_\alpha))$, $\alpha \in \Lambda$, be its projective limit algebra equipped with its projective topology generated by supremum seminorms $\widehat{\|.\|}_\alpha$, $\alpha \in \Lambda$ built from the C^*-supremum norms on functional C^*-algebras $C(\mathfrak{M}(B_\alpha))$. Let

$$\widetilde{\pi}_\alpha : \varprojlim C(\mathfrak{M}(B_\alpha)) \to C(\mathfrak{M}(B_\alpha)),$$

be the natural projection which is a surjective *-homomorphism from $\varprojlim C(\mathfrak{M}(B_\alpha))$ onto $C(\mathfrak{M}(B_\alpha))$, such that

$$\widetilde{\pi}_\alpha = \widetilde{g}_\alpha^\beta \circ \widetilde{\pi}_\beta,$$

for each $\alpha \preceq \beta$, $\alpha, \beta \in \Lambda$

Let now

$$\varphi : B \to \varprojlim C(\mathfrak{M}(B_\alpha)),$$

be such that

$$\widetilde{\pi}_\alpha \circ \varphi = \varphi_\alpha \circ \pi_\alpha,$$

for each $\alpha \in \Lambda$. It is routine to check that φ is a locally *-isometry from B onto $\varprojlim C(\mathfrak{M}(B_\alpha))$. On the other hand, Let $C(\mathfrak{M}(B))$ be the set of all continuous complex-valued functions f on $\mathfrak{M}(B)$ such that $f|_{\mathfrak{M}(B_\alpha)}$ is continuous for all $\alpha \in \Lambda$. It is shown in [11] that $C(\mathfrak{M}(B))$ is naturally endowed with operations and topology turning it to a locally C^*-algebra, and that $\varprojlim C(\mathfrak{M}(B_\alpha))$ is *-isomorphic to $C(\mathfrak{M}(B))$, and from the proof one can see that the aforementioned isomorphism is a locally *-isometry, thus, B is locally *-isometric to $C(\mathfrak{M}(B))$.

So, without a loss of generality we now can say that $C(\mathfrak{M}(B))$ locally *-isometric to $\varprojlim C(\mathfrak{M}(B_\alpha))$, $\alpha \in \Lambda$, and

$$\varphi : B \to C(\mathfrak{M}(B)) \cong \varprojlim C(\mathfrak{M}(B_\alpha)),$$

is a locally *-isometry, and

$$\widetilde{\pi}_\alpha : C(\mathfrak{M}(B)) \cong \varprojlim C(\mathfrak{M}(B_\alpha)) \to C(\mathfrak{M}(B_\alpha)),$$

is the natural projection for all $\alpha \in \Lambda$.

Now, let non-zero $h, t \in \mathfrak{M}(B)$, and let us assume they cannot be separated by continuous functions, Thus, for any $a \in B$,

$$(\varphi(a))(h) = (\varphi(a))(t).$$

But, on the other hand, h and t are continuous on B, thus, for each $a \in B$, there exist indices α_h and α_t in Λ, such that $h_{\alpha_h} \in \mathfrak{M}(B_{\alpha_h})$ and $h_{\alpha_t} \in \mathfrak{M}(B_{\alpha_t})$, where

$$h(a) = h_{\alpha_h}(\pi_{\alpha_h}(a)) \text{ and } t(a) = t_{\alpha_t}(\pi_{\alpha_t}(a)).$$

Let $\beta \in \Lambda$ be such that $\beta \succeq \alpha_h$ and $\beta \succeq \alpha_h$. We can now define $h_\beta, t_\beta \in \mathfrak{M}(B_\beta)$ such that

$$h_\beta(\pi_\beta(a)) = h_{\alpha_h}(g_{\alpha_h}^\beta(\pi_\beta(a)) = h(a) \text{ and } t_\beta(\pi_\beta(a)) = t_{\alpha_t}(g_{\alpha_t}^\beta(\pi_\beta(a)) = t(a).$$

So, we get that

$$\begin{aligned} h(a) &= h_\beta(\pi_\beta(a)) = (\varphi_\beta(\pi_\beta(a)))(h_\beta) = (\varphi(a))(h) \\ &= (\varphi(a))(t) = (\varphi_\beta(\pi_\beta(a)))(t_\beta) = t_\beta(\pi_\beta(a)) = t(a), \end{aligned}$$

for all $a \in B$, thus, $h = t$. □

On the other hand, the following theorem is valid:

Theorem 3.8. *Let (X, F_X) be a functionally Hausdorff filtered space of type Λ, and $C(X)$ be the algebra of all continuous complex-valued functions on the space X, such that their restrictions on each compact $X_\alpha \in F_X$, $\alpha \in \Lambda$, are continuous, with the locally convex topology generated by supremum seminorms on $C(X_\alpha)$, $\alpha \in \Lambda$. Then there exists a family of surjective *-morphisms*

$$g_\alpha^\beta : C(X_\beta) \to C(X_\alpha),$$

, $\alpha \preceq \beta$, $\alpha, \beta \in \Lambda$, and with these morphisms the family of functional unital abelian C^*-algebras $C(X_\alpha)$, $\alpha \in \Lambda$, forms a projective family, and its projective limit $\varprojlim C(X_\alpha)$ with projective topology is a unital abelian locally C^*-algebra of type Λ, which is locally *-isometric to $C(X)$.

Proof. Let $f \in C(X)$ be arbitrary. Denote by

$$f_\alpha = f|_{X_\alpha},$$

$\alpha \in \Lambda$. Because f_α is continuous on Hausdorff compact X_α, it is bounded on X_α, thus $f_\alpha \in C(X_\alpha)$ for all $\alpha \in \Lambda$. We define

$$\pi_\alpha : C(X) \to C(X_\alpha),$$

as $f \mapsto f_\alpha$. One can easily verify that π_α is a surjective *-homomorphism for all $\alpha \in \Lambda$. Now, for every $\alpha \preceq \beta$, $\alpha, \beta \in \Lambda$ we define a mapping

$$g_\alpha^\beta : C(X_\beta) \to C(X_\alpha),$$

as

$$\pi_\beta(f) \mapsto \pi_\alpha(f),$$

for every $f \in C(X)$. One can easily observe that each g_α^β, $\alpha \preceq \beta$, $\alpha, \beta \in \Lambda$ is a surjective *-homomorphism from the C^*-algebra $C(X_\beta)$ with supremum norm onto $C(X_\alpha)$ with supremum norm. Thus, the family $C(X_\alpha)$ with morphisms g_α^β, $\alpha \preceq \beta$, $\alpha, \beta \in \Lambda$, is a projective family. Its projective limit $\varprojlim C(X_\alpha)$, $\alpha \in \Lambda$, with projective topology is an unital functional locally C^*-algebra of type Λ which is locally *-isometric to $C(X)$. □

The following corollary is a version of the Spectral Theorem for locally C^*-algebras.

Corollary 3.9 (Spectral Theorem). *Let B be a unital locally C^*-algebra of type Λ, a be its self-adjoint element, and $LC^*(a)$ be the unital abelian locally C^*-subalgebra of type Λ of B generated by a and 1_B. Then $LC^*(a)$ is locally *-isometric to the functional locally C^*-algebra $C(\mathfrak{M}(LC^*(a)))$ of type Λ with the Arens-Michael decomposition $\varprojlim C(Sp(\pi_\alpha(a)))$, $\alpha \in \Lambda$.*

Proof. Immediately follows from theorems 3 and 4. □

From a categorical prospective we get the following generalization of the Gelfand duality for the category of unital locally C^*-algebras of type Λ with local *-isometries vs. functionally Hausdorff filtered spaces of type Λ with local homeomorphisms.

Theorem 3.10. *Let (X, F_X) and $C(X)$ be as in theorem 4. Then the functor*

$$X \mapsto C(X),$$

is a contravariant category equivalence from the category of functionally Hausdorff space filtered space of type Λ with locally continuous morphisms to the category of unital abelian locally C^*-algebras of type Λ with local homomorphisms.

Proof. Direct functor is established the following way:

Let (X, F_X) and (Y, F_Y) be two functionally Hausdorff filtered spaces of type Λ, and

$$\phi : (X, F_X) \to (Y, F_Y)$$

is a locally continuous, i.e. ϕ is continuous, and

$$\phi_\alpha : X_\alpha \to Y_\alpha$$

is continuous for each $\alpha \in \Lambda$, where

$$\phi_\alpha = \phi|_{X_\alpha},$$

and $\phi_\alpha(X_\alpha) \subset Y_\alpha$. One can observe that

$$C(\phi) : C(Y) \to C(X)$$

will be a homomorphism given by

$$C(\phi)(f_Y) = f_Y \circ \phi,$$

where $f_Y \in C(Y)$, and for each $\alpha \in \Lambda$, $C(\phi_\alpha) : C(Y_\alpha) \to C(X_\alpha)$ will be a homomorphism given by

$$C(\phi_\alpha)(f_{Y_\alpha}) = f_{Y_\alpha} \circ \phi_\alpha,$$

where $f_{Y_\alpha} = f|_{Y_\alpha}$, and $f_{Y_\alpha} \in C(Y_\alpha)$.

Now, we need an inverse functor. Such functor is supposed to assign to each unital abelian locally C^*-algebra B of type Λ the spectrum $\mathfrak{M}(B)$ of B as in theorem 4 above.

Let the mapping

$$\nabla : B_1 \to B_2$$

be a locally continuous unital homomorphism from a unital abelian locally C^*-algebra B_1 of type Λ to a unital abelian locally C^*-algebra B_1 of type Λ. It means that for any Arens-Michael locally *-isometric decomposition of

$$B_1 \cong \varprojlim {}_1B_\alpha,$$

as a projective limit of the projective family of unital abelian C^*-algebras ${}_1B_\alpha$, $\alpha \in \Lambda$, there exists an Arens-Michael locally *-isometric decomposition of

$$B_2 \cong \varprojlim {}_2B_\alpha,$$

as a projective limit of the projective family of unital abelian C^*-algebras ${}_2B_\alpha$, $\alpha \in \Lambda$, where each $\alpha \in \Lambda$,

$$\nabla_\alpha : {}_1B_\alpha \to {}_2B_\alpha$$

is defined as

$$\nabla_\alpha({}_1\pi_\alpha(a)) = {}_2\pi_\alpha(\nabla(a))$$

for each $a \in B_1$, and is a continuous (as a composition of continuous mappings) unital homomorphism of C^*-algebras. Let us notice that for every $b \in B_2$, the function

$$x \mapsto x(b)$$

defines a continuous function from $\mathfrak{M}(B_2)$ to $\mathbb{C}$, such that for each $\alpha \in \Lambda$,

$$x_\alpha \mapsto x_\alpha({}_2\pi_\alpha(b))$$

is continuous, where

$$x_\alpha = x|_{{}_2B_\alpha}.$$

Consider now the function

$$x \mapsto x \circ \nabla.$$

It is a continuous mapping from $\mathfrak{M}(B_2)$ to $\mathfrak{M}(B_1)$, such that for each $\alpha \in \Lambda$,

$$x_\alpha \mapsto x_\alpha \circ \nabla_\alpha$$

is a continuous mapping from $\mathfrak{M}({}_2B_\alpha)$ to $\mathfrak{M}({}_1B_\alpha)$, i.e.

$$x \mapsto x \circ \nabla$$

is locally continuous.

Finally, from theorems 3 and 4 it follows that these two functors are inverses of each other. □

Remark 3.11. *Some elements of theorems 3, 4 and 5 first appeared in the paper [36] of Phillips in the realm of a quasitopological structure on $\mathfrak{M}(B)$.*

Remark 3.12. *Examples of Weidner from [43] (see as well [36]) show that you cannot make the restriction on (X, F_X) any weaker.*

4 Arens-Kaplansky type Theorem for unital abelian real locally C*-algebras

Let X be a topological space. By a **topological involution** on X we understand a homeomorphism

$$h : X \to X,$$

such that

$$h(h(x)) = x,$$

for all $x \in X$.

Let now

$$(X, F) = (X, X_\alpha, \alpha \in \Lambda),$$

be a filtered space of type Λ. A topological involution h on X is called a **locally topological involution** on (X, F_X) iff for each $X_\alpha \in F$, $\alpha \in \Lambda$,

$$h_\alpha = h|_{X_\alpha} : X_\alpha \to X_\alpha,$$

is a topological involution on X_α.

The following example of a motivational example for what follows.

Example 4.1. *Let $C(X)$ be a locally C^*-algebra of a type Λ of all continuous complex-valued functions on a functionally Hausdorff filtered space (X, F_X) of type Λ, such that their restrictions on each Hausdorff compact X_α, $\alpha \in \Lambda$, are continuous, equipped with the supremum seminorms $\|.\|_\alpha$, $\alpha \in \Lambda$, corresponding to $C(X_\alpha)$. Let now h be a locally topological involution on (X, F). We define*

$$C(X, h) = \{f \in C(X) : f(h(x)) = \overline{f}(x), \forall x \in X\}.$$

The algebra $C(X, h)$ is a real *-subalgebra of $C(X)$. One can see that $C(X, h)$ is a real locally C^*-algebra of type Λ with the topology inherited from $C(X)$.

The following theorem is valid:

Theorem 4.2. *Let*

$$(X, F_X) = (X, X_\alpha, \alpha \in \Lambda)$$

be a functionally Hausdorff filtered space of type Λ, and h be a locally topological involution on (X, F_X). Let $C(X)$ be a locally C^*-algebra of a type Λ of all continuous complex-valued functions on a functionally Hausdorff space filtered space (X, F_X) of type Λ, such that their restrictions on each Hausdorff compact X_α, $\alpha \in \Lambda$, are continuous, equipped with the corresponding to $C(X_\alpha)$'s supremum seminorms $\|.\|_\alpha$, $\alpha \in \Lambda$. We define a mapping

$$\Psi : C(X) \to C(X)$$

as

$$\Psi(f)(x) = \overline{f}(h(x)),$$

for any $f \in C(X)$ *and all* $x \in X$. *Then:*

(i) Ψ *is a conjugate-linear involutory* **-automorphism of* $C(X)$ *of order* 2, *and*

$$C(X, h) = \{f \in C(X) : \Psi(f) = f\},$$

is a real **-subalgebra in* $C(X)$;

(ii)

$$C(X) = C(X, h) \dotplus iC(X, h),$$

or, speaking otherwise, each $f \in C(X)$ *has a unique decomposition as* $u + iv$, *with* $u, v \in C(X, h)$;

(iii) let

$$\widehat{\pi}_\alpha : C(X, h) \to C(X_\alpha, h_\alpha),$$

$\alpha \in \Lambda$, *be a natural surjective projection from* $C(X, h)$ *onto* $C(X_\alpha, h_\alpha)$. *Define*

$$\Psi_\alpha : C(X_\alpha) \to C(X_\alpha)$$

as

$$\Psi_\alpha(\widehat{\pi}_\alpha(f))(h_\alpha(\widehat{\pi}_\alpha(x)) = \widehat{\pi}_\alpha(\Psi(f)(x)),$$

for any $f \in C(X)$ *and all* $x \in X$, $\alpha \in \Lambda$. *Then* Ψ *is locally isometric, i.e. for each* $\alpha \in \Lambda$, Ψ_α *is a conjugate-linear involutory* **-automorphism of* $C(X_\alpha)$ *of order* 2, *and and an isometry;*

(iv) there exists an Arens-Michael decomposition of $C(X, h)$ *into a projective limit of a projective family of real unital functional* C^**-algebras*

$$C(X, h) \cong \varprojlim C(X_\alpha, h_\alpha),$$

$\alpha \in \Lambda$, *where each* h_α *is a topological involution on* X_α, *i.e* $C(X, h)$ *is a real locally* C^**-algebra of type* Λ;

(v) let $(C(X))_{\mathbb{R}}$ *(resp.* $(C(X_\alpha))_{\mathbb{R}}$, $\alpha \in \Lambda$*) denote* $C(X)$ *(resp.* $C(X_\alpha)$*) regarded as a real algebra. For* $f \in C(X)$, *we define*

$$P(f) = \frac{1}{2}[f + \Psi(f)].$$

Then P is a continuous linear surjective mapping

$$P : (C(X))_{\mathbb{R}} \to C(X, h),$$

satisfying

$$P^2 = P;$$

(vi) every continuous conjugate-linear involutory **-automorphism of* $C(X)$ *of order* 2 *arises from a locally topological involution on X (a functionally Hausdorff filtered space* (X, F_X) *of type* Λ*) in a manner described above.*

Proof. One can see (i) by direct verification.

To establish (ii), notice that since Ψ is a conjugate-linear involutory *-automorphism of order two,

$$\Psi(f + \Psi(f)) = \Psi(f) + f \text{ and } \Psi(\frac{1}{i}(f - \Psi(f)) = \frac{1}{i}(f - \Psi(f)).$$

From this we get that

$$f = \frac{1}{2}(f + \Psi(f)) + i(\frac{1}{2i}(f - \Psi(f))),$$

where $\frac{1}{2}(f + \Psi(f))$, $\frac{1}{2i}(f - \Psi(f)) \in C(X, h)$. Also, because

$$\Psi(f + ig) = f - ig,$$

where $f, g \in C(X, h)$. Now we get

$$f = \frac{1}{2}((f + ig) + \Psi(f + ig)) \text{ and } g = \frac{1}{2i}((f + ig) - \Psi(f + ig)),$$

which proves the uniqueness of decomposition.

One can easily see (iii) by direct verification.

To show (iv), let

$$C(X) \cong \varprojlim C(X_\alpha),$$

$\alpha \in \Lambda$ be a locally *-isometric Arens-Michael decomposition of $C(X)$ as a projective limit of the projective family of unital abelian functional C^*-algebras $C(X_\alpha)$ with supremum norms, $\alpha \in \Lambda$. Let

$$\pi_\alpha : C(X) \to C(X_\alpha)$$

be the natural projection from $C(X)$ onto $C(X_\alpha)$, and

$$g_\alpha^\beta : C(X_\beta) \to C(X_\alpha)$$

be the connecting surjections for all $\alpha, \beta \in \Lambda$, $\alpha \preceq \beta$. We define

$$\Psi_\alpha : C(X_\alpha) \to C(X_\alpha)$$

as

$$\Psi_\alpha = \pi_\alpha \circ \Psi,$$

for every $\alpha \in \Lambda$. One can easily see that

$$\pi_\alpha \circ \Psi = \Psi_\alpha \circ \pi_\alpha$$

for each $\alpha \in \Lambda$. If the topological involution

$$h_\alpha : X_\alpha \to X_\alpha$$

is defined as

$$h_\alpha = h|_{X_\alpha},$$

let

$$C(X_\alpha, h_\alpha) = \{f_\alpha \in C(X_\alpha) : \Psi_\alpha(f_\alpha) = f_\alpha\}$$

for any $\alpha \in \Lambda$ and arbitrary $f_\alpha \in C(X_\alpha)$. From the fact that Ψ_α is a conjugate-linear *-automorphism of order two on a C^*-algebra $C(X_\alpha)$ it follows that $C(X_\alpha, h_\alpha)$ is a unital real C^*-algebra in the topology inherited from $C(X_\alpha)$, and

$$C(X_\alpha) = C(X_\alpha, h_\alpha) \dotplus iC(X_\alpha, h_\alpha),$$

and

$$C(X_\alpha, h_\alpha) \cap iC(X_\alpha, h_\alpha) = \mathbf{0}_{C(X_\alpha)}.$$

Let

$$\widehat{g}_\alpha^\beta : C(X_\beta, h_\beta) \to C(X_\alpha, h_\alpha)$$

be defined as

$$\widehat{g}_\alpha^\beta = g_\alpha^\beta|_{C(X_\beta, h_\beta)}.$$

From the fact that g_α^β was a surjective *-homomorphism of complex C^*-algebras it follows that $\widehat{g}_\alpha^\beta$ is a surjective *-homomorphism of real C^*-algebras for all $\alpha \preceq \beta$, $\alpha, \beta \in \Lambda$. The family of real C^*-algebras $C(X_\alpha, h_\alpha)$ with the morphisms $\widehat{g}_\alpha^\beta$ forms a projective family, and its projective limit $\varprojlim C(X_\alpha, h_\alpha)$ is locally *-isomorphic to $C(X, h)$, thus, $C(X, h)$ is a real unital locally C^*-algebra of type Λ.

Statement in (v) immediately follows from (i) and (iii).

To show (vi), let Ψ be an arbitrary continuous conjugate-linear involutory *-automorphism of $C(X)$ of order 2, where X is a functionally Hausdorff filtered space (X, F_X) of type Λ. Then there exists a locally *-isometric Arens-Michael decomposition

$$C(X) \cong \varprojlim C(X_\alpha),$$

$\alpha \in \Lambda$, of $C(X)$ as a projective limit of the projective family of unital abelian functional C^*-algebras $C(X_\alpha)$ with supremum norms, $\alpha \in \Lambda$, and each $X_\alpha \subset F_X$. Let

$$\pi_\alpha : C(X) \to C(X_\alpha)$$

be the natural projection from $C(X)$ onto $C(X_\alpha)$, and

$$g_\alpha^\beta : C(X_\beta) \to C(X_\alpha)$$

be the connecting surjections for all $\alpha, \beta \in \Lambda$, $\alpha \preceq \beta$. We define

$$\Psi_\alpha : C(X_\alpha) \to C(X_\alpha)$$

as

$$\Psi_\alpha = \pi_\alpha \circ \Psi,$$

for every $\alpha \in \Lambda$. One can easily see that

$$\pi_\alpha \circ \Psi = \Psi_\alpha \circ \pi_\alpha$$

for each $\alpha \in \Lambda$. From [25] it follows that each Ψ_α, $\alpha \in \Lambda$, arises from a locally topological involution h_α on X_α. Direct verification shows that there exists a locally topological involution h on (X, F_X), such that for each $\alpha \in \Lambda$, $h_\alpha = h|_{X_\alpha}$. Thus

$$\Psi(f)(x) = \overline{f}(h(x)),$$

for any $f \in C(X)$ and all $x \in X$ because for each $\alpha \in \Lambda$,

$$\Psi_\alpha(f_\alpha)(x_\alpha) = \overline{f_\alpha}(h_\alpha(x_\alpha)),$$

for any $f_\alpha \in C(X_\alpha)$ and all $x_\alpha \in X_\alpha$. □

Now we can formulate and prove the Arens-Kaplansky type theorem for real abelian locally C^*-algebras:

Theorem 4.3 (Arens-Kaplansky type Theorem). *Let A be a real unital abelian locally C^*-algebra of type Λ, and*

$$B = A \dotplus iA$$

be its complexification. Then A is a real locally *-isometric to the real locally C^*-subalgebra

$$D = \{d \in C(\mathfrak{M}(B)) : d(h(x)) = \overline{d}, \text{ for any } x \in \mathfrak{M}(B)\},$$

where h is a locally topological involution on the filtered space $\mathfrak{M}(B)$.

Proof. Let

$$A \cong \varprojlim A_\alpha,$$

$\alpha \in \Lambda$, be the Arens-Michael decomposition of A as a projective limit of real unital abelian C^*-algebras A_α, $\alpha \in \Lambda$. Then $B_\alpha = A_\alpha \dotplus iA_\alpha$ is a unital abelian C^*-algebra for each $\alpha \in \Lambda$, and

$$B \cong \varprojlim B_\alpha,$$

$\alpha \in \Lambda$, is the Arens-Michael decomposition of B as a projective limit of unital abelian C^*-algebras B_α, $\alpha \in \Lambda$. Each B_α is isometrically *-isomorphic to $C(\mathfrak{M}(B_\alpha))$, where $\mathfrak{M}(B_\alpha)$ is a Hausdorff compact. From theorem 3 above it follows that B is locally *-isometric to $C(\mathfrak{M}(B))$, where $\mathfrak{M}(B)$ is a functionally Hausdorff filtered space of type Λ

From theorem 6 it follows that there exists a continuous conjugate-linear involutory *-antiautomorphism Ψ of $C(\mathfrak{M}(B))$ of order 2, which generates a continuous conjugate-linear involutory *-antiautomorphism Ψ' of B of order 2. We define

$$\Psi'_\alpha = \pi_\alpha \circ \Psi',$$

where π_α is the natural projection from B onto B_α, $\alpha \in \Lambda$, which is a conjugate-linear involutory antiautomorphism of B_α of order 2. Therefore there exists a family of topological involutions h_α on each $\mathfrak{M}(B_\alpha)$ which generate Ψ'_α for each $\alpha \in \Lambda$, and they in turn generate a locally topological involution h on the space $\mathfrak{M}(B)$, which, as one can directly verify, satisfies the condition of the theorem. □

From a categorical prospective we get the following generalization of Arens-Kaplansky duality for the category of pairs $(C(X), \Psi)$ of unital abelian locally C^*-algebras of type Λ with conjugate-linear involutory antiautomorphisms, with morphisms being local *-isometries, vs. pairs (X, h) of functionally Hausdorff filtered spaces of type Λ and local topological involutions with morphisms being local homeomorphisms.

Theorem 4.4. *Let (X, h) and $C(X, h)$ be as in theorem 6. Then the functor*

$$(X, h) \mapsto C(X, h),$$

is a contravariant category equivalence from the category of pairs of functionally Hausdorff filtered spaces of type Λ with local topological involutions and morphisms being local homeomorphisms to the category of pairs of unital abelian locally C^*-algebras of type Λ with continuous conjugate-linear *-antiautomorphisms of order two with morphisms being local *-isometries.

Proof. Direct functor is established the following way:

Let $((X, F_X),_X h)$ and $((Y, F_Y),_Y h)$ be two pairs of functionally Hausdorff filtered spaces of type Λ local topological involutions, with

$$\phi : ((X, F_X),_X h) \to ((Y, F_Y),_Y h)$$

satisfying the conditions (i) and (ii) below:

(i) ϕ is locally continuous, i.e.

$$\phi : X \to Y$$

is continuous, and

$$\phi_\alpha : X_\alpha \to Y_\alpha$$

is continuous for each $\alpha \in \Lambda$, where

$$\phi_\alpha = \phi|_{X_\alpha},$$

and $\phi_\alpha(X_\alpha) \subset Y_\alpha$;

(ii)

$$\phi(_X h(x)) = {}_Y h(\phi(x)),$$

for each $x \in X$, and thus

$$\phi_\alpha(_X h_\alpha(x_\alpha)) = {}_Y h_\alpha(\phi_\alpha(x_\alpha)),$$

for each $x_\alpha \in X_\alpha$.

One can observe that

$$C(\varphi) : C(Y) \to C(X),$$

given by

$$C(\varphi)(f_Y) = f_Y \circ \phi,$$

where $f_Y \in C(Y)$, will be a homomorphism, and for each $\alpha \in \Lambda$,

$$C(\phi_\alpha) : C(Y_\alpha) \to C(X_\alpha),$$

given by

$$C(\phi_\alpha)(f_{Y_\alpha}) = f_{Y_\alpha} \circ \phi_\alpha,$$

where $f_{Y_\alpha} = f|_{Y_\alpha}$, and $f_{Y_\alpha} \in C(Y_\alpha)$, will be a homomorphism. The restriction of $C(\phi)$ on $C(Y, {}_Y h)$ is the required homomorphism

$$C(\phi)|_{C(Y,{}_Y h)} : C(Y, {}_Y h) \to C(X, {}_X h),$$

such that

$$C(\phi)(f_Y({}_Y h(\phi(x)))) = f_Y \circ \phi({}_X h(x)),$$

where $f_Y \in C(Y)$, and for each $\alpha \in \Lambda$ and $x \in X$, and for each $\alpha \in \Lambda$,

$$C(\phi_\alpha)|_{C(Y_\alpha,{}_Y h_\alpha)} : C(Y_\alpha, {}_Y h_\alpha) \to C(X_\alpha, {}_X h_\alpha),$$

given by

$$C(\phi_\alpha)(f_{Y_\alpha}({}_Y h_\alpha(\phi_\alpha(x_\alpha)))) = f_{Y_\alpha} \circ \phi_\alpha({}_X h_\alpha(x_\alpha)),$$

where $f_{Y_\alpha} = f|_{Y_\alpha}$, and $f_{Y_\alpha} \in C(Y_\alpha)$ and $x_\alpha \in X_\alpha$, will be a homomorphism.

Now, we need an inverse functor. Such a functor is supposed to assign to each pair (B, Ψ) composed of a unital abelian locally C^*-algebra B of type Λ with its continuous conjugate-linear *-automorphism Ψ, the pair $(\mathfrak{M}(B), h)$ composed of the spectrum $\mathfrak{M}(B)$ of B with h being a locally topological involution of $\mathfrak{M}(B)$ as in theorem 6 above.

Let the mapping

$$\nabla : (B_1, \Psi_1) \to (B_2, \Psi_2)$$

be a locally continuous unital homomorphism from a unital abelian locally C^*-algebra B_1 of type Λ to a unital abelian locally C^*-algebra B_1 of type Λ, such that

$$\nabla \circ \Psi_1 = \Psi_2 \circ \nabla.$$

It means that for any Arens-Michael locally *-isometric decomposition of

$$B_1 \cong \varprojlim {}_1B_\alpha,$$

as a projective limit of the projective family of unital abelian C^*-algebras ${}_1B_\alpha$, $\alpha \in \Lambda$, there exists an Arens-Michael locally *-isometric decomposition of

$$B_2 \cong \varprojlim {}_2B_\alpha,$$

as a projective limit of the projective family of unital abelian C^*-algebras ${}_2B_\alpha$, $\alpha \in \Lambda$, such that for each $\alpha \in \Lambda$,

$$\nabla_\alpha : ({}_1B_\alpha, {}_1\Psi_\alpha) \to ({}_2B_\alpha, {}_2\Psi_\alpha)$$

is defined as

$$\nabla_\alpha({}_1\pi_\alpha(a)) = {}_2\pi_\alpha(\nabla(a))$$

for each $a \in B_1$, and is a continuous (as a composition of continuous mappings) unital homomorphism of C^*-algebras, and satisfies the condition

$$\nabla_\alpha \circ {}_1\Psi_\alpha = {}_2\Psi_\alpha \circ \nabla_\alpha.$$

Let us notice that for every $b \in B_2$, the function

$$x \mapsto x(b)$$

defines a continuous function from $\mathfrak{M}(B_2)$ to $\mathbb{C}$, such that for each $\alpha \in \Lambda$,

$$x_\alpha \mapsto x_\alpha({}_2\pi_\alpha(b))$$

is continuous, where

$$x_\alpha = x|_{{}_2B_\alpha}.$$

If now ${}_{\mathfrak{M}(B_2)}h$ is the locally topological involution on $\mathfrak{M}(B_2)$ which generates Ψ_2, then

$${}_{\mathfrak{M}(B_2)}h(x) \mapsto x(\Psi_2(b))$$

defines a continuous function from $\mathfrak{M}(B_2)$ to $\mathbb{C}$, such that for each $\alpha \in \Lambda$,

$${}_{\mathfrak{M}({}_2B_\alpha)}h_\alpha(x_\alpha) \mapsto x_\alpha({}_2\Psi_\alpha({}_2\pi_\alpha(b)))$$

Now consider the function

$$x \mapsto x \circ \nabla \circ \Psi_1.$$

It is a locally continuous mapping from $(\mathfrak{M}(B_2), {}_{\mathfrak{M}(B_2)}h)$ to $(\mathfrak{M}(B_1), {}_{\mathfrak{M}(B_1)}h)$, such that for each $\alpha \in \Lambda$,

$$x_\alpha \mapsto x_\alpha \circ \nabla_\alpha \circ {}_1\Psi_\alpha$$

is a continuous mapping from $(\mathfrak{M}({}_2B_\alpha), {}_{\mathfrak{M}(B_2)}h_\alpha)$ to $(\mathfrak{M}({}_1B_\alpha), {}_{\mathfrak{M}(B_1)}h_\alpha)$ i.e.

$$x \mapsto x \circ \nabla \circ \Psi_1$$

is locally continuous.

Finally, from theorems 5 and 6 it follows that these two functors are inverses of each other. □

5 Gelfand-Naimark type Theorem for unital abelian locally JB-algebras

Let B be a C^*-algebra. Then

$$B_{sa} = \{a \in B : a = a^*\}$$

with symmetric multiplication

$$a \circ b = \frac{1}{2}(ab + ba)$$

is a real Jordan algebra which is a JB-algebra (more precisely, a JC-algebra, see [18] for details). Turumaru showed in [41] that a C^*-algebra B is abelian iff $(B_{sa}, \circ)$ is an abelian Jordan algebra under symmetric multiplication.

We start by proving a version of Turumaru's theorem for locally C^*-algebras.

Theorem 5.1. *An abelian unital locally C^*-algebra (B, τ_B) of type Λ is an abelian if the locally JB-algebra $(B_{sa}, \circ), \tau_{(B_{sa},\circ)}$ is an abelian locally JC-algebra of type Λ. Conversely, if (A, τ_A) is an abelian locally JB-algebra of type Λ, then there exist operations and topology τ_B on its complexification $B = A + iA$, turning (B, τ_B) into an abelian unital locally C^*-algebra (B, τ_B) of type Λ.*

Proof. Let (B, τ_B) be a unital abelian locally C^*-algebra (B, τ_B) of type Λ, and

$$B \cong \varprojlim B_\alpha,$$

$\alpha \in \Lambda$ be its Arens-Michael decomposition as a projective limit of the projective family of unital abelian C^*-algebras B_α, $\alpha \in \Lambda$. It is easy to see that each connecting surjective C^*-morphism

$$g_\alpha^\beta : B_\beta \to B_\alpha,$$

has a property that

$$g_\alpha^\beta((B_\beta)_{sa}, \circ)) = ((B_\alpha)_{sa}, \circ),$$

for each pair $\alpha \preceq \beta$, $\alpha, \beta \in \Lambda$. Thus, the family of JC-algebras $((B_\alpha)_{sa}, \circ)$ is projective, and its projective limit with projective topology is Jordan isomorphic and homeomorphic to the abelian locally JC-algebra $((B_{sa}, \circ), \tau_{(B_{sa},\circ)})$, where

$$\tau_{(B_{sa},\circ)} = \tau_B|_{(B_{sa},\circ)}.$$

Conversely, let $((A.\circ), \tau_A)$ be an abelian unital locally JB-algebra of type Λ and let

$$A \cong \varprojlim A_\alpha,$$

$\alpha \in \Lambda$, be its Arens-Michael decomposition as a projective limit of the projective family of abelian unital JB-algebras $(A_\alpha, \circ)$, $\alpha \in \Lambda$, where the topology τ_A is generated by a saturated separating family of JB-seminorms p_α, such that for each $\alpha \in \Lambda$

$$p_\alpha(x) = \|\pi_\alpha(x)\|_{A_\alpha}$$

for every $x \in A$, where

$$\pi_\alpha : A \to A_\alpha$$

is the natural projection from A onto A_α, $\alpha \in \Lambda$. Let

$$B = \{x + iy : x, y \in A\}$$

be the complexification of A. One can easily see that with the product

$$(x + iy)(z + iw) = (x \circ z - y \circ w) + i(x \circ w + y \circ z)$$

and involution

$$(x + iy)^* = x - iy,$$

$x, y, z, w \in A$ we turn B into an abelian unital associative *-algebra with a unit

$$\mathbf{1}_B = \mathbf{1}_A = \mathbf{1}_A + i\mathbf{0}_A.$$

Let us for each $\alpha \in \Lambda$ define a function

$$\widehat{p}_\alpha : B \to \mathbb{R}$$

as

$$\widehat{p}_\alpha(x + iy) = \sqrt{p_\alpha(x^2 + y^2)},$$

$x, y \in A$. Let $x + iy = a \in B$, where $x, y \in A$. Then

$$\widehat{p}_\alpha(a)^2 = p_\alpha(a^* a) = \widehat{p}_\alpha(a^* a),$$

and clearly

$$\widehat{p}_\alpha(\lambda a) = |\lambda|\, \widehat{p}_\alpha(a)$$

for each $\alpha \in \Lambda$. Also,

$$\widehat{p}_\alpha(ab)^2 = \widehat{p}_\alpha((ab)^* ab) = \widehat{p}_\alpha(b^* a^* ab) = \widehat{p}_\alpha((a^* a)(b^* b)) = \widehat{p}_\alpha(\frac{1}{2}((a^* a)(b^* b)+(b^* b)(a^* a)))$$

$$= p_\alpha((a^* a) \circ (b^* b)) \le p_\alpha((a^* a)p_\alpha((b^* b) = \widehat{p}_\alpha((a^* a)\widehat{p}_\alpha((b^* b) = \widehat{p}_\alpha(a)^2\widehat{p}_\alpha(b)^2,$$

which implies that

$$\widehat{p}_\alpha(ab) \le \widehat{p}_\alpha(a)\widehat{p}_\alpha(b).$$

Now, if $a, b \in B$, and $a = x + iy$, $b = z + iw$, where $x, y, z, w \in A$, then we have

$$a^* b + b^* a = 2x \circ z + 2y \circ w \in A.$$

Also, due to the fact that p_α is a JB-seminorm for each $\alpha \in \Lambda$,

$$p_\alpha(x \circ z + y \circ w)^2 = p_\alpha((x \circ z + y \circ w)^2) \le p_\alpha((x \circ z + y \circ w)^2 + (x \circ w - y \circ z)^2)$$
$$= p_\alpha((x^2 + y^2) \circ (z^2 + w^2)) \le p_\alpha(x^2 + y^2) p_\alpha(z^2 + w^2) = \widehat{p}_\alpha(a)^2 \widehat{p}_\alpha(b)^2.$$

Thus, we get that

$$\widehat{p}_\alpha(a + b)^2 = \widehat{p}_\alpha((a + b)^*(a + b)) = p_\alpha((a + b)^*(a + b)) = \widehat{p}_\alpha(a^* a + (a^* b + b^* a) + b^* b)$$
$$= p_\alpha(a^* a + (2x \circ z + 2y \circ w) + b^* b) \le p_\alpha(a^* a) + p_\alpha(2x \circ z + 2y \circ w) + p_\alpha(b^* b)$$
$$= \widehat{p}_\alpha(a)^2 + 2p_\alpha(x \circ z + y \circ w) + \widehat{p}_\alpha(b)^2 = \widehat{p}_\alpha(a)^2 + 2\widehat{p}_\alpha(a)\widehat{p}_\alpha(b) + \widehat{p}_\alpha(b)^2 = (\widehat{p}_\alpha(a) + \widehat{p}_\alpha(b))^2,$$

which implies

$$\widehat{p}_\alpha(a + b) \le \widehat{p}_\alpha(a) + \widehat{p}_\alpha(b).$$

We have proved that for each $\alpha \in \Lambda$, $\widehat{p}_\alpha$ is a C^*-seminorm on B. Also note, that the family of C^*-seminorms $\widehat{p}_\alpha$, $\alpha \in \Lambda$ is saturated and separating as a consequence of saturation and separability of the family of seminorms p_α, $\alpha \in \Lambda$. Thus (B, τ_B) is a locally C^*-algebra of type Λ, where the topology τ_B is generated by family $\widehat{p}_\alpha$, $\alpha \in \Lambda$. □

One can see that the representation theory of unital abelian locally JB-algebra A is the representation theory of pairs of the form $(B, *)$, where

$$B = A + iA$$

is the unital abelian locally C^*-algebras of complexification of A's and " $*$ " is its involution (which is a continuous order 2 involutory linear antiautomorphisms on B). So, our given unital abelian locally JB-algebra A is Jordan isomorphic and homeomorphic to the Jordan algebra $(B_{sa}, \circ)$ of self-adjoint elements of the locally C^*-algebra B of its complexification with topology on B extended from the topology on A, with symmetric multiplication, which is the set of fixed points under the actions of the " $*$ " on B.

Theorem 5.2. *Each unital abelian locally JB-algebra (A, τ_A) of type Λ is locally *-isometric to the algebra $C^{\mathbb{R}}(\mathfrak{M}(B))$ of all continuous real-valued functions on the functionally Hausdorff space $\mathfrak{M}(B, \tau_B)$ (where*

$$(B, \tau_B) = (A, \tau_A) \dotplus i(A, \tau_A)$$

is a complexification of (A, τ_A)), such that their restrictions on each compact $\mathfrak{M}(B_\alpha)$, $\alpha \in \Lambda$ (where $B_\alpha = A_\alpha + iA_\alpha$), are continuous, equipped with the corresponding to $C^{\mathbb{R}}(\mathfrak{M}(B_\alpha))$'s supremum seminorms $\|.\|_\alpha$, $\alpha \in \Lambda$. This family of seminorms generates on $C^{\mathbb{R}}(\mathfrak{M}(B))$ a locally convex topology $\tau_{C^{\mathbb{R}}(\mathfrak{M}(B))}$, such that $(C^{\mathbb{R}}(\mathfrak{M}(B)), \tau_{C^{\mathbb{R}}(\mathfrak{M}(B))})$ becomes an abelian unital functional locally JB-algebra of type Λ, for which there exists an Arens-Michael decomposition

$$(C^{\mathbb{R}}(\mathfrak{M}(B)), \tau_{C^{\mathbb{R}}(\mathfrak{M}(B))}) \cong \varprojlim C^{\mathbb{R}}(\mathfrak{M}(B_\alpha)),$$

$\alpha \in \Lambda$, into a projective limit of a projective family of functional abelian unital JB-algebras

$$C^{\mathbb{R}}(\mathfrak{M}(B_\alpha)),$$

$\alpha \in \Lambda$.

Proof. Let

$$(B, \tau_B) = (A, \tau_A) \dot{+} i(A, \tau_A)$$

be the abelian unital locally C^*-algebra (see theorem 9 above). If

$$(A, \tau_A) \cong \varprojlim A_\alpha,$$

$\alpha \in \Lambda$, is the Arens-Michael decomposition of (A, τ_A), then one can easily see that there exists an Arens-Michael decomposition of (B, τ_B)

$$(B, \tau_B) \cong \varprojlim B_\alpha,$$

$\alpha \in \Lambda$, such that each

$$B_\alpha = A_\alpha \dot{+} iA_\alpha,$$

$\alpha \in \Lambda$. Indeed, if

$$g_\alpha^\beta : A_\beta \to A_\alpha$$

be the contractive JB-surjection for each $\alpha \preceq \beta$, $\alpha, \beta \in \Lambda$, we can define

$$\widehat{g}_\alpha^\beta : B_\beta \to B_\alpha$$

as

$$\widehat{g}_\alpha^\beta(x_\beta + iy_\beta) = g_\alpha^\beta(x_\beta) + ig_\alpha^\beta(y_\beta),$$

where

$$x_\beta = \pi_\beta(x) \in A_\beta \text{ and } y_\beta = \pi_\beta(y) \in A_\beta,$$

while

$$g_\alpha^\beta(x_\beta) = \pi_\alpha(x) = x_\alpha \in A_\alpha \text{ and } g_\alpha^\beta(y_\beta) = \pi_\alpha(y) = y_\alpha \in A_\alpha,$$

for every $x, y \in A$.

One can see that $\widehat{g}_\alpha^\beta$ is a surjective *-homomorphism, and from the fact that each A_α is a C^*-algebra it follows that they are contractions and thus are continuous.

Now, we apply to (B, τ_B) the theorems 3 and 4 above to get that (B, τ_B) is locally *-isometric to the abelian unital locally C^*-algebra

$$(C(\mathfrak{M}(B)), \tau_{C(\mathfrak{M}(B))}) \cong \varprojlim C(\mathfrak{M}(B_\alpha)),$$

and $(C^{\mathbb{R}}(\mathfrak{M}(B)), \tau_{C^{\mathbb{R}}(\mathfrak{M}(B))})$ is obviously its self-adjoint part. Thus, $(C^{\mathbb{R}}(\mathfrak{M}(B)), \tau_{C(\mathfrak{M}(B))})$ locally Jordan isometric to $\varprojlim C^{\mathbb{R}}(\mathfrak{M}(B_\alpha))$, where each $C^{\mathbb{R}}(\mathfrak{M}(B_\alpha))$ is Jordan isometrically isomorphic to the self adjoint part of $C(\mathfrak{M}(B_\alpha))$ for each $\alpha \in \Lambda$. □

The following corollary is a version of the Spectral Theorem for locally JB-algebras.

Corollary 5.3 (Spectral Theorem). *Let A be a unital locally JB-algebra of type Λ, x be its element, and $LJB(x)$ be the unital locally JB-subalgebra of type Λ of A generated by x and $\mathbf{1}_A$. Then $LJB(x)$ is locally Jordan isometric to the functional locally JB-algebra $C^{\mathbb{R}}(\mathfrak{M}(LC^*(x)))$ of type Λ with the Arens-Michael decomposition $\varprojlim C^{\mathbb{R}}(Sp(\pi_\alpha(a)))$, $\alpha \in \Lambda$.*

Proof. Immediately follows from theorem 4, corollary 2 and theorem 10 above. □

Bibliography

[1] Alfsen, E.M., Shultz, F.W., Størmer, E. *A Gelfand-Naimark theorem for Jordan algebras.* (English) Advances in Math. Vol. 28 (1978), No. 1, pp. 11–56.

[2] Allan, G.R. *On a class of locally convex algebras.* (English) Proc. London Math. Soc. (3) Vol. 17 (1967), pp. 91–114.

[3] Apostol, C. *b^*-algebras and their representation.* (English) J. London Math. Soc. (2) Vol. 3 (1971), pp. 30–38.

[4] Arens, R.F., Kaplansky, I. *Topological representation of algebras.* (English) Trans. Amer. Math. Soc. Vol. 63 (1948), pp. 457–481.

[5] Arens, R.F. *A generalization of normed rings.* (English) Pacific J. Math. Vol. 2 (1952), pp. 455–471.

[6] Beckenstein, E., Narici, L., Suffel, C. *Topological algebras.* (English) North-Holland Mathematics Studies, Vol. 24. Notas de Matemática, No. 60. [Mathematical Notes, No. 60] North-Holland Publishing Co., Amsterdam-New York-Oxford (1977), 370 pp.

[7] Binz, E. *Continuous convergence on $C(X)$.* (English) Lectures Notes in Mathematics, Vol. 469. Springer-Verlag, Berlin-New York (1975), 140 pp.

[8] Dubuc, E.J. *Concrete quasitopoi.* (English) Applications of sheaves (Proc. Res. Sympos. Appl. Sheaf Theory to Logic, Algebra and Anal., Univ. Durham, Durham, 1977), pp. 239–254, Lecture Notes in Math., Vol. 753, Springer, Berlin (1979).

[9] Dubuc, E.J., Porta, H. *Convenient categories of topological algebras, and their duality theory.* (English) J. Pure Appl. Algebra Vol. 1 (1971), No. 3, pp. 281–316.

[10] Dubuc, E.J., Porta, H. *Uniform spaces, Spanier quasitopologies, and a duality for locally convex algebras.* (English) J. Austral. Math. Soc. Ser. A Vol. 29 (1980), No. 1, pp. 99–128.

[11] El Harti, R., Lukács, G. *Bounded and unitary elements in pro-C^*-algebras.* (English) Appl. Categ. Structures Vol. 14 (2006), No. 2, pp. 151–164.

[12] Emch, G.G. *Algebraic methods in Statistical Mechanics and Quantum Field Theory.* (English) Interscience Monographs and Texts in Physics and Astronomy. Vol. XXVI. New York etc.: Wiley-Interscience, a division of John Wiley & Sons, Inc. (1972), 333 pp.

[13] Fragoulopoulou, M. *Topological algebras with involution.* (English) North-Holland Mathematics Studies, Vol. 200. Elsevier Science B.V., Amsterdam (2005), 495 pp.

[14] Gelfand, I.M. *Normierte Ringe.* (German) Rec. Math. [Mat. Sbornik] N. S. Vol. 9 (51), (1941), pp. 3–24.

[15] Gelfand, I.M., Naimark, M.A. *On the imbedding of normed rings into the ring of operators in Hilbert space.* (English) Rec. Math. [Mat. Sbornik] N.S. 1Vol. 2(54) (1943), pp. 197–213.

[16] Gillman, L., Jerison, M. *Rings of continuous functions.* (English) The University Series in Higher Mathematics D. Van Nostrand Co., Inc., Princeton, N.J.-Toronto-London-New York (1960), 300 pp.
[17] Goodearl, K.R. *Notes on real and complex C^*-algebras.* (English) Shiva Mathematics Series, No. 5. Shiva Publishing Ltd., Nantwich (1982), 211 pp.
[18] Hanche-Olsen, H., Størmer, E. *Jordan operator algebras.* (English) Monographs and Studies in Mathematics, Vol. 21. Pitman (Advanced Publishing Program), Boston, MA (1984), 183 pp.
[19] Hong, S.-S., Nel, L.D. *Duality theorems for algebras in convenient categories.* (English) Math. Z. Vol. 166 (1979), No. 2, pp. 131–136.
[20] Inoue, A. *Locally C^*-algebra.* (English) Mem. Fac. Sci. Kyushu Univ. Ser. A Vol. 25 (1971), pp. 197–235.
[21] Katz, A.A., Friedman, O. *On Gelfand-Naimark theorem for projective limits of real C^*- and JB-algebras.* (English), pp. 95-97., in the Ayupov, Sh.A. (ed.), *Abstracts of the International Conference "Operator Algebras and Quantum Probability".* (Russian), Tashkent, Uzbekistan, September 7–10, 2005. “Universitet”, Tashkent, 2005, 252 pp.
[22] Katz, A.A., Friedman, O. *On projective limits of real C^*- and Jordan operator algebras.* (English) Vladikavkaz. Mat. Zh. Vol. 8 (2006), No. 2, pp. 33–38.
[23] Katz, A.A., Friedman, O. *On the Gelfand-Naimark type theorem and the extension of a generating family of C^*-seminorms from a real locally C^*-algebra to its complexification.* (English), 2012, 15 pp. Preprint.
[24] Katz, A.A., Friedman, O., Kushnir, R. *On inductive limits of real Hilbert spaces.* (English), pp. 98–100., in the Ayupov, Sh.A. (ed.), *Abstracts of the International Conference "Operator Algebras and Quantum Probability".* (Russian), Tashkent, Uzbekistan, September 7–10, 2005. “Universitet”, Tashkent, 2005, 252 pp.
[25] Kulkarni, S.H., Limaye, B.V. *Real function algebras.* (English) Monographs and Textbooks in Pure and Applied Mathematics, No. 168. Marcel Dekker, Inc., New York (1992), 186 pp.
[26] Lassner, G. *Über Realisierungen gewisser *-Algebren.* (German) Math. Nachr. Vol. 52 (1972), pp. 161–166.
[27] Li, B.-R. *Real operator algebras.* (English) World Scientific Publishing Co., Inc., River Edge, NJ (2003), 241 pp.
[28] MacLane, S. *Categories for the working mathematician.* (English) 2nd Ed. Graduate Texts in Mathematics, Vol. 5. Springer-Verlag, New York (1998), 314 pp.
[29] Mallios, A. *Topological algebras. Selected topics.* (English) North-Holland Mathematics Studies, Vol. 124. Notas de Matemática [Mathematical Notes], No. 109. North-Holland Publishing Co., Amsterdam (1986), 535 pp.
[30] Michael, E.A. *Locally multiplicatively-convex topological algebras.* (English) Mem. Amer. Math. Soc. (1952), No. 11, 79 pp.
[31] Morris, P.D., Wulbert, D.E. *Functional representation of topological algebras.* (English) Pacific J. Math. Vol. 22 (1967) pp. 323–337.
[32] Murphy, G.J. *C^*-algebras and operator theory.* (English) Academic Press, Inc., Boston, MA, (1990), 286 pp.
[33] Nassopoulos, G.F. *On a comparison of real with complex involutive complete algebras.* (English) J. Math. Sci. (New York) Vol. 96 (1999), No. 6, pp. 3755–3765.
[34] Nassopoulos, G.F. *Spectral decomposition and duality in commutative locally C^*-algebras.* (English) Topological algebras and applications, pp. 303–317, Contemp. Math., No. 427, Amer. Math. Soc., Providence, RI, (2007).
[35] Phillips, N.C. *Inverse limits of C^*-algebras and applications.* (English) Operator algebras and applications, Vol. 1, pp. 127–185, London Math. Soc. Lecture Note Ser., No. 135, Cambridge Univ. Press, Cambridge, (1988).

[36] Phillips, N.C. *Inverse limits of C^*-algebras.* (English) J. Operator Theory Vol. 19 (1988), No. 1, pp. 159–195.
[37] Schmüdgen, K. *Uber LMC***-Algebren.* (German) Math. Nachr. Vol. 68 (1975), pp. 167–182.
[38] Spanier, E. *Infinite symmetric products, function spaces, and duality.* (English) Ann. of Math. (2) Vol. 69 (1959), pp. 142–198 [erratum, p. 733].
[39] Steenrod, N.E. *A convenient category of topological spaces.* (English) Michigan Math. J. Vol. 14 (1967), pp. 133–152.
[40] Trèves, F. *Topological vector spaces, distributions and kernels.* (English) Unabridged republication of the 1967 original. Dover Publications, Inc., Mineola, NY, (2006), 565 pp.
[41] Turumaru, T. *On the commutativity of the C^*-algebra.* (English) Kōdai Math. Sem. Rep. Vol. 3 (1951), p. 51.
[42] Warner, S. *The topology of compact convergence on continuous function spaces.* (English) Duke Math. J. Vol. 25 (1958), pp. 265–282.
[43] Weidner, J. *KK-groups for generalized operator algebras. I, II.* (English) K-Theory, Vol. 3 (1989), No. 1, pp. 57–77, 79–98.
[44] Wenjen, C. *On semi-normed *-algebras.* (English) Pacific J. Math. Vol. 8 (1958), pp. 177–186.
[45] Xia, D.-X. *On semi-normed rings with involution.* (Russian) Izv. Akad. Nauk SSSR. Ser. Mat. Vol. 23 (1959), pp. 509–528.

Rodia I. Hadjigeorgiou and Mohamed Oudadess

Multipliers and strictly real topological algebras

Abstract: This work is an extension of an authors' paper on locally convex algebras. Here no convexity is assumed at all. Using adequate notions, in particular the Gel'fand radical, classical statements on multipliers and strictly real Banach algebras are obtained in the context of topological algebras modulo appropriate conditions; specifically Kaplansky's commutativity result.

Keywords: Left(right) multiplier, multiplier algebra, Gel'fand radical, strictly real algebra, projective limit topological algebra, inductive limit topological algebra

MSC: 46H05, 46H20

1 Introduction and Preliminaries

Multipliers have been considered in commutative Banach and Fréchet m-convex algebras without order (see comments at the beginning of Section 2). Here, the frame is the one of general topological algebras. It is shown, in particular, that *on a topological algebra E without order satisfying the closed graph theorem, every multiplier is continuous* (Theorem 2.1). Also, *every multiplier can be seen as a continuous function on the carrier space* (Theorem 2.2, Theorem 2.4), as *Ju-Kwei Wang* [28, p. 1132, Theorem 2.1] has obtained in commutative Banach algebras without order. Moreover, results in relation with the Gel'fand representation $\mathcal{G}_E$ are given (Proposition 2.3, Corollary 2.7), along with a characterization for a linear operator on E to be a multiplier (Theorem 2.6).

Section 3 is devoted to the study of some properties of strictly real topological (non locally convex) algebras. It is shown that *the multiplier algebra $M(E/ker(\mathcal{G}_E))$ of the quotient algebra $E/ker(\mathcal{G}_E)$, over the Gel'fand radical, is strictly real* (Theorem 3.2); hence *$M(E)$ is also so, if E is semi-simple*, in the sense that $ker(\mathcal{G}_E) = \{0\}$. Now, it is known, by a result of *T. Müldner* [26, p. 291, Theorem 1], that a complete topological algebra E with continuous multiplication is a projective limit of complete metrizable topological algebras $\widetilde{E_\lambda}$. Taking benefit of this, one obtains that *E is strictly real if, and only if, every $\widetilde{E_\lambda}$ is strictly real* (Theorem 3.4, 1)). Furthermore, Kaplansky's commutativity result is obtained, but in our setting the rôle of the Jacobson radical $RadE$ is replaced by the Gel'fand one $ker\mathcal{G}_E$. It is proved that *if E is a complete topological al-*

Rodia I. Hadjigeorgiou, Ivis 60, GR-16562 Athens, Hellas; E-mail: rhadjig@math.uoa.gr
Mohamed Oudadess, Ecole Normale Supérieure B.P. 5118, Takaddoum 10105 Rabat, Morocco; E-mail: oudadessm@yahoo.fr

https://doi.org/10.1515/9783110413557-006

gebra with continuous multiplication, then it is commutative modulo its Gel'fand radical $ker(\mathcal{G}_E)$ (Theorem 3.4, 2)). Finally, the analogues of statements 1) and 2) of Theorem 3.4 are obtained for topological inductive limits (Theorems 3.5 and 3.6).

In all that follows by a *topological algebra* E we mean a real or complex topological vector space, which is also an algebra with separately continuous ring multiplication. The *spectrum* or *Gel'fand space* $\mathcal{M}(E)$ of E, endowed with the *Gel'fand topology*, is the set of continuous characters of E (: non zero scalar-valued multiplicative linear forms of E) which is assumed to be non-empty. The respective *Gel'fand map* of E is given by

$$\mathcal{G}_E : E \longrightarrow \mathcal{C}(\mathcal{M}(E)) : x \longmapsto \mathcal{G}_E(x) \equiv \hat{x} : \mathcal{M}(E) \longrightarrow \mathbf{C}$$
$$: f \longmapsto \hat{x}(f) := f(x).$$

The image of $\mathcal{G}_E$, denoted by $E^\wedge$, is called the *Gel'fand transform algebra* of E and is topologized as a *locally m-convex algebra* by the inclusion $E^\wedge \subseteq \mathcal{C}_c(\mathcal{M}(E))$, where the algebra $\mathcal{C}(\mathcal{M}(E))$ carries the topology "c" of compact convergence in $\mathcal{M}(E)$ [25, p. 19, Example 3.1]. We denote by $Sp_E(x)$ and $r_E(x)$, the *spectrum* and *spectral radius* of $x \in E$, respectively, i.e.,

$$Sp_E(x) = \{\lambda \in \mathbf{C}\backslash\{0\} : \lambda^{-1}x \notin G^q(E)\} \tag{1}$$

and

$$r_E(x) = \sup\{|\lambda| : \lambda \in Sp_E(x)\}, \tag{2}$$

where $G^q(E)$ (or E°) stands for the group of quasi-invertible elements of E. A subalgebra F of E is called *quasi-plane* if

$$F \cap E^\circ = F^\circ,$$

while the respective relation in the unital case $F \cap E^\bullet = F^\bullet$, where $E^\bullet$ denotes the group of invertible elements of E, defines a *plane subalgebra* of E. The spectral radius is a *submultiplicative semi-norm*, if, for instance, E satisfies the *(A-V)-condition*, considered by *H. Arizmendi* and *V. Valov* in [5], that is,

(A–V) $$r_E(x) = \sup_{f \in \mathcal{M}(E)} |\hat{x}(f)| \equiv sup|\hat{x}|(\mathcal{M}(E)), \quad x \in E.$$

In this context, one also obtains

$$ker(\mathcal{G}_E) = ker(r_E) \tag{3}$$

and

$$r_E(xy) = 0, \ \ for\ every\ \ y \in E,\ \ implies\ that\ \ r_E(x) = 0. \tag{4}$$

If, for any $x \in E$, $Sp_E(x) \subseteq \mathbf{R}$, E is called a *strictly real topological algebra*. Since for every $x \in E$ one has $\hat{x}(\mathcal{M}(E)) \subseteq Sp_E(x)$, we obtain that

(1.5) *a strictly real topological algebra E owns real continuous characters, hence E is a real algebra.*

Furthermore, recalling that for an algebra morphism $\phi : E \longrightarrow F$ one has (cf. *A. Mallios* [25, p. 49, Proposition 1.1]

(1.6) $$Sp_F(\phi(x)) \subseteq Sp_E(x), \text{ for every } x \in E,$$

we conclude that

(1.7) *any quotient algebra of a strictly real algebra is strictly real too.*

A topological algebra E is called *topologically spectral* if its spectrum $\mathcal{M}(E)$ is a *spectral set*, in the sense that

$$Sp_E(x) = \hat{x}(\mathcal{M}(E)),$$

for every $x \in E$ [13, p. 51, (1.3)]. We say that E is *without order* if

$$xE = Ex = \{0\} \text{ implies that } x = 0. \tag{8}$$

Topological algebras without order are also called *proper* by *W. Ambrose* [4], *Taqdir Husain* [17, p. 6] and Marina Haralampidou, Lourdes Palacios and Carlos J. Signoret Poillon[15, p. 1932 after (1.1)]. They are still named *faithful* by *B. E. Johnson* [18], *R. Larsen* [22] and *K. B. Laursen and M. Neumann* [24, p. 301].

Examples of topological algebras without order are:
1) every unital topological algebra,
2) a topological algebra with approximate identity,
3) a semi-simple topological algebra (: the Gel'fand map is 1-1),
4) a semi-prime topological algebra (: without non zero nilpotent ideals),
5) a topological algebra with an orthogonal basis.

See e.g. *Ju-Kwei Wang* [28, p. 1132, §2] and *Pietro Aiena* [2, p. 2, §1] for Banach algebras, *L. A. Khan, N. Mohammed, and A. B. Thaheem* [21, p. 630, §2], *R. A. Fontenot* [10] for topological algebras.

Now, by a *left multiplier* of E, respectively *right multiplier* of E, we mean a map $T : E \longrightarrow E$ such that for every $x, y \in E$

$$T(xy) := T(x)y, \text{ respectively } T(xy) := xT(y). \tag{9}$$

In the case one has

$$T(x)y := xT(y), \quad x, y \in E \tag{10}$$

then T is called a *multiplier* of E. The concept of multipliers of a Banach algebra was introduced by *Helgason* [16] as follows: In a semi-simple Banach algebra A of continuous functions over its *"maximal ideal space"* ($\equiv$ topological spectrum of A) $\mathcal{M}(A)$, a function h on $\mathcal{M}(A)$ satisfying $hA^\wedge \subseteq A^\wedge$ is a multiplier of A. The notion of a multiplier was generalized by *Ju-Kwei Wang* [28] for a commutative Banach algebra without order as in (10). The term *"centralizer"* instead of *"multiplier"* has been used by *C. N. Kellogg* [20] and *J. G. Wendel* [29]. The set of all multipliers on E is denoted by $M(E)$, and by $L_m(E)$ (resp. $R_m(E)$) the set of all left (resp. right) multipliers of E. By $\mathbf{L}_s(E)$ we denote the algebra of all linear operators of E into itself equipped with the topology "s" of simple convergence. For each $x \in E$, consider the multiplication operators $l_x, r_x : E \longrightarrow E$ defined by $l_x(y) := xy$ *and* $r_x(y) := yx$, for every $y \in E$. Obviously l_x is a left multiplier and r_x a right multiplier of E, respectively. It is evident that $R_m(E) \cap L_m(E) \subseteq M(E)$. Considering $T \in M(E)$, one notes that $(T(xy))z = (xy)T(z) = x(yT(z)) = x(T(y)z) = (xT(y))z$, for $x, y, z \in E$. If E *is without order*, then one has $T(xy) = xT(y)$, that is, $T \in R_m(E)$. Similarly, we show that $T \in L_m(E)$. Thus, we conclude (cf. *Wai-Mee Ching and James S. W. Wong* [9, p. 388, Proposition 1]

(1.11) *in a topological algebra E without order one has*

(1.11.1) $$R_m(E) \cap L_m(E) = M(E);$$

in other words a multiplier T of a topological algebra E without order satisfies

(1.11.2) $$x\,T(y) = T(x)\,y = T(xy).$$

In the case E *is a unital commutative topological algebra,* there are no other multipliers but the multiplication operator l_x by an element x of E. This is due to the fact that every multiplier $T \in M(E)$ on such an algebra satisfies $T(x) = e\,T(x) = T(e)\,x = l_{T(e)}(x)$, for all $x \in E$, with e the unit of E. Thus $T = l_{T(e)}$, so, *we can identify E with* $M(E)$.

In the context of a *topological algebra E without order*, we note the following:

(I) Assuming that $T \in M(E)$, $x, y, z \in E$ and $\lambda, \mu \in \mathbf{C}$, then

$$\begin{aligned} z\,T(\lambda x + \mu y) &= T(z)\,(\lambda x + \mu y) = \lambda\,T(z)\,x + \mu\,T(z)\,y \\ &= \lambda z\,T(x) + \mu z\,T(y) = z\,(\lambda\,T(x) + \mu\,T(y)). \end{aligned} \tag{12}$$

Since E is without order, one gets that $T(\lambda x + \mu y) = \lambda\,T(x) + \mu\,T(y)$, that is, *T is linear.* Hence $M(E) \subseteq \mathbf{L}(E)$.

(II) If T, S are two multipliers, then one has quite algebraically that $T \circ S \in M(E)$, *for* $T, S \in M(E)$, *with* $T \circ S = S \circ T$, therefore $M(E)$ *is a commutative subalgebra of* $\mathbf{L}(E)$, called the *multiplier algebra of E*.

(III) Given $T \in M(E)$, if T^{-1} exists, then $T^{-1} \in M(E)$; that is $M(E)$ *is a plane subalgebra of* $\mathbf{L}(E)$. (The term *inverse closed subalgebra* is still in use by *P. Aiena* [2, p. 2, II)]; cf. *Wai-Mee Ching and James S. W. Wong* [9, p. 388, Proposition 3] for normed algebras, *Pietro Aiena* [2, p. 2, II)] for Banach algebras without order). So, for the spectrum of a $T \in M(E)$, which has an inverse T^{-1}, we get (cf. [13, p. 52, Lemma 2.1]) $Sp_{\mathbf{L}(E)}(T) = Sp_{M(E)}(T)$.

In case E *is commutative*, then for every $x \in E$, we have $l_x \in M(E)$, and, since *E is without order, the left regular representation* $l : E \longrightarrow \mathbf{L}_s(E) : x \longmapsto l(x) := l_x$ *is a continuous bijection onto* $l(E) \equiv \{l_x,\ x \in E\}$. Thus E *coincides with the subset* $\{l_x,\ x \in E\}$ *of the multiplier algebra* $M(E)$. Moreover, for all $x \in E$ and $T \in M(E)$, we have $(l_x \circ T)(y) = l_x(T(y)) = x\,T(y) = T(x)\,y = l_{T(x)}(y)$, $y \in E$, thus $T \circ l_x = l_x \circ T = l_{T(x)}$, showing that $E = l(E)$ *is an ideal of the multiplier algebra* $M(E)$ (cf. *P. Aiena* [3, p. 194, Theorem 4.3] and/or *K.B. Laursen M. Neumann* [24, p. 302, Proposition 4.1.1 (taken from R. Larsen p. 318)] for faithful commutative Banach algebras).

(IV) Let $(T_\delta)_\delta \subseteq M(E)$ be a net which converges to $T \in \mathbf{L}_s(E)$. Then for every $x, y \in E$, we have $T_\delta(x)\,y \underset{\delta}{\longrightarrow} T(x)\,y$, and $x\,T_\delta(y) \underset{\delta}{\longrightarrow} x\,T(y)$. Since $T_\delta \in M(E)$, we get $T_\delta(x)\,y = x\,T_\delta(y)$, which implies that $T(x)\,y = x\,T(y)$. Thus $T \in M(E)$, that is, $M(E) \subseteq \mathbf{L}_s(E)$ *is closed.*

(V) *Every multiplier* T *has a closed graph:* Assume that $x, y \in E$ and $(x_\delta) \subseteq E$, with $\lim_\delta x_\delta - x$ and $lim_\delta T(x_\delta) = y$. Then, for any $z \in E$, one has

$$\begin{aligned} y\,z &= (\lim_\delta T(x_\delta))\,z = \lim_\delta (T(x_\delta)\,z) = \lim_\delta (x_\delta\,T(z)) \\ &= (\lim_\delta x_\delta)\,T(z) = x\,T(z) = T(x)\,z. \end{aligned} \tag{13}$$

Since E is without order, one has $y = T(x)$, thus T has closed graph. See also *T. Husain* [17, p. 7, Theorem 2.2] for proper topological algebras with continuous multiplication.

Based on the identities $x\,T(y) = T(x\,y)$ and $T(y)\,x = T(y\,x)$, for all $x, y \in E$, one has that, the set $M(E)$ of multipliers may be described as the commutant in $\mathbf{L}(E)$ of all operators of multiplication (on the right or on the left) by the elements of E.

Denoting by $\mathcal{M}(E)$ the *algebra of continuous multipliers of* E and by $\mathcal{L}_s(E)$ the algebra of continuous linear operators of E into itself with the topology "s" of simple convergence, we have, as the preceding, that $\mathcal{M}(E)$ *is a commutative closed plane subalgebra of* $\mathcal{L}_s(E)$. If E *is commutative, then, algebraically,* $E \equiv l(E)$, *viz.* E *becomes an ideal of the algebra of continuous multipliers* $\mathcal{M}(E)$. See *L. A. Khan, N. Mohammad, A. B. Thaheem* [21, p. 630, Theorem 2.1] for proper topological algebras, Taqdir Husain [17, p. 6, §2] for proper topological algebras with continuous multiplication, *R. C. Busby* [8], *B. E. Johnson* [18], *Pietro Aiena* [2, p. 2, §1], *C. N. Kellogg* [20, p. 122, Theorem 2.1], *K. B. Laursen-M. Mbekhta* [23, p. 130, Multipliers], *Ju-Kwei Wang* [28, p. 1133] for Banach algebras without order, *Jong-Kwang Yoo* [30, p. 769] for commutative semi-simple Banach algebras, *K. B. Laursen and M. Neumann* [24, p. 301 and 302] for faithful Banach algebras.

2 Multipliers of topological algebras

In the classical theory of multipliers, *every multiplier is a bounded linear operator on a commutative Banach algebra without order* (cf. *Ju-Kwei Wang* [28, p. 1132, Theorem 2.1], *B. E. Johnson* [18], *B. E. Johnson* [19], *R. C. Busby* [8]). The above result has been proved for *a complete metrizable locally m-convex algebra with a uniformly bounded one-sided approximate identity or a complete metrizable algebra without order* by *A. L. Khan* [21, p. 631, Theorem 2.3 (a) and (b)]. Here we extend it to topological algebras without order satisfying the closed graph theorem.

Theorem 2.1. *In a topological algebra E without order every multiplier T is linear. If, in addition, E satisfies the closed graph theorem, T is continuous, that is $T \in \mathcal{M}(E)$.*

Proof. The linearity of T was proved by (12), while its continuity results from the closed graph theorem in view of (13). □

The following result, generalizing a relevant one owed to *Ju-Kwei Wang* [28, p. 1135, Theorem 3.1], proves that *every multiplier of E can be realized as a continuous complex-valued function on $\mathcal{M}(E)$*. See also *P. Aiena* [3, p. 201, Theorem 4.14] for commutative semi-simple Banach algebras. A relevant result has been given by *T. Husain* [17, p. 8, Theorem 2.5] for a commutative semi-simple topological algebra with continuous multiplication.

Theorem 2.2. *Let E be a topological algebra, and $T \in M(E)$ a multiplier of E. Then, there is a unique continuous function φ_T on $\mathcal{M}(E)$, such that*

$$\widehat{T(x)}(f) = \varphi_T(f)\hat{x}(f), \tag{1}$$

for all $x \in E$, and $f \in \mathcal{M}(E)$.

Proof. Taking an arbitrary element $f \in \mathcal{M}(E)$, choose a point $x \in E$ with $f(x) \neq 0$, and set

$$\varphi_T(f) := \frac{\widehat{T(x)}(f)}{\hat{x}(f)}. \tag{2}$$

We note that, φ_T is independent of the choice of x and defines a continuous function on $\mathcal{M}(E)$: Indeed, since T is a multiplier, for $x,\ y \in E$ with $f(x) \neq 0 \neq f(y)$, we have

$$\begin{aligned} x\,T(y) = T(x)\,y \quad &\Longrightarrow \quad f(x\,T(y)) = f(T(x)\,y) \\ &\Longleftrightarrow \quad f(x)f(T(y)) = f(T(x))f(y) \\ &\Longleftrightarrow \quad \hat{x}(f)\widehat{T(y)}(f) = \widehat{T(x)}(f)\hat{y}(f) \\ &\Longleftrightarrow \quad \frac{\widehat{T(x)}(f)}{\hat{x}(f)} = \frac{\widehat{T(y)}(f)}{\hat{y}(f)}, \end{aligned} \tag{3}$$

implying the independence of φ_T from $x \in E$. Thus, we get

$$\widehat{T(x)}(f) = \varphi_T(f)\hat{x}(f), \ \textit{for all}\ x \in E,\ f \in \mathcal{M}(E), \tag{4}$$

even if, $f(x) = 0$: Indeed, if $f(x) = 0$, and $f(y) \neq 0$, for some $y \in E$, then we have $\widehat{T(x)}(f)\hat{y}(f) = \hat{x}(f)\widehat{T(y)}(f) = 0$, implying that $\widehat{T(x)}(f) = 0$. Therefore, (2) holds identically. The continuity of φ_T is immediate, since if $f_\delta \underset{\delta}{\longrightarrow} f$, then, one has $f_\delta(x) \underset{\delta}{\longrightarrow} f(x)$, for every $x \in E$, much more for $T(x)$. Hence $f_\delta(T(x)) \underset{\delta}{\longrightarrow} f(T(x))$; that is, $\widehat{T(x)}(f_\delta) \underset{\delta}{\longrightarrow} \widehat{T(x)}(f)$, implying $\varphi_T(f_\delta) \underset{\delta}{\longrightarrow} \varphi_T(f)$. For the uniqueness of φ_T, let ψ_T be another complex-valued function on $\in \mathcal{M}(E)$, for which $\widehat{T(x)}(f) = \psi_T(f)\,\hat{x}(f),\ f \in \mathcal{M}(E)$. Then by (1), $(\varphi_T(f) - \psi_T(f))\hat{x}(f) = 0,\ \textit{for all}\ x \in E$, which yields $\varphi_T = \psi_T$. □

The function φ_T which corresponds to a multiplier T, by the previous theorem, is called *Helgason-Wang function of* T. For the multiplication operators l_x, r_x we have

$$\varphi_{l_x}(f) = \frac{\widehat{(l_x)(y)}(f)}{\hat{y}(f)} = \frac{\widehat{xy}(f)}{\hat{y}} = \frac{f(x)f(y)}{f(y)} = \hat{x}(f),$$

for $x,\ y \in E,\ f \in \mathcal{M}(E)$ *with* $f(y) \neq 0$, that is, one has

$$\varphi_{l_x} = \hat{x} \ \ or \ \ \varphi_{r_x} = \hat{x},\ \ x \in E. \tag{5}$$

Denoting by

$$E_m := \{h \in \mathcal{C}_c(\mathcal{M}(E)) : h\,E^\wedge \subseteq E^\wedge\} \subseteq \mathcal{C}_c(\mathcal{M}(E)), \tag{6}$$

we note that *every multiplier T on E defines*, by the relation (1), *a continuous complex function* φ_T *on* $\mathcal{M}(E)$, *such that* $\widehat{T(x)} = \varphi_T\,\hat{x},\ x \in E$, that is $\varphi_T \in E_m$. In the following, we prove the opposite (cf. *F. T. Birtel* [7, p. 815, Preliminaries]).

Proposition 2.3. *Let E be a semi-simple topological algebra. Then, every* $h \in E_m$ defines a multiplier T on E, *by the relation*

$$\widehat{T(x)} = h\,\hat{x},\ \ x \in E. \tag{7}$$

Proof. Indeed, for $x,\ y \in E$, one has $\widehat{y\,T(x)} = \hat{y}\,\widehat{T(x)} = \hat{y}\,h\,\hat{x} = h\,\hat{y}\,\hat{x} = \widehat{T(y)}\,\hat{x} = \widehat{T(y)\,x}$. Thus, by the semi-simplicity of E, $yT(x) = T(y)x$; that is, T is a multiplier on E. □

Now, summarizing the two preceding results, we have the next.

Theorem 2.4. *In a semi-simple topological algebra E, there is an algebra isomorphism of the algebra* $M(E)$ *onto the algebra* E_m *given by* (cf. (1))

$$\varphi : M(E) \longrightarrow E_m : T \longmapsto \varphi(T) := \varphi_T. \tag{8}$$

Regarding *the algebra* $E_m \subseteq \mathcal{C}_c(\mathcal{M}(E))$, we note that it *is a quasi-plane subalgebra of* $\mathcal{C}_c(\mathcal{M}(E))$ as the following result proves.

Proposition 2.5. E_m *is a quasi-plane subalgebra of* $\mathcal{C}_c(\mathcal{M}(E))$.

Proof. Assuming that $h \in E_m \cap (\mathcal{C}_c(\mathcal{M}(E)))^\circ$, one has to prove that $h \in (E_m)^\circ$: Indeed, there exists $\phi \in \mathcal{C}_c(\mathcal{M}(E))$, such that $h \circ \phi = h + \phi - h\,\phi = 0$. Since $h \in E_m$, there exists $x \in E$ with $h\,\hat{x} \in E^\wedge$, hence $0 = h\hat{x} + \phi\hat{x} - h\,\phi\hat{x} = h\hat{x} + \phi\hat{x} - h\hat{x}\phi = \hat{y} + \phi\hat{x} - \hat{y}\phi$, with $h\hat{x} = \hat{y}$. So $\phi(\hat{y} - \hat{x}) = \hat{y}$, that is $\phi\hat{z} = \hat{y} \in E^\wedge$, where $\hat{z} = \hat{y} - \hat{x}$, thus $\phi \in E_m$, proving the assertion. □

A characterization of multipliers for semi-simple topological algebras is given in the following theorem, generalizing a relevant one by *P. Aiena* [3, p. 202, Theorem 4.15] for commutative semi-simple Banach algebras. See also *T. Husain* [17, p 10, Theorem 2.9] for commutative semi-simple topological algebras with continuous multiplication.

Theorem 2.6. *Let E be a topological algebra and T a linear operator on E* (viz. $T \in \mathbf{L}(E)$). *Moreover, assume the following assertions:*

1) $T \in M(E)$.

2) $T(ker(f)) \subseteq ker(f)$, for all $f \in \mathcal{M}(E)$.

Then, 1)$\Longrightarrow$2). If, in addition, E is semi-simple, then 2)$\Longrightarrow$1), as well.

Proof. Assuming $T \in M(E)$ and $f \in \mathcal{M}(E)$, we have for every $x \in ker(f)$, $\widehat{T(x)}(f) = \varphi_T(f)\hat{x}(f) = 0$. Thus, $\widehat{T(x)}(f) = f(T(x)) = 0$, that is, $T(x) \in ker(f)$.

Conversely, if 2) holds true, let $x,\ y \in E$, and an arbitrary $f \in \mathcal{M}(E)$. Suppose that either x or y belongs to $ker(f)$, say $x \in ker(f)$. Since $T(x) \in ker(f)$, we get

$$(\widehat{T(x)y - xT(y)})(f) = 0. \tag{9}$$

Now, assume that $x,\ y \notin ker(f)$. Since $E/ker(f) \cong \mathbf{C}$, there exist $\lambda, \mu \in \mathbf{C}$, $z \notin ker(f)$ and $u, u \in ker(f)$, such that $x = \lambda z + u$, $y = \mu z + v$. Since $T(u)$, $T(v) \in ker(f)$, we obtain $\widehat{xT(y)}(f) = \widehat{T(x)y}(f) = \lambda\hat{z}(f)\,\mu\widehat{T(z)}(f)$. Hence, one obtains $(\widehat{T(x)y - xT(y)})(f) = 0$, therefore (9) is satisfied for every $f \in \mathcal{M}(E)$ and $x,\ y \in E$. The semi-simplicity of E implies $T(x)y = xT(y)$, for all $x,\ y \in E$, hence $T \in M(E)$. □

Since $ker(\mathcal{G}_E) = \bigcap_{f \in \mathcal{M}(E)} ker(f)$, one immediately has the next.

Corollary 2.7. *In a topological algebra E, every multiplier T of E satisfies $T(ker(\mathcal{G}_E)) \subseteq ker(\mathcal{G}_E)$.*

Considering the spectrum $\mathcal{M}(M(E))$ of $M(E) \subseteq \mathbf{L}_s(E)$, we describe the relationship between $\mathcal{M}(M(E))$ and $\mathcal{M}(E)$. For this we recall the *algebraic identification of E with the ideal $l(E)$ of $M(E)$ for a commutative topological algebra E without order*. We know that *the spectrum of the ideal $l(E) \equiv E$ in $M(E)$ is, within a homeomorphism, the complement in $\mathcal{M}(M(E))$ of its closed subset*

$$h_{M(E)}(E) = \{\psi \in \mathcal{M}(M(E)) : \psi|_E = 0\}, \tag{10}$$

(cf. *Rodia I. Hadjigeorgiou* [11, p. 314, Theorem 2.1] and/or [12, p. 392, (1.3)]. That is,

$$\mathcal{M}(l(E) \equiv \{l_x,\ x \in E\}) \cong C\,h_{M(E)}(E) \tag{11}$$

is, within a homeomorphism, an open subset of $\mathcal{M}(M(E))$. Obviously, we also have the set-theoretic decomposition

$$\mathcal{M}(M(E)) = h_{M(E)}(E) \cup C\,h_{M(E)}(E), \tag{12}$$

which entails that

$$\textit{if } \psi \in \mathcal{M}(M(E)) \textit{ either } \psi|_E = 0 \textit{ or } \psi|_E = f, \textit{ for a unique } f \in \mathcal{M}(E). \tag{13}$$

In fact, there exists a homeomorphism from $\mathcal{M}(E)$ onto $C\,h_{M(E)}(E)$ as follows in the next result. See also *P. Aiena* [2, p. 4] or [3, p. 203-204, Theorem 4.16 and the comments after] for commutative faithful Banach algebras.

Theorem 2.8. *Let E be a commutative topological algebra without order. Then there exists a map*

(2.14) $\theta : \mathcal{M}(E) \longrightarrow \mathcal{M}(M(E)) : f \longmapsto \theta(f) \equiv \psi : M(E) \longrightarrow \mathbf{C}, : T \longmapsto \psi(T) := \frac{f(T(x))}{f(x)},\ x \in E,$

which defines a homeomorphism of $\mathcal{M}(E)$ onto the open subset $C\,h_{M(E)}(E)$ of $\mathcal{M}(M(E))$.

Proof. It is clear that ψ is independent of the choice of $x \in E$ (cf. (3)), and it is a continuous linear form on $M(E)$ with

$$\psi(T \circ S) = \frac{f(T(S(x)))}{f(x)} = \frac{\psi(T)f(S(x))}{f(x)} = \psi(T)\psi(S),$$

that is $\psi \in \mathcal{M}(M(E))$. Moreover, $\psi|_E = f$, since for all $y \in E$ we have

$$\psi(l_y) = \frac{f(l_y(x))}{f(x)} = \frac{f(yx)}{f(x)} = f(y). \tag{15}$$

So for $\psi(l_x) \neq 0$, for some $x \in E$, $\psi \in C\,h_{M(E)}(E)$. We shall show that ψ is unique : Indeed, suppose there is another $\chi \in \mathcal{M}(M(E))$ with the same restriction on E, that is $\chi(l_y) = f(y)$, $y \in E$. Choosing any $z \in E$, such that $f(z) \neq 0$, if $T \in M(E)$, then (cf. also (10) and (15)) $\chi(T)f(z) = \chi(T)\chi(l_z) = \chi(T \circ l_z) = \chi(l_{T(z)}) = \psi(l_{T(z)}) = f(T(z)) = \psi(T)f(z)$, hence $\psi = \chi$. Now, if $\sigma \in C\,h_{M(E)}(E)$, that is $\sigma \in \mathcal{M}(M(E))$ with $\sigma(l_y) \neq 0$ for some $y \in E$, the relation $f(x) = \sigma(l_x)$, $x \in E$, defines a continuous character f of E, such that $\sigma|_E = f$ and, as before, σ is unique. Hence, θ is a bijection onto $C\,h_{M(E)}(E)$. It is also continuous, since if $f_\delta \underset{\delta}{\longrightarrow} f$, then $f_\delta(x) \underset{\delta}{\longrightarrow} f(x)$, for every $x \in E$, let alone for $T(x)$, therefore, by (2.14), $\psi_\delta(T) \underset{\delta}{\longrightarrow} \psi(T)$, for every $T \in \mathcal{M}(M(E))$. The inverse map is continuous as well, since for $\psi_\delta \underset{\delta}{\longrightarrow} \psi$ on $E = l(E) \equiv \{l_x,\ x \in E\}$, we have $\psi_\delta(l_x) \underset{\delta}{\longrightarrow} \psi(l_x)$, thus, in view of (15), $f_\delta(x) \underset{\delta}{\longrightarrow} f(x)$ for every $x \in E$. □

Based on the previous theorem, we get $\mathcal{M}(E) \cong C\,h_{M(E)}(E)$, so the relation (12) is written

$$\mathcal{M}(M(E)) = h_{M(E)}(E) \cup \mathcal{M}(E), \tag{16}$$

where the Gel'fand topology of $\mathcal{M}(E)$ coincides with the Gel'fand topology induced by $\mathcal{M}(M(E))$, regarded as a subset of $\mathcal{M}(M(E))$. Considering the Gel'fand transform $\widehat{T}$ of an element $T \in M(E)$, we remark that

$$\widehat{T}\,|_{\mathcal{M}(E)} = \varphi_T, \quad T \in M(E), \tag{17}$$

and thus we use both notations $\widehat{T}$ and φ_T for the *Helgason-Wang* function on $\mathcal{M}(E)$.

The results examined in this section remain true if we consider continuous multipliers of the given topological algebra E, *or, much more, if we take a topological algebra E satisfying the closed graph theorem*, so that any multiplier becomes continuous, that is $M(E) = \mathcal{M}(E)$.

3 Strictly real topological algebras

The assertion of Corollary 2.7 has been proved by *R. Hadjigeorgiou and M. Oudadess* [19, Remark 8, (2)] in the context of strictly real Banach algebras using the *topological Jacobson radical* $RadE$ of E, in place of the *"Gel'fand radical"* $ker(\mathcal{G}_E)$ of E. We know that in commutative Banach algebras $RadE = \bigcap_{f \in \mathcal{M}(E)} ker(f) = ker(r_E)$ and r_E is a submultiplicative semi-norm of E, facts which are basically needed in proof. Thus, in the general context of topological algebras, we consider $ker(\mathcal{G}_E)$ in place of $RadE$, and assume that r_E is a submultiplicative semi-norm of E to get the above result following this process. The latter is valid, for example, if E satisfies the (A-V)-condition, so that one has the identification $ker(\mathcal{G}_E) = ker(r_E)$(cf. (3)). So we have the next.

Proposition 3.1. *Let E be a strictly real topological algebra without order satisfying the (A-V)-condition. If $T \in M(E)$, then $T(ker(\mathcal{G}_E)) \subseteq ker(\mathcal{G}_E)$.*

Proof. Considering $x \in ker(\mathcal{G}_E)$, we have to show $T(x) \in ker(\mathcal{G}_E) = ker(r_E)$, i.e. $r_E(T(x)) = 0$. It suffices to prove that $r_E(T(x)y) = r_E(xT(y)) = 0$, for every $y \in E$. Since E is strictly real and $r_E(xT(y)) \leq r_E(x)r_E(T(y))$, with $r_E(x) = 0$, one has the assertion. □

Theorem 3.2. *Let E be a commutative strictly real topological algebra without order. Then, one has*

1) *The multiplier algebra $M(E/ker(\mathcal{G}_E))$ is strictly real.*
2) *If, moreover, E is semi-simple, the algebra $M(E)$ is strictly real.*

Proof. 1) Considering the algebra $F \equiv E/ker(\mathcal{G}_E)$, we note that it is a semi-simple (hence commutative, since it is identified with its Gel'fand transform algebra) topological algebra being also strictly real (cf. (1.6)). Hence (Theorem 2.4), every multiplier $T \in M(F) \equiv M(E/ker(\mathcal{G}_E))$ coincides with a continuous function φ_T on the spectrum $\mathcal{M}(F) \equiv \mathcal{M}(E/ker(\mathcal{G}_E)) \cong h_E(ker(\mathcal{G}_E))$ of $F \equiv E/ker(\mathcal{G}_E)$, which is real valued since F has real characters (cf. (1.5) and (2)). Since $M(F)$ is algebraically identical with F_m and F_m is topologically spectral, as a semi-simple (hence commutative) quasi-plane (Proposition 2.5) l.m.c. subalgebra of the advertibly complete l.m.c. algebra $\mathcal{C}_c(\mathcal{M}(F))$ (cf. [13, p. 52, Lemmas 2.2, 2.3 and p. 54, Proposition 2.11]), one obtains $Sp_{M(F)}(T) = Sp_{F_m}(\varphi_T) = \widehat{\varphi_T}(\mathcal{M}(F_m)) = \varphi_T(\mathcal{M}(F)) \subseteq \mathbf{R}$, yielding 1). Assertion 2) derives from 1). □

Completion preserves strict reality under an appropriate condition for the completion of a given topological algebra. Namely, we have.

Theorem 3.3. *Let E be a strictly real topological algebra, whose completion $\widetilde{E}$ is a topologically spectral algebra. Then $\widetilde{E}$ is strictly real.*

Proof. Assuming $z \in \widetilde{E}$, then $z = \lim_{\delta} x_\delta$, $(x_\delta) \subseteq E$. By the hypothesis for $\widetilde{E}$, one has $Sp_{\widetilde{E}}(z) = \hat{z}(\mathcal{M}(\widetilde{E})) = \{\chi(z) = \lim_{\delta} \chi(x_\delta),\ \chi \in \mathcal{M}(\widetilde{E})\}$. Since E is strictly real, every $\chi(x_\delta) \in \hat{x}_\delta(\mathcal{M}(E)) \subseteq Sp_E(x_\delta) \subseteq \mathbf{R}$, thus $\chi(z) \in \mathbf{R}$, $\chi \in \mathcal{M}(\widetilde{E})$, hence $Sp_{\widetilde{E}}(z) \subseteq \mathbf{R}$. □

Strict reality is preserved when taking projective limits as proved by *R. Hadjigeorgiou and M. Oudadess* [19, Proposition 13] in the context of complete l.m.c. algebras. We remind that every complete locally m-convex algebra E is represented as a projective limit of Banach algebras $\widetilde{E}_\alpha$, within a topological algebraic isomorphism $E = \varprojlim_{\alpha} \widetilde{E}_\alpha$, known as the *Arens-Michael decomposition of E corresponding to a local basis* $\mathcal{N} = \{U_\alpha\}$ *of E* (cf. *A. Mallios* [25, p. 85, §3 Definition 3.1]). The latter has been generalized by *T. Müldner* [26, p. 291, Theorem 1] for topological algebras with continuous multiplication, as follows:

(3.1) *every complete topological algebra with continuous multiplication E is isomorphic to a projective limit of Fréchet (:complete and metrizable) topological algebras* $\widetilde{E}_\lambda$

(3.1.1) $$E = \varprojlim_{\lambda} \widetilde{E}_\lambda .$$

We recall some basic points of this representation : First, we note that (ibid., p. 291, Lemma; for its proof see *M. Oudadess* [27, Lemma 2.3] and/or *Mati Abel* [1, p. 11, Lemma 2.1])

(3.2) *given a local basis* $\mathcal{N} = \{U\}$ *of closed balanced neighbourhoods of zero in a topological algebra E with continuous multiplication, then*

(3.2.1) $$N = \bigcap_{U \in \mathcal{N}} U \text{ is a closed ideal of } E.$$

Now, for complete E, there is a family $\{\mathcal{N}_\lambda = \{U_n^\lambda\}_{n=1}^{\infty}\}_{\lambda\in\Lambda}$, where each $\mathcal{N}_\lambda$ defines in E a topology by the F-semi-norm p_λ. Setting $N_\lambda = \bigcap_{n=1}^{\infty} U_n^\lambda$, $E_\lambda = E/N_\lambda$, $\pi_\lambda : E \longrightarrow E_\lambda : x \longmapsto x_\lambda := x + N_\lambda$ the canonical surjection, we have by (3.2) that N_λ it is p_λ-closed ideal of E, and E_λ a metrizable topological algebra, equipped by the F-norm $\|x_\lambda\|_\lambda = p_\lambda(x)$. $\widetilde{E}_\lambda$ stands for the completion of E_λ.

Theorem 3.4. *Let E be a complete topological algebra with continuous multiplication. Then, the following assertions hold true:*

1) E is strictly real if, and only if, every factor $\widetilde{E}_\lambda$ *is a strictly real Fréchet topological algebra.*

2) If E is strictly real, then it is commutative modulo its Gel'fand radical $ker(\mathcal{G}_E)$ (viz. *"almost commutative"*).

Proof. 1) It is immediate from the relation $Sp_E(x) = \bigcup_{\lambda} Sp_{\widetilde{E}_\lambda}(x_\lambda)$. 2) Assuming $x = (x_\lambda)_\lambda$, $y = (y_\lambda)_\lambda \in E$, then $xy - yx = (x_\lambda y_\lambda - y_\lambda x_\lambda)_\lambda$. Since $\widetilde{E}_\lambda$ is, by 1), strictly real, it has (cf. (1.5)) real characters, thus $x_\lambda y_\lambda - y_\lambda x_\lambda \in ker(\mathcal{G}_{\tilde{E}_\lambda})$. By the relation $ker(\mathcal{G}_E) = \bigcap_\lambda \pi_\lambda^{-1}(ker(\mathcal{G}_{\tilde{E}_\lambda}))$ one obtains $xy - yx \in ker(\mathcal{G}_E)$. □

Strict reality is a property that is preserved when taking inductive limits as showed by *R. Hadjigeorgiou and M. Oudadess* [19, Propositions 14 and 15] in the context of strict inductive limits of Banach algebras. Given an inductive system of topological algebras $(E_\alpha, f_{\beta\alpha})$, with respect to a directed index set I, let

$$E = \varinjlim_\alpha (E_\alpha, f_{\beta\alpha}) = \bigcup_{\alpha\in I} f_\alpha(E_\alpha) \tag{3.3}$$

be the corresponding *inductive limit topological algebra*, where the canonical maps $f_\alpha : E_\alpha \longrightarrow E,\ \alpha \in I$, and the connecting maps $f_{\beta\alpha} : E_\alpha \longrightarrow E_\beta,\ \alpha \le \beta$ are continuous algebra morphisms (cf. *A. Mallios* [25, p. 109-113ff], while E is topologized by the *final vector space topology defined by the canonical maps* $f_\alpha,\ \alpha \in I$ (ibid., p. 115, Lemma 2.2). Canonical and connecting maps are related by

$$f_\beta \circ f_{\beta\alpha} = f_\alpha, \quad \textit{for any } \alpha \le \beta \in I, \tag{3.4}$$

therefore one gets

$$f_\alpha(E_\alpha) \subseteq f_\beta(E_\beta), \quad \textit{with } \alpha \le \beta \in I; \tag{3.5}$$

thus, (3.3) may be considered as the *union of an increasing family* $(f_\alpha(E_\alpha))_{\alpha\in I}$ *of topological algebras*. The spectrum of an element x of an inductive limit topological algebra $E = \varinjlim_\alpha E_\alpha$ is given by

$$Sp_E(x) = \bigcap_{\alpha\in I(x)} Sp_{f_\alpha(E_\alpha)}(x), \tag{3.6}$$

where $I(x) = \{\alpha \in I : x \in f_\alpha(E_\alpha)\}$, following an analogous argument with that one given by *A. Arosio* [6, p. 338, Theorem 1, b)]; cf. also (3.5) and (1.6). Now, we extend results in [19] for the general context of an inductive limit topological algebra. The proof of the first one is actually obvious by (3.6).

Theorem 3.5. *Let $(E_\alpha, f_{\beta\alpha})$ be an inductive system of topological algebras and $E = \varinjlim_\alpha E_\alpha$ the respective inductive limit topological algebra. Then E is strictly real if every factor E_α, with $\alpha \in I$, is strictly real.*

Theorem 3.6. *Let $(E_\alpha, f_{\beta\alpha})$ be an inductive system of strictly real topological algebras with respect to a totally ordered index set I, and $E = \varinjlim_\alpha E_\alpha$ the respective inductive limit topological algebra, which is, moreover, complete having continuous multiplication. Then E is commutative modulo its Gel'fand radical $ker(\mathcal{G}_E)$.*

Proof. By Theorem 3.5, E is strictly real, hence, by hypothesis and Theorem 3.4, it fulfils the assertion. □

Bibliography

[1] Abel, M. *Representations of topological algebras by projective limit of Fréchet algebras*, Communications in Mathematics and applications 1(3), (2012), pp. 9–15.
[2] Aiena, P. *Fredholm, Riesz and local spectral theory of multipliers*, Extr. Math. 9(1994), 1–20.
[3] Aiena, P. *Fredholm and Local Spectral Theory with Applications to Multipliers*, Kluwer, Dordrecht, 2004.
[4] Ambrose, W. *Structure theorems for a special class of Banach algebras*, Trans. Amer. Math. Soc. 57(1945), 364–386.
[5] Arizmendi, H., Valov, V. *Some characterizations of Q-algebras*, Comment. Math. 39(1999), 11–21.
[6] Arosio, A. *Locally convex inductive limits of normed algebras*, Rend. Sem. Mat. Univ. Padova 51(1974), 333–359.
[7] Birtel, F. T. *On a commutative extension of a Banach algebra*, Proc. Amer. Math. Soc. 13(1962), 815–822.
[8] Busby, R. C. *Double centralizers and extensions of C^*-algebras*, Trans. Amer. Math. Soc. 132(1968), 79–99.
[9] Ching, W.-M. and Wong, J. S. W. *Multipliers and H^* algebras*, Pacific J. Math. 22(1967), 387–395.
[10] Fontenot, R. A. *The double centralizer algebra as a linear space*, Proc. Amer. Math. Soc. 53(1975), 99–103.
[11] Hadjigeorgiou, R. I. *Spectral Geometry of Topological Algebras*, Doctoral Thesis, Univ. of Athens, 1995.
[12] Hadjigeorgiou, R. I. *Spectral geometry of non-local topological algebras*, Bull. Belg. Math. Soc. 11(2004), 391–407.
[13] Hadjigeorgiou, R. I. *On some more characterizations of Q-algebras*, Proc. International Conference on "*Topological Algebras and Their Applications*" (ICTAA 4), Oaxaca, Mexico, 2002. Contemp. Math. 341(2004). (*H. Arizmendi–C. Bosch–L. Palacios*, Eds). pp. 49–61.
[19] Hadjigeorgiou, R. I., and Oudadess, M. *Strictly real locally convex algebras*, Frontiers in Sci. and Engineering, Vol. 2(2008), 1–12.
[15] Haralampidou, M., Palacios, L., and Signoret Poillon, C. J. *Multipliers on locally H^*-algebras*, Rocky Mountain J. Math., 43(2013), no 6, 1931–1940.
[16] Helgason, S. *Multipliers of Banach algebras*, Ann. of Math. 64(1956), 240–254.
[17] Husain, T. *Multipliers of topological algebras*, Dissertationes Math. (Rozprawy Mat.) 285(1989).
[18] Johnson, B. E. *An introduction to the theory of centralizers*, Proc. London Math. Soc. 14(1964), 299–320.
[19] Johnson, B. E. *Continuity of centralisers on Banach algebras*, J. London Math. Soc. 41(1966), 639–640.
[20] Kellogg, C. N. *Centralizers and $H*$-algebras*, Pacific J. Math. 17(1966), 121–129.
[21] Khan, L. A., Mohammad, N., and Thaheem, A. B. *Double multipliers on topological algebras*, Internat. J. Math. & Math. Sci. 22(1999), 629–636.
[22] Larsen, R. *An Introduction to the Theory of Multipliers*, Grundlehren der mathematischen Wissenschaften, Springer-Verlag, 1969.
[23] Laursen, K. B. and Mbekhta, M. *Closed range multipliers and generalized inverses*, Studia Math. 107(1993), 127–135.
[24] Laursen, K. B. and Neumann, M. *An introduction to local spectral theory*, Oxford University Press Inc., New York, 2000.

[25] Mallios, A. *Toplogical Algebras. Selected topics*, North-Holland, Amsterdam, 1986.
[26] Müldner, T. *Projective limits of topological algebras*, Colloq. Math. 33(1973), 291–294.
[27] Oudadess, M. *Structure of topological algebras and a lemma of Müldner*, Frontiers in Sci. and Engineering (to appear).
[28] Wang, J.-K. *Multipliers of commutative Banach algebras*, Pacific J. Math. 11(1961), 1131–1149.
[29] Wendel, J. G. *Left centalizers and isomorphisms of group algebras*, Pacific J. Math. 2(1952), 251–261.
[30] Yoo, J.-K. *Spectral properties of multipliers*, Comm. Korean Math. Soc. 14(1999), 769–776.

Marina Haralampidou, Lourdes Palacios*, and Carlos Signoret

Multipliers in some perfect locally m-pseudo-convex algebras

Abstract: In this paper we describe the multiplier algebra $M(E)$ when E is a perfect complete locally m-pseudo-convex algebra with an approximate identity and with complete generalized Arens-Michael k_α-normed factors. This generalizes a previous result for locally m-convex algebras that was recently obtained by the authors.

Keywords: Proper algebra, generalized Arens-Michael decomposition, multiplier algebra, perfect projective system, perfect locally m-pseudo-convex algebra

MSC: Primary 46H05; Secondary 46H10

1 Introduction and Preliminaries

This paper is devoted primarily to provide a generalization of a result obtained by the authors in [6, Theorem 2.8] for locally m-convex algebras, to the more general case of locally m-pseudo-convex ones. The generalization relies on the automatic continuity of multipliers in various settings, that are clearly stated, and in the fact that the classical Arens-Michael decomposition of locally m-convex algebras is extended to locally m-pseudo-convex ones in a very similar way. Throughout this paper, by a *topological algebra* we mean an algebra (over the field of complex numbers $\mathbb{C}$) endowed with a Hausdorff topology that makes the multiplication *separately continuous.*

Definition 1.1. *A mapping $T : E \to E$ is called a* left (right) multiplier *on E if $T(xy) = T(x)y$ (resp. $T(xy) = xT(y)$) for all $x, y \in E$; it is called a* two-sided multiplier *on E if it is both a left and a right multiplier.*

In the sequel, a two-sided multiplier will be called in short, a *multiplier*. We employ the symbol $M(E)$ for the set of all multipliers on E. It is known that if E is a *proper algebra*, namely $xE = \{0\}$ implies $x = 0$ or $Ex = \{0\}$ implies $x = 0$, where 0 denotes the zero element of E , then any two-sided multiplier on E is automatically a linear mapping (see [7, p. 20, Theorem 1.1.1]), so $M(E)$ is a subalgebra of the algebra $L(E)$ of all linear operators on E.

Marina Haralampidou, Department of Mathematics, University of Athens, Panepistimioupolis, Athens, 15784, Greece
***Corresponding author: Lourdes Palacios,** Universidad Autónoma Metropolitana Iztapalapa, Mexico City 09340, Mexico
Carlos Signoret, Universidad Autónoma Metropolitana Iztapalapa, Mexico City 09340, Mexico

https://doi.org/10.1515/9783110413557-007

Let E be a proper algebra. For $x \in E$, due to the associativity of the multiplication on E, the operators l_x and r_x given by $l_x(y) = xy$ and $r_x(y) = yx$ $(y \in E)$, respectively, define an embedding from E to $M(E)$ in such a way that E is identified with a two-sided ideal of $M(E)$ (see [7, p. 23, Theorem 1.1.4] and [5, p. 1933, Proposition 2.2]).

Definition 1.2. *An* approximate identity *in a topological algebra E is a net $(e_\delta)_{\delta\in\Delta}$ such that for each $x \in E$ we have*

$$x - xe_\delta \underset{\delta}{\to} 0 \text{ and } x - e_\delta x \underset{\delta}{\to} 0 \text{ for all } x \in E.$$

Note that every algebra with an approximate identity is proper.

Definition 1.3. *A k-*seminorm *on E, with $k \in (0, 1]$, is a function*

$$p : E \to \mathbb{R}^+ \cup \{0\}$$

such that, for each $x, y \in E$,

$$p(x + y) \leq p(x) + p(y), \tag{1}$$

$$\text{for each } \lambda \in \mathbb{C},\ p(\lambda x) \leq |\lambda|^k p(x). \tag{2}$$

If, in addition, the function satisfies

$$p(xy) \leq p(x)p(y), \tag{3}$$

then the k-seminorm is called submultiplicative.

A k-seminorm p is also called a pseudo-seminorm *and k is called the* homogenity index *of p (see [3, p. 189]). Occasionally, we shall employ the symbol k_p to indicate that index, corresponding to p. A pseudo-seminorm p is a* pseudo-norm *if $p(x) = 0$ implies $x = 0$.*

If p is a k-seminorm (k-norm) on a linear space E, then the resulting topological linear space $E = (E, p)$ is called a k-seminormed (k-normed) linear space. *A topological algebra whose topology is induced by a k-seminorm p is called a* k-seminormed (k-normed) algebra.

A locally pseudo-convex space *E is a topological linear space equipped with a family $\mathcal{P} = (p_\alpha)_{\alpha\in\Lambda}$ of pseudo-seminorms on E which define its topology. If each $p_\alpha \in \mathcal{P}$ is a k-seminorm, then E is called a* locally k-convex space.

A locally pseudo-convex algebra *E is a topological algebra such that its underlying topological linear space is locally pseudo-convex. If its underlying topological linear space is locally k-convex, then E is called a* locally k-convex algebra. *E is called a* locally m-pseudo-convex algebra *(or* locally m-(k-convex) algebra*) if p_α is submultiplicative for each $\alpha \in \Lambda$.*

If $p_1, p_2, \ldots, p_n$ are k_{p_j}-seminorms on E and $k = \min\limits_{1\leq j\leq n}\{k_{p_j}\}$, then the function $q = p_1 \vee p_2 \vee \ldots \vee p_k$ defined as

$$q(x) = \max_{1\leq j\leq n}\{p_1^{\frac{k}{k_{p_1}}}(x), p_2^{\frac{k}{k_{p_2}}}(x), \ldots, p_n^{\frac{k}{k_{p_n}}}(x)\}$$

is also a pseudo-seminorm on E (with homogenity index k). We say that the family of pseudo-seminorms $\mathcal{P} = (p_\alpha)_{\alpha\in\Lambda}$ is saturated *if* $p_{\alpha_1} \vee p_{\alpha_2} \vee \ldots \vee p_{\alpha_n} \in \mathcal{P}$ *for each finite family* $\{p_{\alpha_1}, p_{\alpha_2}, \ldots, p_{\alpha_n}\} \subset \mathcal{P}$. *If* $\mathcal{P}$ *is any family of pseudo-seminorms on E, then the family*

$$\tilde{\mathcal{P}} = (p_{\alpha_1} \vee p_{\alpha_2} \vee \ldots \vee p_{\alpha_n})_{\{p_{\alpha_1}, p_{\alpha_2}, \ldots, p_{\alpha_n} \in \mathcal{P}\}}$$

is saturated (it is called the saturated closure *of* $\mathcal{P}$*). It can be proved that* $\mathcal{P}$ *and* $\tilde{\mathcal{P}}$ *induce the same topology on E (see [3, p. 191, Lemma 4.3.8 (a)]).*

Recall that a topological algebra is said to be an *F-algebra* if it is complete and metrizable.

Example 1.4. *[3, p. 209] Let* (k_n) *be a sequence of real numbers such that* $0 < k_n \leq 1$. *The set C of all* $\mathbb{C}$*-valued continuous functions* $x = x(t)$ *on* $\mathbb{R}$ *is an algebra over* $\mathbb{C}$ *under pointwise operations. Set*

$$\|x\|_n = \sup_{|t|\leq n} |x(t)|^{k_n} .$$

Then, $\|\cdot\|_n$ is a submultiplicative k_n-seminorm and $(C, (\|\cdot\|_n)_{n\in\mathbb{N}})$ is a locally m-pseudo-convex algebra. In fact, C is an F-algebra. *Of course, if* $k_n = 1$ *for every* $n \in \mathbb{N}$, *then* $(C, (\|\cdot\|_n)_{n\in\mathbb{N}})$ is a locally m-convex algebra.

Example 1.5. *[3, p. 209] We denote by* $l^k = l^k(\mathbb{C})$ *the algebra of all sequences* $(x^{(m)})_{m=1}^{\infty}$ *of elements of the complex field* $\mathbb{C}$ *such that*

$$p(x) = \sum_{m=1}^{\infty} \left|x^{(m)}\right|^k < \infty$$

under coordinatewise operations. Consider the cartesian power

$$A = (l^k)^\omega = l^k \times l^k \times \ldots..$$

For $y = (y_1, y_2, \ldots) \in A$ $(y_n \in l^k)$, *we set*

$$p_n(y) = p(y_n) = \sum_{m=1}^{\infty} \left|y_n^{(m)}\right|^k ,$$

where $y_n = (y_n^{(m)})_{m=1}^{\infty}$. *Then, it is easy to check that* $(A, \{p_n : n = 1, 2, \ldots\})$ *is a locally m-(k-convex) algebra. Moreover, it is an F-algebra too.*

Example 1.6. *Let* $\mathcal{C} = \{f : \mathbb{R} \to \mathbb{R} : f \text{ is continuous}\}$ *and let* $\mathcal{K}$ *be the family of all compact subsets of* $\mathbb{R}$. *For each* $K \in \mathcal{K}$, *choose a real number* $\rho_K \in (0, 1]$ *and define the* ρ_K*-seminorm on* $\mathcal{C}$ *by*

$$p_K(f) = \sup_{x\in K} |f(x)|^{\rho_K} .$$

Then, $(\mathcal{C}, (p_K)_{K\in\mathcal{K}})$ *is a locally m -pseudo-convex algebra which is not an F-algebra.*

2 Locally *m*-pseudo-convex algebras

We need some propositions in order to obtain the desired generalization. The next one can be found in [3, p. 192, Proposition 4.3.11].

Proposition 2.1. Let $(E, \mathcal{P})$ and $(F, \mathcal{Q})$ be two locally m-pseudo-convex spaces where $\mathcal{P} = (p_a)_{\alpha\in\Lambda}$ is a saturated family of k_α-seminorms on E defining its topology, and $\mathcal{Q} = (q_\beta)_{\beta\in\Gamma}$ is a family of k_β-seminorms on F defining its topology. Then a linear transformation $T : E \to F$ is continuous if and only if for each $q_\beta \in \mathcal{Q}$, there exists a $p_\alpha \in \mathcal{P}$ and a constant $C = C_{\alpha,\beta} > 0$ such that

$$q_\beta(T(x)) \leq C \cdot p_\alpha(x)^{\frac{k_\beta}{k_\alpha}} \text{ for each } x \in E. \tag{1}$$

Here, k_β and k_α are the homogenity indexes of q_β and p_α, respectively.

In the case where $E = (E, p)$ is a k_p-seminormed space and $F = (F, q)$ is a k_q-seminormed space, relation (2.1) takes the form (see [3, p. 140])

$$q(T(x)) \leq C \cdot p(x)^{\frac{k_q}{k_p}} \text{ for each } x \in E. \tag{2}$$

We denote by $\mathcal{L}(E, F)$ the space of all continuous linear transformations from (E, p) to (F, q) as above. For $T \in \mathcal{L}(E, F)$, we set

$$\|T\| = \inf\{C > 0 : q(T(x)) \leq C \cdot p(x)^{\frac{k_q}{k_p}} \text{ for each } x \in E\}. \tag{3}$$

>From this, it can be easily derived the following proposition.

Proposition 2.2. Relation (3) defines a k_q-seminorm in $\mathcal{L}(E, F)$. Hence $\mathcal{L}(E, F)$ is a k_q-seminormed space.

Much more can be said in the particular case of the algebra of bounded linear operators $\mathcal{L}(E)$ on the k_p-seminormed space (E, p) (see [3, p. 143, Proposition 3.5.7]).

Proposition 2.3. Let (E, p) be a k_p-seminormed space. Then $\mathcal{L}(E)$, the algebra of all bounded linear operators on E, is a k_p-seminormed algebra under the k_p-seminorm

$$\|T\| = \sup_{p(x)\leq 1} p(T(x)).$$

Moreover, if p is a k_p-norm in E, then so is $\|\cdot\|$ in $\mathcal{L}(E)$, and if (E, p) is a complete k_p-norm in E, then so is $\|\cdot\|$ in $\mathcal{L}(E)$.

It is known that multipliers in proper Banach algebras are bounded (see [7, p. 20, Theorem 1.1.1]). For the present case, we have the following proposition.

Proposition 2.4. If (E, p) is a proper k_p-Banach algebra (complete k_p-normed) and T is a multiplier on E, then T is bounded. Moreover, $M(E)$ is a closed commutative subalgebra of $\mathcal{L}(E)$, hence a commutative k_p -Banach algebra.

Proof. The proof goes as Larsen's just taking into account that the Closed Graph Theorem holds in the case of k_p-Banach algebras (see, for instance [9, p. 50, Theorem 2.15]). □

Proposition 2.5. If $(E, \mathcal{P})$ is a locally m-pseudo-convex algebra, then the algebra $\mathcal{L}(E)$ is also a locally m-pseudo-convex algebra under the topology defined by the family of pseudo-seminorms $\dot{\mathcal{P}} = (\dot{p})_{p\in\mathcal{P}}$, where, for $T \in \mathcal{L}(E)$, we define

$$\dot{p}(T) = \sup_{p(x)\leq 1} p(T(x)).$$

Proof. See [3, p. 143, the proof of Proposition 3.5.7, and p. 197, Corollary 4.4.6]). □

Now we turn to the generalization of the classical Arens-Michael decomposition to locally m-pseudo-convex algebras. See [1], [2] and [3, p. 202, Theorem 4.5.3].

Let $(E, \mathcal{P})$ be a Hausdorff complete locally m-pseudo-convex algebra, where $\mathcal{P} = (p_\alpha)_{\alpha\in\Lambda}$ is a saturated family of k_α-seminorms in E. Let

$$\rho_\alpha : E \to E/\ker(p_\alpha) \equiv E_\alpha$$

be the respective quotient maps, defined by $\rho_\alpha(x) = x + \ker(p_\alpha) \equiv x_\alpha$, $\alpha \in \Lambda$. Then, the relation

$$\dot{p}_\alpha(x_\alpha) := p_\alpha(x),\ x \in E,\ \alpha \in \Lambda$$

defines on E_α an algebra k_α-norm, so that E_α is a k_α-normed algebra and the morphisms ρ_α, $\alpha \in \Lambda$, are continuous. If $\tilde{E}_\alpha$, $\alpha \in \Lambda$ denotes the completion of E_α (with respect to $\dot{p}_\alpha$), then $\tilde{E}_\alpha$ is a k_α-Banach algebra. Λ is endowed with a partial ordering by putting

$$\alpha \leq \beta \text{ if and only if } p_\alpha(x) \leq p_\beta(x) \text{ for every } x \in E.$$

Note that the family is directed in this way due to the fact that $\mathcal{P}$ is a saturated family. We have $\ker(p_\beta) \subseteq \ker(p_\alpha)$ (if $\alpha \leq \beta$) and hence the continuous (onto) morphism

$$f_{\alpha\beta} : E_\beta \to E_\alpha : x_\beta \mapsto f_{\alpha\beta}(x_\beta) = x_\alpha,\ \ \alpha \leq \beta,$$

is well defined. Moreover, $f_{\alpha\beta}$ is extended to a continuous morphism

$$\tilde{f}_{\alpha\beta} : \tilde{E}_\beta \to \tilde{E}_\alpha,\ \ \alpha \leq \beta.$$

Thus, $(E_\alpha, f_{\alpha\beta})$, $(\tilde{E}_\alpha, \tilde{f}_{\alpha\beta})$, $\alpha, \beta \in \Lambda$ with $\alpha \leq \beta$, are projective systems of k_α-normed (resp. k_α -Banach) algebras, and

$$E \cong \varprojlim E_\alpha \cong \varprojlim \tilde{E}_\alpha$$

within topological algebra isomorphisms (*Generalized Arens-Michael decomposition for locally m-pseudo-convex algebras*).

Theorem 2.6. *Let $(E, \mathcal{P})$ be a Hausdorff complete locally m-pseudo-convex algebra, where*

$$\mathcal{P} = (p_\alpha)_{\alpha\in\Lambda}$$

is a saturated family of k_α-seminorms in E, and suppose that E has an approximate identity $(e_\delta)_{\delta\in\Delta}$. Suppose that each factor

$$E_\alpha = E/\ker p_\alpha$$

in the generalized Arens-Michael decomposition of E is complete. Then, each (two-sided) multiplier T of E is continuous, viz. $M(E)$ is a subalgebra of $\mathcal{L}(E)$.

Proof. Note that, due to Proposition 2.4, multipliers on k_α-Banach algebras are continuous; the proof can readily be adapted from [5, p. 1934, Theorem 3.1]. □

Remark 2.7. *As a by-product of the proof of this theorem, it is easily seen that, under the given hypothesis,* $\ker(p_\alpha)$ *is T-invariant for each $T \in M(E)$ and for each $\alpha \in \Lambda$.*

Remark 2.8. *If ϕ is the isomorphism $E \longrightarrow \varprojlim E_\alpha$ given by $\phi(x) = (x_\alpha)_{\alpha\in\Lambda}$, then for each $\alpha \in \Lambda$, $\rho_\alpha = f_\alpha \circ \phi$. Therefore,* $\ker p_\alpha = \ker \rho_\alpha = \ker(f_\alpha \circ \phi)$.

Next we present the notion of a *perfect projective system* as it appeared in [4, p. 199, Definition 2.7].

Definition 2.9. *A projective system $\{(E_\alpha, f_{\alpha\beta})\}_{\alpha\in\Lambda}$ of topological algebras is called* perfect *if the restrictions to the projective limit algebra*

$$E = \varprojlim E_\alpha = \{(x_\alpha) \in \prod_{\alpha\in\Lambda} E_\alpha : f_{\alpha\beta}(x_\beta) = x_\alpha, \text{ if } \alpha \le \beta \in \Lambda\}$$

of the canonical projections $\pi_\alpha : \prod_{\alpha\in\Lambda} E_\alpha \to E_\alpha$, $\alpha \in \Lambda$, namely, the continuous algebra morphisms

$$f_\alpha = \pi_\alpha\,|_{E=\varprojlim E_\alpha} : E \to E_\alpha, \quad \alpha \in \Lambda,$$

are onto maps. The resulting projective limit algebra $E = \varprojlim E_\alpha$ is then called a perfect (topological) algebra.

Definition 2.10. *In the sequel, by the term* perfect locally m-pseudo-convex algebra *we mean a locally m-pseudo-convex algebra $(E, (p_\alpha)_{\alpha\in\Lambda})$ for which the respective generalized Arens-Michael projective system $\{(E_\alpha, f_{\alpha\beta})\}_{\alpha\in\Lambda}$ is perfect.*

Since locally m-convex algebras are locally m-pseudo-convex ones, the examples and facts about perfect algebras given in [6] are pertinent here.

Example 2.11. *Let Λ be any infinite non-countable directed index set and, for each $\alpha \in \Lambda$, let E_α be a k_α-Banach algebra. Let $E = \prod_{\alpha\in\Lambda} E_\alpha$. Then E is a perfect locally m-pseudo-*

convex algebra for which each factor in its generalized Arens-Michael decomposition (namely, E_α) is complete.

Let $(E_\alpha, f_{\alpha\beta})$ be the generalized Arens-Michael projective system just defined for E, so that $E \cong \varprojlim E_\alpha$ algebraically and topologically.

Now we state our main Theorem.

Theorem 2.12. *Let $(E, \mathcal{P})$ be a complete locally m-pseudoconvex Hausdorff algebra, where $\mathcal{P} = (p_\alpha)_{\alpha\in\Lambda}$ is a saturated family of k_α-seminorms in E; suppose E has an approximate identity $(e_\delta)_{\delta\in\Delta}$ and that each factor $E_\alpha = E/\ker p_\alpha$ in its generalized Arens-Michael decomposition is complete. Then:*

(a) *E is perfect,*

(b) *$M(E) \cong \varprojlim M(E_\alpha)$ within a topological algebra isomorphism.*

Proof. The first assertion follows directly from the hypothesis that the factors are complete. We proceed to prove the second assertion.

By the perfectnes of E, each f_β is surjective, so each time we have an element $x_\beta \in E_\beta$, we can choose an element $\omega \in E$ such that $\omega_\beta = x_\beta$, and consequently $\omega_\alpha = x_\alpha = f_{\alpha\beta}(x_\beta)$, whenever $\alpha \leq \beta$.

For each $\alpha \leq \beta$, we define the map $h_{\alpha\beta} : M(E_\beta) \to M(E_\alpha)$ by

$$[h_{\alpha\beta}(T_\beta)](x_\alpha) = f_{\alpha\beta}(T_\beta(x_\beta))$$

As in [6, Lemma 2.6], we can prove that, under the given hypothesis, $\ker f_{\alpha\beta}$ is T_β-invariant for each $T_\beta \in M(E_\beta)$, that is, $T_\beta(\ker f_{\alpha\beta}) \subseteq \ker f_{\alpha\beta}$, if $\alpha \leq \beta$. So the map $h_{\alpha\beta}$ is a well-defined multiplicative linear mapping.

Let us see that the map $h_{\alpha\beta}$ is continuous. Since $f_{\alpha\beta} : E_\beta \to E_\alpha$ is a continuous mapping between k_α-normed algebras, by relation (2.2), there exists a constant $C > 0$ such that

$$\dot{p}_\alpha(f_{\alpha\beta}(y_\beta)) \leq C\,(\dot{p}_\beta)^{\frac{k_\alpha}{k_\beta}}(y_\beta)$$

for each $y_\beta \in E_\beta$. In particular,

$$\dot{p}_\alpha(f_{\alpha\beta}(T_\beta(x_\beta))) \leq C\,(\dot{p}_\beta)^{\frac{k_\alpha}{k_\beta}}(T_\beta(x_\beta))$$

for each $x_\beta \in E_\beta$.

Taking the supremum on the right hand of this relation, and since $M(E_\beta)$ is a k_β-Banach algebra (see Proposition 2.4), we get

$$\begin{aligned}\dot{p}_\alpha(f_{\alpha\beta}(T_\beta(x_\beta))) &\leq C\,(\dot{p}_\beta)^{\frac{k_\alpha}{k_\beta}}(T_\beta(x_\beta)) \leq \\ &\leq C \sup_{\dot{p}_\beta(x_\beta)\leq 1} \{(\dot{p}_\beta)^{\frac{k_\alpha}{k_\beta}}(T_\beta(x_\beta))\} \leq C\,(\|T_\beta\|_\beta)^{\frac{k_\alpha}{k_\beta}}\end{aligned} \tag{4}$$

for every $x_\beta \in E_\beta$ with $\dot{p}_\beta(x_\beta) \leq 1$. Here $\|\cdot\|_\beta$ is the k_β-norm on the multiplier algebra $M(E_\beta)$. Since

$$f_{\alpha\beta}(T_\beta(x_\beta)) = [h_{\alpha\beta}(T_\beta)](x_\alpha), \alpha \leq \beta$$

(hence $\dot{p}_\alpha(x_\alpha) \leq \dot{p}_\beta(x_\beta)$), we get by (4),

$$\dot{p}_\alpha([h_{\alpha\beta}(T_\beta)](x_\alpha)) \leq C\,(\|T_\beta\|_\beta)^{\frac{k_\alpha}{k_\beta}} \tag{5}$$

for every $x_\alpha \in E_\alpha$ with $\dot{p}_\alpha(x_\alpha) \leq 1$.
Taking now the supremum in (5), we have

$$\sup_{\dot{p}_\alpha(x_\alpha)\leq 1} \dot{p}_\alpha([h_{\alpha\beta}(T_\beta)](x_\alpha)) \leq C(\|T_\beta\|_\beta)^{\frac{k_\alpha}{k_\beta}}.$$

Thus

$$\|h_{\alpha\beta}(T_\beta)\|_\alpha \leq C\,(\|T_\beta\|_\beta)^{\frac{k_\alpha}{k_\beta}},$$

namely, each $h_{\alpha\beta}$ is continuous.

By the previous argumentation, we have the family of topological algebras $M(E_\alpha)$ and the family of multiplicative continuous linear mappings

$$h_{\alpha\beta} : M(E_\beta) \to M(E_\alpha), \alpha \leq \beta \in \Lambda.$$

Actually, it is readily seen that they form a projective system of k_α-Banach algebras $\{(M(E_\alpha), h_{\alpha\beta})\}_{\alpha\in\Lambda}$ and we can take its inverse limit, $\varprojlim M(E_\alpha)$.

Take $T \in M(E)$. Due to Remark 1, T induces a well-defined map $T_\alpha : E_\alpha \to E_\alpha$ such that $T_\alpha \circ \rho_\alpha = \rho_\alpha \circ T$ for each $\alpha \in \Lambda$. In fact, T_α is a multiplier in E_α; moreover, $(T_\alpha)_{\alpha\in\Lambda} \in \varprojlim M(E_\alpha)$. So, we can define the map

$$\Phi : M(E) \longrightarrow \varprojlim M(E_\alpha)$$

by $\Phi(T) = (T_\alpha)_{\alpha\in\Lambda}$.

The proof that Φ is actually a multiplicative linear isomorphism, as well as that it is a continuous map, is carried out in the same way as in the case of locally m-convex algebras and we just refer the reader to [6, Th. 2.8].

Here we shall prove that Φ is an open map. Let us recall that the topology of $M(E)$ can be given by the set of seminorms $(\overline{p}_\alpha, \alpha \in \Lambda)$ defined as $\overline{p}_\alpha(T) = \sup_{p_\alpha(x)\leq 1} p_\alpha(T(x))$ for each $T \in M(E)$. Further, the topology of $M(E_\alpha)$ can be given by the norm $\|\cdot\|_\alpha$ defined as $\|S\|_\alpha = \sup_{\dot{p}(x)\leq 1} \dot{p}(S(x))$ for each $S \in M(E_\alpha)$, where, as usual, $\dot{p}$ is the induced norm in E_α given by $\dot{p}(x_\alpha) = \dot{p}(x + \ker p_\alpha) = p_\alpha(x)$. The topology of $\varprojlim M(E_\alpha)$ can be defined by the local base consisting of neighborhoods $V = \bigcap_{i=1}^{n} h_{\alpha_i}^{-1}(V_{\alpha_i})$, where V_{α_i} is a basic neighborhood in $M(E_{\alpha_i})$.

Let $\varepsilon > 0$ be given and let $V_\alpha = \{S \in M(E_\alpha) : \|S\|_\alpha < \varepsilon\}$ and

$U_\alpha = \{T \in M(E) : \overline{p}_\alpha(T) < \varepsilon\}$. We claim that

$$T \in U_\alpha \Longleftrightarrow T_\alpha \in V_\alpha. \tag{6}$$

Indeed,

$$T \in U_\alpha \Longleftrightarrow \overline{p}_\alpha(T) < \varepsilon \Longleftrightarrow \sup_{p_\alpha(x)\le 1} p_\alpha(T(x)) < \varepsilon$$

$$\Longleftrightarrow \sup_{\dot{p}(x_\alpha)\le 1} \dot{p}\,((T(x))_\alpha) < \varepsilon \Longleftrightarrow \sup_{\dot{p}(x_\alpha)\le 1} \dot{p}(T_\alpha(x_\alpha)) < \varepsilon \Longleftrightarrow \|T_\alpha\|_\alpha < \varepsilon$$

$$\Longleftrightarrow T_\alpha \in V_\alpha.$$

Note that for each $\alpha \in \Lambda$ we have $h_\alpha \circ \phi = g_\alpha$; this implies that $g_\alpha \circ \phi^{-1} = h_\alpha$, for each $\alpha \in \Lambda$. Therefore, in order to prove that ϕ is an open mapping (equivalently, that ϕ^{-1} is a continuous mapping), and since h_α is continuous for each $\alpha \in \Lambda$, we only have to show that the topology on $M(E)$ coincides with the initial topology σ given by the family of mappings $\{g_\alpha\}_{\alpha\in\Lambda}$.

For, let V_α be a basic neighborhood in $M(E_\alpha)$, say

$$V_\alpha = \{S \in M(E_\alpha) : \|S\|_\alpha < \varepsilon_\alpha\},$$

and take the basic neighborhood (in the original topology in $M(E)$)

$$U_\alpha = \{T \in M(E) : \overline{p}_\alpha(T) < \varepsilon_\alpha\}$$

just as above. Then $g_\alpha^{-1}(V_\alpha)$ is a basic neighborhood in $M(E)$ according to the topology σ and, applying relation (6), we have:

$$\begin{aligned} g_\alpha^{-1}(V_\alpha) &= \{T \in M(E) : g_\alpha(T) \in V_\alpha\} = \\ &= \{T \in M(E) : T_\alpha \in V_\alpha\} = \{T \in M(E) : T \in U_\alpha\}. \end{aligned}$$

So, $g_\alpha^{-1}(V_\alpha) = U_\alpha$ and we are done. This completes the proof. □

Note: The locally m-convex algebras considered in [6, Lemma 2.6, and in Theorem 2.8], are, actually, perfect (see 1 of Theorem 2.12). Moreover, the proof that Φ is open, appeared in the proof of Theorem 2.12, amends also an oversight in the final part of the proof of Theorem 2.8 in [6]. We thank Luis Roberto Hernández for having pointed out the oversight in the proof.

Acknowledgment: We would like to thank Professor Mati Abel for his valuable comments and encouragement in writing this paper.

Bibliography

[1] Abel, M. *Projective limits of topological algebras*, Tartu Ül. Toimetised, 836(1989), 3–27 (in Russian).

[2] Abel, M. *Representation of topological algebras by projective limit of Fréchet algebras*, Communications in Mathematics and Applications, No. 1, 3(2012), 9–15. RGN Publications.

[3] Balachandran, V.K. *Topological Algebras*, North-Holland Math. Studies 185, Elsevier, Amsterdam, 2000.

[4] Haralampidou, M. *The Krull nature of locally C^*-algebras*, Function Spaces (Edwardsville, IL, 2002), 195–200, Contemp. Math., 328, Amer. Math. Soc., Providence, RI, 2003.

[5] Haralampidou, M., L. Palacios, C. Signoret *Multipliers in locally convex*-algebras*, Rocky Mountain J. Math., No. 6, 43(2013), 1931–1940.

[6] Haralampidou, M., Palacios, L., Signoret, C. *Multipliers in perfect locally m-convex algebras*, Banach J. Math. Anal. 9(2015) no.1, 137–143.

[7] Larsen, R. *The multiplier problem*, Lectures Notes in Math. No. 105, Springer-Verlag, Berlin, 1969.

[8] Mallios, A. *Topological Algebras. Selected Topics*, North-Holland, Amsterdam, 1986.

[9] Rudin, W. *Functional Analysis*, McGraw Hill Book Company, 1973.

Marina Haralampidou and Reyna María Pérez-Tiscareño

Wedderburn structure theorems for two-sided locally m-convex H^*-algebras

Abstract: We give the first and the second structure theorems for certain two-sided locally m-convex H^*-algebras, which constitute a generalization of two-sided H^*-algebras considered by P. Saworotnow.

Keywords: Two-sided locally (resp. locally m-)convex H^*-algebra, proper algebra, left-right projection, left (resp. right) selfadjoint element, invariant ideal, discrete two-sided locally convex H^*-algebra, orthocomplemented pseudo-H-algebra.

MSC: Primary 46H05, 46H10, 46H20, 46K05.

1 Preliminaries

A *left* (resp. *right*) *locally convex H^*-algebra* is a complex algebra E equipped with a family $(p_\alpha)_{\alpha\in\Lambda}$ of seminorms such that each p_α, $\alpha \in \Lambda$ arises from a positive semi-definite (pseudo-)inner product (denoted by $\langle\cdot,\cdot\rangle_\alpha$) so that the induced topology makes E into a locally convex topological algebra. Moreover, for each $x \in E$, there is an element x^l (resp. x^r) in E such that $\langle xy, z\rangle_\alpha = \langle y, x^l z\rangle_\alpha$ (resp. $\langle yx, z\rangle_\alpha = \langle y, zx^r\rangle_\alpha$) for any $y, z \in E$ and $\alpha \in \Lambda$.

The locally m-convex case is also employed, in the obvious way.

A left and right locally (m-)convex H^*-algebra is simply called a *two-sided locally (m-)convex H^*-algebra*. The latter topological algebras were studied in [7], and our concern here is to give two of the Wedderburn structure theorems.

In the sequel, for such a topological algebra we shall employ either the symbol $(E, (p_\alpha)_{\alpha\in\Lambda})$ or $(E, (\langle\cdot,\cdot\rangle_\alpha)_{\alpha\in\Lambda})$. We denote by $\mathcal{A}_r(S)$ the right annihilator of a (non empty) subset S of a $\mathbb{C}$-algebra E. If $\mathcal{A}_r(E) = (0)$ we say that E is *right preannihilator* or *proper*. Hereafter, (0) will stand for the set or yet the ideal containing only the zero element of an algebra.

For a proper Hausdorff two-sided locally convex H^*-algebra E the mappings $l\colon E \to E\colon x \mapsto x^l$, $r\colon E \to E\colon x \mapsto x^r$ are involutions (see [7, p. 412, Corollary 1.3 and p. 413, Lemma 1.4]). In that case, we say that l (resp. r) is the *left* (resp. *right*) *involution* of E. If $x = x^l$ (resp. $x = x^r$), then x is called *left* (resp. *right*) *selfadjoint*.

Marina Haralampidou, Department of Mathematics, University of Athens, Panepistimioupolis, Athens 15784, HELLAS (GREECE). e-mail: mharalam@math.uoa.gr
Reyna María Pérez-Tiscareño, Departament of Mathematics, Autonomous Metropolitan University (UAM) Iztapalapa (Ixtapalapa) Campus, Av. San Rafael Atlixco No.186 Col.Vicentina, 09340 Mexico DF, MEXICO. e-mail: reynapt@xanum.uam.mx

https://doi.org/10.1515/9783110413557-008

Let $(E,(p_\alpha)_{\alpha\in\Lambda})$ be a locally m-convex algebra. Consider the algebra $E_\alpha = E/ker(p_\alpha)$, $\alpha \in \Lambda$, being a normed one, under the norm

$$\dot{p}_\alpha([x]_\alpha) = \dot{p}_\alpha(x + ker(p_\alpha)) := p_\alpha(x),\ [x]_\alpha \equiv x_\alpha \in E_\alpha.$$

Then $\{(E_\alpha, \dot{p}_\alpha), f_{\alpha\beta}\}_{\alpha\in\Lambda}$, (where $f_{\alpha\beta} : E_\beta \to E_\alpha$ with $f_{\alpha\beta}([x]_\beta) := [x]_\alpha$, $\alpha \leq \beta$ in Λ) is a projective system of normed algebras. If in particular, E is a two-sided locally m-convex H^*-algebra, we consider the map

$$(\cdot,\cdot)_\alpha : E_\alpha \times E_\alpha \to \mathbb{C},\ ([x]_\alpha,[y]_\alpha) \mapsto ([x]_\alpha,[y]_\alpha)_\alpha := \langle x, y\rangle_\alpha \tag{1}$$

where $\langle\cdot,\cdot\rangle_\alpha$ is the positive semi-definite inner product from which p_α arises, so it defines the algebra E_α, $\alpha \in \Lambda$, as a vector space with an inner product. Moreover, the E_α's turn to be two-sided H^*-algebras (not necessarily complete) (see [7, p. 419, (2.4) and Lemma 2.4]). In the sequel, we shall freely employ the symbol x_α in place of $[x]_\alpha$, $\alpha \in \Lambda$. Concerning the projective system $\{(E_\alpha, f_{\alpha\beta})\}_{\alpha\in\Lambda}$ we note that since

$$f_{\alpha\beta}((x_\beta)^l) = f_{\alpha\beta}([x^l]_\beta) = [x^l]_\alpha = (x_\alpha)^l$$

(resp. $f_{\alpha\beta}((x_\beta)^r) = (x_\alpha)^r$), *the connecting mappings* $f_{\alpha\beta}$, $\alpha \leq \beta$ *are compatible with the correspondences* l *and* r (l, r are involutions when E is proper and Hausdorff). So, if $(E,(p_\alpha)_{\alpha\in\Lambda})$ is a Hausdorff locally m-convex algebra, we get the Arens-Michael embedding.

$$E \underset{\to}{\subset} \underset{\leftarrow}{\lim} E_\alpha \underset{\to}{\subset} \underset{\leftarrow}{\lim} \widetilde{E}_\alpha = \widetilde{E}$$

within topological algebra isomorphisms (see [8, pp. 85-88]). Here, $\widetilde{E_\alpha}$ (resp. $\widetilde{E}$) stands for the completion of E_α (resp. E).

If, in particular, E is complete, then we get the so-called *Arens - Michael decomposition*

$$E \underset{\to}{=} \underset{\leftarrow}{\lim} E_\alpha \underset{\to}{=} \underset{\leftarrow}{\lim} \widetilde{E_\alpha}$$

within topological algebra isomorphisms.

2 Existence of idempotents

An element x in an algebra E is called *idempotent* (or *projection*) if $x = x^2$. In the sequel, we shall consider only nonzero projections. The set of the projections of E is denoted by $Id(E)$. The statements, in this section, hold true, if we interchange "left" by "right".

Lemma 2.1. *Let $(E,(p_\alpha)_{\alpha\in\Lambda})$ be a proper Hausdorff two-sided locally m-convex H^*-algebra. Then each left selfadjoint element $x \in E$ satisfies*

$$\sup_{p_\alpha(y)=1} |\langle xy, y\rangle_\alpha| = \sup_{p_\alpha(y)=1} p_\alpha(xy), \quad \textit{for every} \quad \alpha \in \Lambda. \tag{1}$$

Proof. We have $E \subset \underset{\rightarrow}{lim} \underset{\leftarrow}{E_\alpha}$, where each E_α is a two-sided H^*-algebra, not necessarily complete (see [7, p. 419, Theorem 2.5]) with inner-product given by (1). Now, following the proof of Lemma 2.1 in [4, p. 141] and [1, p. 374, Lemma 3.1], we obtain the analogous (1) for the factors E_α, $\alpha \in \Lambda$. For the sake of completeness we present the details:

The Cauchy-Schwarz inequality, $|(x_\alpha y_\alpha, y_\alpha)_\alpha| \le p_\alpha(x_\alpha y_\alpha)p_\alpha(y_\alpha)$, for each $\alpha \in \Lambda$, and every $x_\alpha, y_\alpha \in E_\alpha$), implies

$$\sup_{p_\alpha(y_\alpha)=1} |(x_\alpha y_\alpha, y_\alpha)_\alpha| \le \sup_{p_\alpha(y_\alpha)=1} p_\alpha(x_\alpha y_\alpha), \quad \text{for every } \alpha \in \Lambda.$$

To prove the opposite inequality, we consider $y_\alpha' s$ such that $x_\alpha y_\alpha \neq 0$ (if $x_\alpha y_\alpha = 0$, then $(x_\alpha y_\alpha, y_\alpha)_\alpha = p_\alpha(x_\alpha y_\alpha) = 0$). Now, for every $y_\alpha, z_\alpha \in E_\alpha$, and using the left selfadjointness of x_α, we have

$$\begin{aligned}
&(x_\alpha(y_\alpha + z_\alpha), y_\alpha + z_\alpha)_\alpha - (x_\alpha(y_\alpha - z_\alpha), y_\alpha - z_\alpha)_\alpha \\
&= (x_\alpha y_\alpha, y_\alpha + z_\alpha)_\alpha + (x_\alpha z_\alpha, y_\alpha + z_\alpha)_\alpha - ((x_\alpha y_\alpha, y_\alpha - z_\alpha)_\alpha - (x_\alpha z_\alpha, y_\alpha - z_\alpha)_\alpha) \\
&= 2(x_\alpha y_\alpha, z_\alpha)_\alpha + 2(x_\alpha z_\alpha, y_\alpha)_\alpha \\
&= 2(x_\alpha y_\alpha, z_\alpha)_\alpha + 2\overline{(z_\alpha, x_\alpha y_\alpha)_\alpha} \\
&= 2(x_\alpha y_\alpha, z_\alpha)_\alpha + 2\overline{(x_\alpha y_\alpha, z_\alpha)_\alpha} \\
&= 4Re((x_\alpha y_\alpha, z_\alpha)_\alpha).
\end{aligned}$$

Moreover, if we consider $y_\alpha + z_\alpha = 0$ and $y_\alpha - z_\alpha = 0$, then $y_\alpha = z_\alpha = 0$ and hence $x_\alpha y_\alpha = 0$ which is a contradiction. So, $y_\alpha + z_\alpha \neq 0$ or $y_\alpha - z_\alpha \neq 0$.

Suppose $y_\alpha + z_\alpha \neq 0$ and $y_\alpha - z_\alpha = 0$ (analogously, if $y_\alpha - z_\alpha \neq 0$ and $y_\alpha + z_\alpha = 0$).

We have $|(x_\alpha(\frac{y_\alpha+z_\alpha}{p_\alpha(y_\alpha+z_\alpha)}), \frac{y_\alpha+z_\alpha}{p_\alpha(y_\alpha+z_\alpha)})_\alpha| \le \sup_{p_\alpha(u_\alpha)=1} |(x_\alpha u_\alpha, u_\alpha)_\alpha|$ so,

$$|(x_\alpha(y_\alpha + z_\alpha), y_\alpha + z_\alpha)_\alpha| \le p_\alpha^2(y_\alpha + z_\alpha) \sup_{p_\alpha(u_\alpha)=1} |(x_\alpha u_\alpha, u_\alpha)_\alpha|.$$

Hence,

$$\begin{aligned}
4Re((x_\alpha y_\alpha, z_\alpha)_\alpha) &\le \sup_{p_\alpha(u_\alpha)=1} |(x_\alpha u_\alpha, u_\alpha)_\alpha| \{p_\alpha^2(y_\alpha + z_\alpha) + p_\alpha^2(y_\alpha - z_\alpha)\} = \\
&\quad 2 \sup_{p_\alpha(u_\alpha)=1} |(x_\alpha u_\alpha, u_\alpha)_\alpha| (p_\alpha^2(y_\alpha) + p_\alpha^2(z_\alpha)).
\end{aligned}$$

The rest of the assertion is proved in a similar way.

Now, suppose that $y_\alpha + z_\alpha \neq 0$ and $y_\alpha - z_\alpha \neq 0$. By the previous argument, we have

$$|(x_\alpha(y_\alpha + z_\alpha), y_\alpha + z_\alpha)_\alpha| \le p_\alpha^2(y_\alpha + z_\alpha) \sup_{p_\alpha(u_\alpha)=1} |(x_\alpha u_\alpha, u_\alpha)_\alpha|$$

and

$$|(x_\alpha(y_\alpha - z_\alpha), y_\alpha - z_\alpha)_\alpha| \le p_\alpha^2(y_\alpha - z_\alpha) \sup_{p_\alpha(u_\alpha)=1} |(x_\alpha u_\alpha, u_\alpha)_\alpha|$$

$$4Re((x_\alpha y_\alpha, z_\alpha)_\alpha) \le \sup_{p_\alpha(u_\alpha)=1} |(x_\alpha u_\alpha, u_\alpha)_\alpha| \{p_\alpha^2(y_\alpha + z_\alpha) + p_\alpha^2(y_\alpha - z_\alpha)\} =$$

$$2 \sup_{p_\alpha(u_\alpha)=1} |(x_\alpha u_\alpha, u_\alpha)_\alpha| (p_\alpha^2(y_\alpha) + p_\alpha^2(z_\alpha)).$$

If we take $z_\alpha = x_\alpha y_\alpha / p_\alpha(x_\alpha y_\alpha)$, then

$$4Re((x_\alpha y_\alpha, x_\alpha y_\alpha / p_\alpha(x_\alpha y_\alpha))_\alpha) \leq$$

$$2 \sup_{p_\alpha(u_\alpha)=1} |(x_\alpha u_\alpha, u_\alpha)_\alpha| \{p_\alpha^2(y_\alpha) + p_\alpha^2(x_\alpha y_\alpha / p_\alpha(x_\alpha y_\alpha))\}.$$

Then, for every y_α such that $p_\alpha(y_\alpha) = 1$, and since $p_\alpha^2(x_\alpha y_\alpha / p_\alpha(x_\alpha y_\alpha)) = 1$ and $(x_\alpha y_\alpha, x_\alpha y_\alpha / p_\alpha(x_\alpha y_\alpha)_\alpha) \in \mathbb{R}$, we have

$$4(x_\alpha y_\alpha, x_\alpha y_\alpha)_\alpha / p_\alpha(x_\alpha y_\alpha) = 4Re((x_\alpha y_\alpha, x_\alpha y_\alpha / p_\alpha(x_\alpha y_\alpha))_\alpha) \leq$$

$$4 \sup_{p_\alpha(u_\alpha)=1} |(x_\alpha u_\alpha, u_\alpha)_\alpha|$$

Hence, the opposite inequality

$$\sup_{p_\alpha(y_\alpha)=1} p_\alpha(x_\alpha y_\alpha) \leq \sup_{p_\alpha(u_\alpha)=1} |(x_\alpha u_\alpha, u_\alpha)_\alpha|$$

holds true, as well. □

Lemma 2.2. *Let $(E, (p_\alpha)_{\alpha \in \Lambda})$ be a proper Hausdorff two-sided locally convex H^*-algebra. Then E contains a non zero left (right) selfadjoint element.*

Proof. For $x \in E$, $x \neq 0$, the left (right) selfadjoint element $x^l x$ (xx^r) is not zero (see [7, p. 412, Corollary 1.3]). □

For the next result, we follow the proof of Lemma 2.3 in [4, p. 142].

Lemma 2.3. *Let $(E, (\langle \cdot, \cdot \rangle_\alpha)_{\alpha \in \Lambda})$ be a proper Hausdorff two-sided locally m-convex H^*-algebra and y a left selfadjoint element of E satisfying the conditions:*

$$0 \leq \langle yz, z \rangle_\alpha \leq \langle z, z \rangle_\alpha \text{ for every } z \in E, \alpha \in \Lambda \tag{2}$$

and

$$\sup_{p_\alpha(z)=1} |\langle yz, z \rangle_\alpha| = 1 \text{ for every } \alpha \in \Lambda. \tag{3}$$

Then y^2 satisfies (2) *and* (3), *as well.*

Proof. Since $y = y^l$,

$$\langle y^2 z, z \rangle_\alpha = \langle yz, y^l z \rangle_\alpha = \langle yz, yz \rangle_\alpha \geq 0,$$

for every $z \in E$ and $\alpha \in \Lambda$. Namely, y^2 satisfies the first inequality of (2). Moreover, by (1) and (3), we have that $\sup_{p_\alpha(z)=1} p_\alpha(yz) = 1$ so,

$$\sup_{p_\alpha(z)=1} \langle y^2 z, z \rangle_\alpha = \sup_{p_\alpha(z)=1} p_\alpha^2(yz) = 1,$$

i.e., y^2 satisfies (3), as well. Thus, $\langle y^2 z, z\rangle_\alpha \leq 1$, for every $z \in E$ such that $p_\alpha(z) = 1$ and hence $\langle y^2 \frac{z}{p_\alpha(z)}, \frac{z}{p_\alpha(z)}\rangle_\alpha \leq 1$, for every $z \in E$ such that $p_\alpha(z)$ is non-zero. (If $p_\alpha(z) = 0$, then since E is locally m-convex, we have that $\langle y^2 z, z\rangle_\alpha = \langle z, z\rangle_\alpha = 0$). So, $\langle y^2 z, z\rangle_\alpha \leq \langle z, z\rangle_\alpha$ for every $z \in E$ (see the proof of Lemma 2.3 in [4, p. 142]), i.e. (2) is fulfilled for y^2. □

We also have the analogous version of the last lemma, for right selfadjoints. In that case, the respective of (2) and (3) are the following.

$$0 \leq \langle zy, z\rangle_\alpha \leq \langle z, z\rangle_\alpha, \quad \text{for every } z \in E, \alpha \in \Lambda. \tag{4}$$

$$\sup_{p_\alpha(z)=1} |\langle zy, z\rangle_\alpha| = 1, \quad \text{for every } \alpha \in \Lambda. \tag{5}$$

We observe that an element x of a proper Hausdorff two-sided locally convex H^*-algebra E that satisfies

$$\langle xy, y\rangle_\alpha \geq 0, \text{ for all } y \in E, \alpha \in \Lambda,$$

is left selfadjoint. In fact, $0 \leq \langle xy, y\rangle_\alpha = \langle y, x^l y\rangle_\alpha = \langle x^l y, y\rangle_\alpha$, for all $y \in E$ and $\alpha \in \Lambda$. So, $\langle (x - x^l)y, y\rangle_\alpha = 0$, for all $y \in E$, and hence $(x - x^l)E = (0)$. Properness of E yields now $x = x^l$.

For the next result see also Lemma 1.1 in [5, p. 65], the proof of which is modified here, appropriately.

Lemma 2.4. *Let $(E, (p_\alpha)_{\alpha\in\Lambda})$ be a proper Hausdorff two-sided locally m-convex H^*-algebra that satisfies the property:*

(lbs) There exists some nonzero element $w \in E$ of the form $w = h^l h$, $h \in E$ with $\sup_{\alpha\in\Lambda}(p_\alpha(w)) < +\infty$.

Then there exists $x \in E$ such that

$$x = y^2, \quad \textit{for some non zero left selfadjoint } y \in E. \tag{6}$$

$$0 \leq \langle xz, z\rangle_\alpha \leq \langle z, z\rangle_\alpha, \quad \textit{for every } z \in E, \alpha \in \Lambda, \tag{7}$$

and

$$\sup_{p_\alpha(z)=1} |\langle xz, z\rangle_\alpha| = 1, z \in E, \alpha \in \Lambda. \tag{8}$$

Proof. We consider $w = h^l h$ as in (lbs). By employing also the Cauchy-Schwarz inequality, we get

$0 \leq \langle hz, hz\rangle_\alpha = \langle h^l hz, z\rangle_\alpha \leq p_\alpha(h^l h) p_\alpha(z)^2$, for all $z \in E$, $\alpha \in \Lambda$ and hence

$$\sup_{p_\alpha(z)=1} \langle h^l hz, z\rangle_\alpha \leq p_\alpha(h^l h) \leq \sup_\alpha \; p_\alpha(h^l h) < +\infty.$$

Moreover, $\sup_{p_\alpha(z)=1} \langle h^l hz, z\rangle_\alpha \neq 0$, otherwise $\langle hz, hz\rangle_\alpha = 0$, for every $z \in E$ with $p_\alpha(z) = 1$. Hence $\langle hz, hz\rangle_\alpha = 0$, for every $0 \neq z \in E$. So, $p_\alpha(hz) = 0$ for every $z \in E$, and by Hausdorff, $hz = 0$, for all $z \in E$. Thus, by properness (see also [2, p. 452, Theorem 1.3]) $h = 0$ that yields $w = h^l h = 0$, a contradiction.

Now, consider $k = \sup_{p_\alpha(z)=1} \langle h^l hz, z\rangle_\alpha$ and $y = \frac{1}{k} h^l h$. Then

$$\sup_{p_\alpha(z)=1} \langle yz, z\rangle_\alpha = \frac{1}{k} \sup_{p_\alpha(z)=1} \langle h^l hz, z\rangle_\alpha = 1, \text{ for every } z \in E \text{ and } \alpha \in \Lambda.$$

Moreover, $\langle yz, z\rangle_\alpha \leq 1$, for every $z \in E$ such that $p_\alpha(z) = 1$ and hence $\langle yz, z\rangle_\alpha \leq \langle z, z\rangle_\alpha$, for every $z \in E$, $\alpha \in \Lambda$ (see at the end of the proof of Lemma 2.3).

The previous argument shows that y satisfies the conditions of Lemma 2.3 so, the same is true for y^2. □

We have an analogous lemma if we consider a similar *(rbs) property* on the right. The next results are proved for appropriate topological algebras with the property *(lbs)*, while analogous versions still hold, if we consider the *(rbs)* property.

For the proofs of the next two lemmas we refer to [4, p. 143, Lemmas 2.5, 2.6].

In the sequel, by an *(lbs) two-sided locally (m-)convex H^*-algebra* we mean a two-sided locally (m)-convex H^*-algebra satisfying the *(lbs)* property.

Lemma 2.5. *Let $(E, (p_\alpha)_{\alpha\in\Lambda})$ be a proper (lbs) two-sided locally convex H^*-algebra. Consider $x \in E$ such that* (6) *and* (7) *are fulfilled. Then, for every $z \in E$, $(\langle x^n z, z\rangle_\alpha)_{n\in\mathbb{N}}$ is a decreasing sequence of non negative numbers.*

Lemma 2.6. *Let $(E, (p_\alpha)_{\alpha\in\Lambda})$ be a proper Hausdorff complete (lbs) two-sided locally m-convex H^*-algebra. If $x \in E$ satisfies* (6), (7) *and* (8) (see Lemma 2.4), *then the sequence $(x^n z)_{n\in\mathbb{N}}$ converges, for every $z \in E$.*

For the next definition see also [9, p. 367].

Definition 2.7. *Let $(E, (p_\alpha)_{\alpha\in\Lambda})$ be a proper Hausdorff two-sided locally convex H^*-algebra. By the term left (right) projection in E, we mean a projection $x \in E$ which is left (right) selfadjoint. If x is both left and right selfadjoint, then it is simply called a left-right projection.*

The proof of the next result goes as in the proof of [4, p. 144, Theorem 2.7] where an involution is used instead of a left involution. For the shake of completeness, we refer it here.

Theorem 2.8. *Let $(E, (p_\alpha)_{\alpha\in\Lambda})$ be a proper Hausdorff complete (lbs) two-sided locally m-convex H^*-algebra with a left continuous involution. Then E has a non zero left projection.*

Proof. Consider $x \in E$ as in Lemma 2.4. Then, by Lemma 2.6, (x^n) is convergent, say to x_0. Thus,

$$x_0^l = \lim_n (x^n)^l = \lim_n x^n = x_0$$

and since $x^{2n} \longrightarrow x_0^2$, we get $x_0^2 = x_0$. Moreover, by Lemma 2.3, x^2 satisfies (2) and (3).

If we apply the preceding argument to x^2, which is left selfadjoint (see the comment after (5)), we conclude that (2) and (3) are satisfied for every x^m with $m = 2^k$, $k \in \mathbb{N}$. Thus,

$$1 = \sup_{p_\alpha(z)-1} |\langle x^m z, z\rangle_\alpha| = \sup_{p_\alpha(z)=1} p_\alpha(x^m z), \quad z \in E$$

for $m = 2k$, $k \in \mathbb{N}$, $\alpha \in \Lambda$ and by the submultiplicativity of p_α, $p_\alpha(x^m) \geq 1$. Now, by the continuity of p_α, we get

$$p_\alpha(x_0) = p_\alpha(\lim_m x^m) = \lim_m p_\alpha(x^m) \geq 1,$$

for all $\alpha \in \Lambda$. Thus $x_0 \neq 0$. □

Corollary 2.9. Let $(E, (p_\alpha)_{\alpha\in\Lambda})$ be a proper Hausdorff complete (lbs) two-sided locally m-convex H^*-algebra such that $E = \overline{[E^2]}$ *(the closed two-sided ideal generated by* E^2*)* and $(y^l - y)^l = (y^l - y)^r$, for every $y \in [E^2]$. Then E has a non zero left projection.

Proof. Apply Theorem 2.8 and [7, p. 421, Theorem 3.2]. □

Definition 2.10. *Let E be a proper Hausdorff two-sided locally convex H^*-algebra. A (left, right) ideal I of E is said to be* invariant *if $I^r = I^l$, where I^l (resp. I^r) denotes the image of I under the involution l (resp. r).*

Theorem 2.11. *Let $(E, (p_\alpha)_{\alpha\in\Lambda})$ be a proper Hausdorff two-sided locally convex H^*-algebra and $e \in E$ a left projection. Then, eE is invariant if and only if e is a left-right projection.*

Proof. Since eE is invariant, for any $x \in E$, we get

$$xe^r = (ex^r)^r \in (eE)^r = (eE)^l.$$

Thus, there exists $y \in E$ such that $xe^r = (ey)^l = y^l e$, and hence $Ee^r \subseteq Ee$. Since $xe = (ex^l)^l$, we can argue as before to get $Ee \subseteq Ee^r$. The previous argument yields $Ee = Ee^r$. Hence, since e^r is idempotent and $e \in Ee = Ee^r$, we have $ee^r = e$. Thus, $e^r = e$.

Conversely, if e is a left-right projection and $y \in (eE)^l$, then $y = (ex)^l$, for some $x \in E$ and since

$$(ex)^l = x^l e^l = x^l e^r = (e(x^l)^r)^r \in (eE)^r,$$

we get $(eE)^l \subset (eE)^r$. Similarly, $(eE)^r \subset (eE)^l$, and the assertion follows. □

As a realization of the last theorem we have the next example, that assures also the existence of invariant ideals.

Example 2.12. *Take $\rho \in \mathbb{R}$. Consider the set $\{t_i \in \mathbb{R}, i \in J$ with $1 \leq t_i \leq \rho\}$, which obviously is non-empty. The set*

$$E = \{x : J \times J \to \mathbb{C} : \sum_{i,j} t_i |x(i,j)|^2 t_j \text{ converges}\} \tag{9}$$

becomes an algebra under the following operations:

$$\begin{aligned}(x+y)(i,j) &= x(i,j) + y(i,j),\\ (\lambda x)(i,j) &= \lambda(x(i,j)) \equiv \lambda x(i,j),\\ (xy)(i,j) &= \textstyle\sum_k x(i,k)y(k,j) \text{ “matrix multiplication”}.\end{aligned}$$

Let $y : J \times J \to \mathbb{C}$ and $(k,l) \in J \times J$ be such that $y(k,l) = 1$ and $y(i,j) = 0$ elsewhere. For each element $(k,l) \in J \times J$, the respective function is denoted by y_{kl}. Clearly, each one of them are elements in E and $\lambda y_{kl} \in E$, for every $\lambda \in \mathbb{C}$. We note that if $x \in E$, then $x = \sum_{k,l} x(k,l) y_{kl}$. A family of pseudo-inner products on E is defined by

$$\langle x, y\rangle_\alpha = s_\alpha \sum_{i,j} t_i x(i,j)\overline{y}(i,j) t_j, \tag{10}$$

where $x = x(i,j), y = y(i,j)$ and $(s_\alpha)_{\alpha\in\Lambda}$ is a family of numbers such that $s_\alpha \geq 1$, for every $\alpha \in \Lambda$. Moreover, the right (resp. left) adjoint of $x \in E$ is given by

$$x^r(i,j) = (\tfrac{t_i}{t_j})\overline{x}(j,i), \text{ resp. } x^l(i,j) = (\tfrac{t_j}{t_i})\overline{x}(j,i).$$

In this context, E is a proper two-sided locally convex H^*-algebra (see [7, p. 416, Example 1.10]).

Note that, for each $k \in J$, y_{kk} is a left-right projection and thus by the previous theorem, $y_{kk}E$ is an invariant ideal, for each $k \in J$.

Remark 2.13. *If, in the previous example, we consider on E the pointwise multiplication (namely, $(xy)(i,j) = x(i,j)y(i,j)$) instead of the “matrix” one, then E is an algebra which the right (resp. left) adjoint of $x \in E$ is*

$$x^r(i,j) = \overline{x}(i,j), \quad \text{resp. } x^l(i,j) = \overline{x}(i,j)$$

Indeed (see also (9) *and* (10)*) we have*

$$\begin{aligned}\langle y, zx^r\rangle_\alpha &= s_\alpha \sum_{i,j} t_i y(i,j)\overline{zx^r}(i,j)t_j = s_\alpha \sum_{i,j} t_i y(i,j)\overline{z}(i,j)\overline{x^r}(i,j)t_j\\ &= s_\alpha \sum_{i,j} t_i y(i,j)\overline{z}(i,j)x(i,j)t_j = s_\alpha \sum_{i,j} t_i yx(i,j)\overline{z}(i,j)t_j = \langle yx, z\rangle_\alpha.\end{aligned}$$

Similarly, $\langle y, x^l z\rangle_\alpha = \langle xy, z\rangle_\alpha$.
Since for every $x \in E$, $x^r = x^l$, E is an example of a (two-sided) locally convex H^*-algebra in which all ideals are trivially invariant.

Discrete algebras were employed by P.P. Saworotnow in the context of Banach algebras (see [9, p. 368, Definition]). In the non normed case, we set the next.

Definition 2.14. *A proper Hausdorff two-sided locally convex H^*-algebra E is said to be discrete, if each invariant ideal in E contains an invariant ideal of the form eE with e a left projection.*

The following result is a direct consequence of Theorem 2.11.

Theorem 2.15. *If E is a discrete two-sided locally convex H^*-algebra, then each invariant right ideal of E has a left-right projection.*

3 Structure theory

We remind that a *pseudo-H-space* is a locally convex space E, whose topology is defined by a family $(\langle\cdot,\cdot\rangle_\alpha)_{\alpha\in\Lambda}$ of positive semi-definite (:pseudo-)inner products. The *orthogonal* $S^\perp$ of a nonempty subset S of E is given by

$$S^\perp := \{x \in E : \langle x, y\rangle_\alpha = 0 \quad \text{for every } y \in S, \alpha \in \Lambda\}$$

being a closed linear subspace of E.

A *pseudo-H-algebra* is a pseudo-H-space and an algebra (which is locally convex) with separately continuous multiplication (or even locally m-convex); see [2, p. 456, Definition 3.1]. The topology of a pseudo-H-algebra E is defined by a family $(p_\alpha)_{\alpha\in A}$ of seminorms so that $p_\alpha(x) = \langle x, x\rangle_\alpha^{1/2}$ for every $x \in E$. Such a topological algebra is still denoted by $(E, (p_\alpha)_{\alpha\in\Lambda})$ or yet by $(E, (\langle\cdot,\cdot\rangle_\alpha)_{\alpha\in\Lambda})$. A closed (left) ideal I in a pseudo H-algebra E is *orthocomplemented*, if $E = I \oplus^\perp I^\perp$ ($I^\perp$ is named the *orthocomplement* of I). Besides, E is called *orthocomplemented*, if every closed (left) ideal in E has an orthocomplement of the same type.

We say that $x, y, x \neq y$, in a pseudo-H-algebra E, are *(algebraically) orthogonal*, if $xy = yx = 0$. If moreover, x, y are orthogonal elements (viz. $\langle x, y\rangle_\alpha = 0$, for all $\alpha \in \Lambda$), then x, y are called *doubly orthogonal*. An element $x \in E$ is *primitive*, if it can not be expressed as the sum of two (non zero) doubly orthogonal projections. A family $(x_i)_{i\in K} \subseteq E$ such that for every $i \neq j$ $(i, j \in K)$, $x_i x_j = 0$ is called a *family (algebraically) orthogonal elements*, and a maximal family of doubly orthogonal primitive left (right, left-right) projections is called a *left (right, left-right) family of axes*.

Let E be a proper Hausdorff two-sided locally m-convex H^*-algebra. We consider the following properties (cf also [4, p. 141]):

(H_l) Each closed left ideal I in E with $I \subseteq Ex$, for some $x \in Id(E)$, has an orthocomplement in Ex. Namely, $Ex = I \oplus^{\perp} I^{\perp}$.

(In the sequel, this property will be also considered in the context of pseudo-H-algebras).

(H_l') Each closed left ideal $I \subseteq Ex$, for some $x \in Id(E)$, is invariant and orthocomplemented in Ex. Namely, $I^l = I^r$ and $Ex = I \oplus^{\perp} I^{\perp}$.

Similarly, by employing the right ideal xE, we can define the properties (H_r) and (H_r').

If E is a proper Hausdorff two-sided locally m-convex H^*-algebra, satisfying the (H_l') property, and $I \subseteq Ex$ is a closed left ideal, for some $x \in Id(E)$, then $I^{\perp}$ is also an invariant closed left ideal. By [7, p. 420, Lemma 2.7], $I^{\perp} \subseteq Ex$ is a closed left ideal in E. So, by the (H_l') property, $I^{\perp}$ is invariant and orthocomplemented.

Clearly, (H_l') implies (H_l). Moreover, if $x^r = x^l$, for every $x \in E$ we get that E is then a locally convex H^*-algebra. Besides, the properties (H_l') and (H_l) are equivalent, and (H_l) coincides with the property (H) given in [4, p. 141].

To shorten the terminology, we set the next definitions.

Let E be a proper Hausdorff two-sided locally m-convex H^*-algebra. Then

(1) E is called a *left* (resp. *right*) *pre-Ambrose algebra*, if it has the (H_l) (resp. (H_r)) property.

(2) E is called an (H_l') (resp. (H_r')) *pre-Ambrose algebra* if it has the (H_l') (resp. (H_r')) property.

(3) E is named a *left* (resp. *right*) *Ambrose algebra*, if it is a complete left (resp. right) pre-Ambrose algebra having its left and right involutions continuous.

Lemma 3.1. *Let $(E, (p_\alpha)_{\alpha \in \Lambda})$ be a pseudo H-algebra, $x \in Id(E)$ and $I \equiv Ex$ the respective (closed) left ideal. Suppose that $I = \bigoplus^{\perp}_{1 \le i \le n} I_i$ with I_i a (non zero) left ideal, for each i $(1 \le i \le n)$. Moreover, if $x = \sum_{i=1}^n x_i$, $x_i \in I_i$, then the following hold true.*

(1) $I_i = Ex_i$, $1 \le i \le n$.

(2) If E is Hausdorff, then the x_i's are (non zero) doubly orthogonal projections.

(3) If, in particular, E is a proper Hausdorff two-sided locally convex H^*-algebra and x is right selfadjoint, then x_i is right selfadjoint, for each i.

Proof. For (1) and (2), see the proof of Lemma 3.1 in [3, p. 1180]. Now, suppose that E is as in (3). By (2), $x_i = x_i x = x x_i$, $1 \le i \le n$ and hence $x_i^r = x x_i^r = x_i^r x$, $1 \le i \le n$. Moreover, by (1), $\langle y x_i, z x_k \rangle_\alpha = 0$, for any $y, z \in E$, $\alpha \in \Lambda$, $i \neq k$. Thus, since E is proper and Hausdorff, $x_k x_i^r = 0$, $i \neq k$. Hence $x_i^r = \sum_{k=1}^n x_k x_i^r = x_i x_i^r$ and since r is a right involution, we get $x_i = x_i x_i^r$, for any $1 \le i \le n$. Therefore (3) is fulfilled. □

An analogous to the last result holds true when we consider the (closed) right ideal xE.

Lemma 3.2. *Let $(E, (p_\alpha)_{\alpha \in \Lambda})$ be a pseudo-H-algebra and $x \in Id(E)$. Then the following statements hold:*

(i) If Ex is a minimal closed (left) ideal, then x is primitive.

(ii) If moreover, E is Hausdorff, satisfies the (H_l) property and x is primitive, then Ex is a minimal closed ideal.

Proof. See the proof of Lemma 3.3 in [3, p. 1181] and the remarks after the (H_l) property. □

Theorem 3.3. *Let $(E, (p_\alpha)_{\alpha \in \Lambda})$ be a Hausdorff two-sided locally m-convex H^*-algebra. If E satisfies the (H_l) (resp. (H_r)) property and $x \in Id(E)$, then*

$$x = \sum_{i=1}^{n} x_i,$$

with x_i $(1 \le i \le n)$ non zero doubly orthogonal primitive projections. If E is moreover proper, and x is right (resp. left) selfadjoint, then x_i is right (resp. left) selfadjoint, for each i.

Proof. We follow the proof of Theorem 3.4 in [3, p. 1182]. According to this, $J = Ex$ can be written as a finite sum $J = \underset{1 \le i \le n}{\oplus^\perp} I_i$ with I_i minimal closed left ideals of E. Thus, by Lemmas 3.1 and 3.2, $x = \sum_{i=1}^{n} x_i$ with x_i, $(1 \le i \le n)$ non zero doubly orthogonal primitive projections.

Now, if x is right selfadjoint, then by Lemma 3.1, each x_i $(1 \le i \le n)$ is right selfadjoint, as well. □

Corollary 3.4. Every (lbs) complete right pre-Ambrose algebra with continuous left involution contains a left family of axes.

Proof. By Theorem 2.8, we have a non zero left selfadjoint projection. The assertion follows from Theorem 3.3 and Zorn's lemma. □

An analogous result holds true when interchange "left" by "right".

Corollary 3.5. Let $(E, (p_\alpha)_{\alpha \in \Lambda})$ be a proper Hausdorff two-sided locally m-convex H^*-algebra. If E satisfies the (H_l') (resp. (H_r')) property and $x \in Id(E)$, then

$$x = \sum_{i=1}^{n} x_i$$

with x_i $(1 \le i \le n)$ non zero doubly orthogonal primitive projections. If moreover, x is right (resp. left) selfadjoint, then x_i is left-right selfadjoint for each i.

Proof. Since the (H_l') property implies the (H_l) one, by Theorem 3.3, we have the first part of the assertion.

Also, based on the proof of Theorem 3.3 and due to the (H_l') property, we conclude that the factors I_i, in $J = \bigoplus_{1\leq i\leq n}^{\perp} I_i$, are invariant, as well.

Now, if x is right selfadjoint, by Lemma 3.1 each x_i is right selfadjoint too. Due to the invariant character of $I_i = Ex_i$, for any $1 \leq i \leq n$, x_i is a left-right projection, for any $1 \leq i \leq n$ (see Theorem 2.11). □

Corollary 3.6. Let $(E, (p_\alpha)_{\alpha\in\Lambda})$ be a complete (lbs) (H_r') pre-Ambrose algebra with continuous left involution. Then E contains a left-right family of axes.

Proof. By Theorem 2.8, E has a non-zero left projection, say x. Thus, due to the analysis of x (see Corollary 3.5), E contains left-right selfadjoint idempotents. Applying now a proof analogous to that of Theorem 3.1 in [4, p. 144], we get the desired left-right family of axes. Here, the Zorn's lemma is also applied.

□

Corollary 3.7. Let $(E, (p_\alpha)_{\alpha\in\Lambda})$ be a complete (lbs) (H_r') pre-Ambrose algebra such that $E = \overline{[E^2]}$ and $(y^l - y)^l = (y^l - y)^r$, for every $y \in [E^2]$. Then E contains a left-right family of axes.

Proof. Immediate from Corollary 2.9 and the proof of Corollary 3.6. □

Corollary 3.8. Every discrete complete (H_r') pre-Ambrose algebra contains a left-right family of axes.

Proof. By Theorem 2.15, there is a non zero left-right projection. The assertion follows from Corollary 3.5, by applying a proof analogous to that of Corollary 3.6. □

The next is a key result in the proof of Theorem 3.10. For this we need some more terminology. In [10], M.F. Smiley introduced and studied *right H^*-algebras*. Such an algebra is a Banach algebra E, the underlying linear space being a Hilbert one. Moreover, for each $x \in E$, there is an $x^* \in E$ such that $\langle xy, z\rangle = \langle x, zy^*\rangle$ for every $y, z \in E$. In our notation $*$ stands for the right involution r. A *left H^*-algebra* is defined analogously.

Lemma 3.9. *Let E be a proper left (resp. right) H^*-(Banach) algebra. Then E is a two-sided H^*-algebra and for any $x \in E$, $\langle x, x\rangle = \langle x^l, x^r\rangle$.*

Proof. Assume that E is a left H^*-algebra. We consider $0 \neq x \in E$ and let N be the closed linear subspace of E, spanned by x. Since N is closed, $E = N \oplus N^\perp$ and $N^\perp$ is a closed space. Take $N_1 = \{y \in E : y^l \in N^\perp\}$. Since the left involution is continuous, N_1 is a closed linear subspace of E. Moreover, $N_1^\perp$ is non zero. Otherwise, $E = N_1$. Thus $x^l \in N_1$ and hence $x \in N^\perp$, that yields a contradiction.

Claim that $N_1^\perp$ is one-dimensional. Indeed, take $r, s \in N_1^\perp, r \neq 0$. We note that $\langle x, r^l\rangle \neq 0$, otherwise, $\langle \mu x, r^l\rangle = 0$, for every $\mu \in \mathbb{C}$, that yields $r^l \in N^\perp$ so, $r \in N_1$ and hence $r = 0$, a contradiction. Therefore, for $\lambda = \frac{\langle x, s^l\rangle}{\langle x, r^l\rangle} \in \mathbb{C}$, we get $\langle \lambda r - s, t\rangle = 0$, for every $t \in N_1$. Now, $\lambda r - s \in N_1$ since for every $\mu \in \mathbb{C}$,

$$\langle(\lambda r - s)^l, \mu x\rangle = \overline{\lambda}\langle r^l, \mu x\rangle - \overline{\mu}\langle s^l, x\rangle = \overline{\mu}\overline{\frac{\langle x, s^l\rangle}{\langle x, r^l\rangle}}\langle r^l, x\rangle - \overline{\mu}\langle s^l, x\rangle = 0.$$

Thus, $\langle \lambda r - s, \lambda r - s\rangle = 0$ and hence $\lambda r - s = 0$, that yields the assertion.

Now, we note that there exist $u \in N_1^\perp$ such that $\langle x^l, u\rangle \neq 0$. Otherwise, $\langle x^l, u\rangle = 0$, for every $u \in N_1^\perp$ and $\langle(\lambda x + w)^l, u\rangle = 0$, for every $w \in N^\perp, u \in N_1^\perp$ and hence $E^l \subset N_1^{\perp\perp} = N_1$ (since N_1 is closed). So, there exists $y \subset N_1$ such that $x^l = y$. The left involution has period 2, thus $x = y^l \in N^\perp$, which is a contradiction.

Put $0 \neq \kappa = \langle x^l, u\rangle$ with $u \in N_1^\perp$ and $\kappa_1 = \langle x, x\rangle$. Then

$$u' \equiv \frac{\kappa_1}{\overline{\kappa}} u \in N_1^\perp$$

and

$$\langle x^l, u'\rangle = \frac{\overline{\kappa_1}}{\kappa}\langle x^l, u\rangle = \overline{\kappa_1} = \langle x, x\rangle.$$

We shall prove that $u' = x^r$. Take $y, z \in E$, then $y^l z = \lambda x + v$ with $\lambda \in \mathbb{C}$ and $v \in N^\perp$ and hence $z^l y = \overline{\lambda} x^l + v^l$. It follows that

$$\langle yx, z\rangle = \langle x, y^l z\rangle = \langle x, \lambda x + v\rangle = \langle x, \lambda x\rangle = \overline{\lambda}\langle x, x\rangle = \overline{\lambda}\langle x^l, u'\rangle =$$
$$\langle \overline{\lambda} x^l, u'\rangle = \langle \overline{\lambda} x^l + v^l, u'\rangle = \langle z^l y, u'\rangle = \langle y, z u'\rangle.$$

The previous argumentation assures that E is finally a two-sided H^*-algebra.

An analogous proof is applied when E is a right H^*-(Banach) algebra.

□

The next result holds true when interchanging "left" by "right".

Theorem 3.10. *Let $(E, (\langle\cdot,\cdot\rangle_\alpha)_{\alpha\in\Lambda})$ be a proper Hausdorff complete left locally m-convex H^*-algebra with continuous left involution. Then E is a two-sided locally m-convex H^*-algebra with continuous left and right involutions.*

Proof. By the Arens-Michael decomposition, $E = \varprojlim \widetilde{E_\alpha}$ and each $\widetilde{E_\alpha}$ *is a proper left H^*-algebra* (see the proof of Theorem 2.6 in [7, p. 419]). Moreover, by Lemma 3.9, each $\widetilde{E_\alpha}$, $\alpha \in \Lambda$ is a two-sided H^*- algebra and by Corollary 2.3 in [7, p. 418], E is a two-sided locally convex H^*-algebra. Finally, since $\widetilde{E_\alpha}$ is a proper two-sided H^*-algebra (and thus with continuous left and right involutions) Theorem 2.6 in [7, p. 419] completes the assertion. □

Definition 3.11. *Let E be a proper Hausdorff two-sided locally m-convex H^*-algebra. We shall say that*

(i) E has the (LPH)-property, *if*

the closed subalgebra of E generated by a left-selfadjoint element $x \in E$ is an (rbs)-algebra and it has the (H_l) property.

(ii) E has the (LPH')-property, if

the closed subalgebra of E generated by a left-selfadjoint element $x \in E$ is an (rbs)-algebra and it has the (H_l') property .

By using right-selfadjoint elements, we define the analogous (RPH) and (RPH') properties.

For the next result we follow, with appropriate adjustments, the proof of Theorem 3.3 in [4, p. 145]. For the terminology applied, see also the comments before Lemma 3.1. Moreover, in what follows, a subalgebra or an ideal I in a proper Hausdorff (left, right, two-sided) locally convex H^*-algebra is called *left (right) selfadjoint*, if $I^l = I$ (resp. $I^r = I$). See also Definition 2.10. Besides, an analogous to the next statement holds when we interchange "left" by "right".

Theorem 3.12. (First structure theorem) *Let $(E, (p_\alpha)_{\alpha\in\Lambda})$ be an (rbs) left Ambrose algebra, having the (LPH)-property. Take a right family of axes, say $(x_i)_{i\in K}$ and consider the left ideal $L \equiv \underset{i\in K}{\oplus}{}^{\perp} Ex_i$ and the right ideal $R \equiv \underset{i\in K}{\oplus}{}^{\perp} x_iE$. Then*

$$\overline{\underset{i\in K}{\oplus}{}^{\perp} Ex_i}^{\perp} = (0) \text{ and } \overline{\underset{i\in K}{\oplus}{}^{\perp} x_iE}^{\perp} = (0). \tag{1}$$

Moreover, if

$$E = \overline{L} + \overline{L}^{\perp} \text{ and } E = \overline{R} + \overline{R}^{\perp}, \text{ then}$$

$$E = \overline{\underset{i\in K}{\oplus}{}^{\perp} Ex_i} = \overline{\underset{i\in K}{\oplus}{}^{\perp} x_iE} \tag{2}$$

and

$$E = \overline{\underset{i,j}{\oplus}{}^{\perp} x_iEx_j} \equiv \overline{\underset{i,j}{\oplus}{}^{\perp} E_{ij}}. \tag{3}$$

Proof. In view of Corollary 3.4, a family $(x_i)_{i\in K}$, as in the statement, does exist. Claim that $L^\perp = (0)$, otherwise there is $0 \neq y \in L^\perp$. We consider $x = y^ly \in L^\perp$, which is left selfadjoint. By Corollary 1.3 in [7, p. 412], x is non zero. We denote by A the closed subalgebra generated by x.

From $x^n = x^{n-1}x \in L^\perp$, for all $n \in \mathbb{N}$, we get that $A \subseteq L^\perp$. Moreover, by the continuity of the left and the right involutions, A is left and right selfadjoint. So, A is a proper Hausdorff complete two-sided locally m-convex H^*-algebra with both one-sided involutions continuous (see also Lemma 1.11 in [7, p. 417]). Therefore, by the

(LPH)-property and in view of Theorems 2.8 and 3.3, there exists a right projection, say $y_0 \in A$ such that $y_0 = \sum_{j=1}^{n} y_j$ with y_j $(1 \leq j \leq n)$ doubly orthogonal primitive right projections in A.

Since $A \subseteq L^\perp$, $y_0 \in L^\perp$ and $y_j = y_j y_0$, for every j $(1 \leq j \leq n)$. Thus $L^\perp$ contains at least a primitive right selfadjoint projection which we denote by y_1 and this is doubly orthogonal with the x_i's. In fact, since $wx_i \in L$ and $zy_1 \in L^\perp$, for every $w, z \in E$, we have $\langle wx_i, zy_1\rangle_\alpha = 0$ for every $w, z \in E, \alpha \in M, i \in K$. So, $\langle x_i, y_1\rangle_\alpha = 0, \alpha \in M, i \in K$ and $0 = \langle w, zy_1x_i^r\rangle_\alpha = \langle wx_iy_1^r, z\rangle_\alpha$. By Hausdorff, $zy_1x_i = 0$ and $wx_iy_1 = 0$ and hence $y_1x_i = x_iy_1 = 0, i \in K$. This contradicts the maximality of (x_i). Thus, $L^\perp = (0)$ and $\overline{L}^\perp = (0)$. Similarly for the ideal R.

The proof of (3) is similar to that for (4) in [4, p. 145, Theorem 3.3]. □

We give now a variation of the first structure theorem for certain two-sided locally m-convex H^*-algebras. The benefit here is that we employ a left-right family of axes, succeeding thus an analysis with at least "fewer terms". We only note that in view of Theorem 2.8, Corollary 3.5 and the (LPH') property, the topological algebra A (the closed subalgebra generated by x, see the proof of Theorem 3.12) has a right projection, say y_0, from which we succeed doubly orthogonal (H)-primitive left-right projections of A, and in turn, $L^\perp$ contains at least a primitive left-right projection. The rest of the proof is quite similar to that of Theorem 3.12, and we omit it.

We stretch the fact that Corollaries 3.6, 3.7 and 3.8 refer to contexts in which a left-right family of axes exists.

Theorem 3.13. (First structure theorem-Second version) *Let $(E, (p_\alpha)_{\alpha\in\Lambda})$ be a proper Hausdorff complete two-sided locally m-convex H^*-algebra, having the (LPH') property and continuous (left, right) involutions. Suppose E has a left-right family of axes, say $(x_i)_{i\in K}$. Consider the left ideal $L \equiv \bigoplus_{i\in K}^{\perp} Ex_i$ (resp. the right ideal $R \equiv \bigoplus_{i\in K}^{\perp} x_iE$). Then* (1) *holds. If, moreover*

$$E = \overline{L} + \overline{L}^\perp \text{ and } E = \overline{R} + \overline{R}^\perp,$$

then (2) *and* (3) *hold true, as well.*

For the second structure theorem we employ the following notation. Let E be a topological algebra, $x \in E$ and $J = Ex$. We denote by $\mathcal{RL}(J)$ the closed two-sided ideal generated by J. In particular, if $x \in Id(E)$, $\mathcal{RL}(J)$ is the closed linear subspace generated by $JE = ExE$. Namely, $\mathcal{RL}(J) = \overline{[ExE]}$.

Lemma 3.14. *Let E be a proper Hausdorff two-sided locally m-convex H^*-algebra with continuous left and right involutions. If $x \neq 0$ is a left-right projection of E, such that the closed left ideal $J = Ex$ is minimal closed, then $\mathcal{RL}(J)$ is left-right selfadjoint and thus, an algebra of the same type as E. Moreover, $\mathcal{RL}(J)$ as closed two-sided ideal, is minimal.*

Proof. Since $[ExE]^r = [(ExE)^r]$, we have $\overline{[ExE]^r} = \overline{[ExE]}$ and by the continuity of the right involution, the ideal $\mathcal{RL}(J)$ is right selfadjoint. Similarly, $\mathcal{RL}(J)$ is left selfadjoint. So, by Lemma 1.11 in [7, p. 417], $\mathcal{RL}(J)$ is an algebra of the same type as E.

Now, we suppose that there exists a closed two-sided ideal I such that $I \subseteq \mathcal{RL}(J)$ and we denote by $(IJ)_l$ the left ideal of E generated by IJ. Then

$$\overline{(IJ)_l} = \overline{[IJ]} \subseteq I \cap J \subseteq J. \tag{4}$$

Since $\overline{[IJ]}$ is a closed left ideal and J is a minimal closed left ideal, we get that, either $\overline{[IJ]} = J$ or $\overline{[IJ]} = (0)$. So, if we consider the first case, then by (4), $I \cap J = J$ and hence $J \subseteq I$. Thus, $\mathcal{RL}(J) \subseteq I$ which implies that $I = \mathcal{RL}(J)$.

In the second case, we have $\overline{[IExE]} = (0)$. Furthermore, $I\mathcal{RL}(J) = I\overline{[ExE]} \subseteq \overline{[IExE]} = (0)$ and since $I^r \subseteq \mathcal{RL}(J)^r$, we get $II^r \subseteq I\mathcal{RL}(J)^r$ and hence $II^r = (0)$. So, by Corollary 1.3 in [7, p. 412], $I = (0)$. □

Concerning the last lemma, we note that the closed left ideal J is minimal closed, if E has the (H_l) property and x is primitive (see Lemma 3.2).

In the rest of the paper, by a *topologically simple algebra* we mean a topological algebra without nonzero proper closed two-sided ideals.

Corollary 3.15. Let E be a proper Hausdorff two-sided locally m-convex H^*-algebra with continuous left and right involutions, and x a primitive left-right projection in E. We consider the minimal closed left ideal $J = Ex$. If $\mathcal{RL}(J)$ is algebraically complemented, then it is an algebra of the same type as E being moreover, topologically simple.

Proof. See Lemma 3.14 and [4, p. 147, Lemma 3.5]. □

We note that Lemma 3.14 and Corollary 3.15 are also true for the closed two-sided ideal $\mathcal{RL}(xE)$.

Lemma 3.16. *Let E be an (H_l') pre-Ambrose algebra with continuous left and right involutions and $(x_i)_{i\in K}$ a left-right family of axes. Then the elements of the family $\mathcal{J} = \{\mathcal{RL}(J_i)\}$ consisting of all minimal closed two-sided ideals of E, where $J_i = Ex_i$ $(i \in K)$, are mutually orthogonal* (viz. their elements are pairwise orthogonal).

Proof. See Lemma 3.14 and Lemma 3.2 that, in our context, holds true when the (H_l) property is replaced by the (H_l') one. For the mutual orthogonality of the $\mathcal{RL}(J_i)'$s apply a proof analogous to that of Lemma 3.7 in [4, p. 148]. □

In the context of the previous lemma, we also get

$$\mathcal{RL}(J_i)\mathcal{RL}(J_k) = (0) \quad \text{for } \mathcal{RL}(J_i) \neq \mathcal{RL}(J_k).$$

Theorem 3.17. (Second structure theorem) *Let $(E, (p_\alpha)_{\alpha\in\Lambda})$ be an (H_l') left and right Ambrose algebra, having the (LPH')-property. If $(x_i)_{i\in K}$ is a left-right family of axes in E such that the closed two-sided ideal*

$$I \equiv \overline{\bigoplus_{i\in K}^{\perp} \mathcal{RL}(Ex_i)}$$

(see Lemma 3.16) is orthocomplemented, then

$$E = \overline{\bigoplus_{i\in K}^{\perp} \mathcal{RL}(Ex_i)}. \tag{5}$$

Namely, E is a topological orthogonal direct sum of proper Hausdorff complete two-sided locally m-convex H^*-algebras with continuous left and right involutions. Moreover, each $\mathcal{RL}(Ex_i)$, $i \in K$ is a minimal closed two-sided ideal in E. In particular, if $\mathcal{RL}(Ex_i)$, $i \in K$ is orthocomplemented in E, then it is topologically simple.

Proof. Following the proof of Theorem 3.13 (note that the proof is independent of the form of I), we have that $I^\perp = (0)$. Thus, the result follows from Lemma 3.14 and [4, p. 147, Lemma 3.5]. □

We note that (5) is valid for a two-sided locally m-convex H^*-algebra having a left-right family of axes $(x_i)_{i\in K}$ with $E = \overline{\bigoplus_{i\in K}^{\perp} Ex_i}$. See also Theorem 3.13. Indeed, $Ex_i \subseteq \mathcal{RL}(Ex_i)$ and $\bigoplus_{i\in K}^{\perp} Ex_i \subseteq \bigoplus_{i\in K}^{\perp} \mathcal{RL}(Ex_i)$.

Corollary 3.18. Let E be an orthocomplemented (rbs) (H_l') Ambrose algebra, having the (LPH')-property. Then E is a topological orthogonal direct sum of all minimal closed two-sided ideals, each one of which is a proper Hausdorff, topologically simple, complete two-sided locally m-convex H^*-algebra with continuous left and right involutions.

Proof. By Corollary 3.6 (which holds by interchanging "left" by "right"), E contains a left-right family of axes $(x_i)_{i\in K}$. So, by Lemma 3.16, the minimal closed two-sided ideals $\mathcal{RL}(Ex_i)$ are mutually orthogonal. Consider the family $\mathcal{J} = \{K_\alpha\}_{\alpha\in A}$ of all minimal closed two-sided ideals in E, which are mutually orthogonal. Moreover, by [4, p. 147, Lemma 3.5], the algebras K_α are topologically simple. Mutually orthogonality of the K_α's yields $\sum_{\alpha\in A} K_\alpha = \bigoplus_{\alpha\in A}^{\perp} K_\alpha$. Furthermore,

$$\sum_{i\in K} \mathcal{RL}(Ex_i) \subseteq \sum_{\alpha\in A} K_\alpha$$

and by Theorem 3.17, $E = \overline{\bigoplus_{\alpha\in A}^{\perp} K_\alpha}$. Finally, Lemma 4.2 in [7, p. 423] completes the proof. □

Acknowledgment: The second author was supported by Consejo Nacional de Ciencia y Tecnología (CONACYT).

Bibliography

[1] Ambrose, W. *Structure theorems for a special class of Banach algebras*, Trans. Amer. Math. Soc. 57 (1945), 364–386.
[2] Haralampidou, M. *On locally convex H^*-algebras*, Math. Japon. 38 (1993), 451–460.
[3] Haralampidou, M. *On Ambrose algebras*, Math. Japon. 38 (1993), 1175–1187.
[4] Haralampidou, M. *Structure theorems for Ambrose algebras*, Period. Math. Hung. 31 (1995), 139–154.
[5] Haralampidou, M. *Matrix representations of Ambrose Algebras*, Contemp. Math. 341 (2004), 63–71.
[6] Haralampidou, M. *Wedderburn decompositions of pseudo-H-algebras*, Contemp. Math., 547, Amer. Math. Soc., Providence, RI, 2011, 97–102.
[7] Haralampidou, M., Pérez-Tiscareño, R.M. *On two-sided locally convex H^*-algebras*, Mediterr. J. Math. 10 (2013), 411–424.
[8] Mallios, A. *Topological Algebras. Selected Topics.* North-Holland, Amsterdam, 1986.
[9] Saworotnow, P.P. *On two-sided H^*-algebras*, Pacific J. Math. 16 (1966), 365–370.
[10] Smiley, M.F. *Right H^*-algebras*, Proc. Amer. Math. Soc. vol. 4 (1953), 1–4.

Alexander Ya. Helemskii

Homologically best modules in classical and quantized functional analysis

Abstract: In the present paper we give a survey of some recent, as well of relevant old results concerning three approaches of what we call projective, as well as injective and flat, modules in functional analysis. These are the so-called relative, topological and metric approaches. It turns out that a certain categorical-general scheme contains, as particular cases, all these versions of projectivity and injectivity. In this scheme, the notion of a free object comes to the forefront, and in the best categories, that we call freedom-loving, projective objects are exactly retracts of free objects. We give a full description of relatively, topologically and metrically free "quantum" (operator) modules.

Keywords: projective and injective modules, quantized functional analysis

MSC: 46L85

1 Homologically best modules

By 'homologically best modules' we mean projective, injective and flat modules; they form the three pillars of the whole building of homological algebra. Functional-analytic versions of these notions are also very important. In analysis, however, there exist several different approaches to each one of these notions, corresponding to different functional-analytic versions of the problem of extension and lifting of operators.

Every type of these 'best modules' has what could be called classical and what could be called operator, or quantized, version. Each of them, in its turn, has completed, or Banach, and non-completed versions. Mostly, but not always, I shall write in this paper about completed quantized modules, and mostly about the projectivity.

Saying about abstract operator spaces (we shall write, for brevity, '*quantum spaces*') and *completely bounded linear and bilinear operators*, we shall follow the terminology of the textbook of Effros and Ruan [4]. In this paper, *quantum algebra* is a unital algebra and simultaneously a quantum space, such that the bilinear operator of multiplication is completely contractive. A *quantum module* (left, right or two-sided) over such an algebra is a unital module in the algebraic sense and simultaneously quantum space such that the bilinear operator of left or/and right outer multiplication is completely contractive.

Alexander Ya. Helemskii, Faculty of Mechanics and Mathematics, Moscow State University. e-mail: helemskii@rambler.ru

https://doi.org/10.1515/9783110413557-009

By *morphism* between quantum modules we mean a morphism in algebraic sense which is also a completely bounded operator.

We begin with the oldest, at least in the classical setting, version of projectivity.

Definition 1. A quantum module P is called *relatively projective* (some people say traditionally projective, and some just projective) if for every morphism $\tau : Y \to X$ that has a right inverse completely bounded operator (generally speaking, not morphism), and for an arbitrary morphism $\varphi : P \to X$ there exists a morphism $\psi : P \to Y$, making the diagram

$$\begin{array}{ccc} & & Y \\ & \overset{\psi}{\nearrow} & \downarrow{\scriptstyle \tau} \\ P & \xrightarrow{\varphi} & X \end{array}$$

commutative.

(Such a ψ is called a *lifting* of φ across τ). It is clear that there is abundance of completely bounded operators that are liftings of φ, but the problem is to find a lifting that is a morphism, i.e. an operator, respecting the outer multiplication.

The classical prototype of this definition for (just) Banach modules is obvious.

In fact, there are plenty of relatively projective modules, sufficient to build a rich homological theory [7]. Here is an illustration.

Proposition 1 (cf. [7][12][10]). Let A be a separable C^*-algebra with its standard quantum norm. Then all its left closed ideals are relatively projective as quantum (as well as classical) left A-modules.

The condition, concerning separability, is not necessary. However, it can not be just omitted. For example, maximal ideals in the algebra $l_\infty = C(\beta\mathfrak{B}N)$, not corresponding to points of $\mathfrak{B}N$, are not projective (cf. [7]).

It is worth noting that the quantization of modules can make non-projective modules into projective and vice versa.

Example 1 [11]. Let H be an infinite-dimensional Hilbert space. Consider the algebra $\mathcal{B}(H)$ in the standard form, that is the operator algebra

$$A := \{a \mathbin{\dot{\otimes}} \mathbb{I}_H; a \in \mathcal{B}(H)\}$$

acting on $H \mathbin{\dot{\otimes}} H$. (Here $\dot{\otimes}$ is the symbol of the Hilbert tensor product). Then the spatial Banach A-module $H \mathbin{\dot{\otimes}} H$ is not projective. However, if we equip the latter with the column quantization, and the C^*-algebra A with the standard quantization, it becomes projective as a quantum module.

Example 2. Let H be as above. Consider it as a spatial module over $\mathcal{N}(H)$, the Banach algebra of trace-class operators. Then it is known long ago that it is projective as a (just) left Banach module [7]. However, O. Yu. Aristov made the following observation (see [10]). Let us quantize our algebra by the known recipe, prepared by Blecher and Le

Merdy [2]. (In this way they succeeded to make $\mathcal{N}(H)$, up to a topological isomorphism, a uniformly closed operator algebra (!)) Namely, let us represent $\mathcal{N}(H)$ as the quantum algebra $\max H \overset{h}{\otimes} \max \bar{H}$, where $\bar{H}$ is a complex-conjugate space of H, max is the symbol of the maximal quantization, and $\overset{h}{\otimes}$ is the symbol of the Haagerup tensor product. Then, as Aristov has shown, our H, equipped with the column quantization, fails to be a projective quantum $\mathcal{N}(H)$-module.

One of advantages of the relative projectivity is that this property can be equivalently expressed in the language of derivations.

Recall that, if X and Y are left quantum modules over a quantum algebra A, then the quantum space $\mathcal{CB}(X, Y)$ of completely bounded operators between X and Y becomes a quantum bimodule over A with the outer multiplication, defined by $[a \cdot \varphi](x) = a \cdot [\varphi(x)]$ and $[\varphi \cdot a](x) = \varphi(a \cdot x)$.

Proposition 2. *A left quantum A-module P is relatively projective if, and only if for every quantum A-module Y every completely bounded derivation of A with values in $\mathcal{CB}(P, Y)$ is inner.*

We turn to another type of projectivity. Outwardly, it seems to be even more natural than the relative projectivity. However, demanding a very strong condition, in general it is not so useful.

Let $\tau : F \to E$ be an operator between (complete) quantum spaces. We shall call it *completely open*, if there is $C > 0$ such that for every $n \in \mathfrak{B}N$ and $x \in M_n(E)$ there exists $y \in M_n(F)$ with $\tau_n(y) = x$ and $\|y\| \le C\|x\|$. Here and thereafter $\tau_n : M_n(F) \to M_n(E)$ is the n-th amplification of τ.

Definition 2. A quantum module P is called *topologically projective*, if for every completely open morphism $\tau : Y \to X$, and for an arbitrary morphism $\varphi : P \to X$ there exists a morphism $\psi : P \to Y$, making the diagram

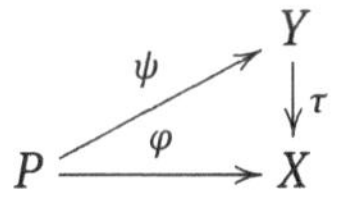

commutative.

This definition provides much less projective modules than its ‘relative’ counterpart. Compare the above-formulated result about ideals in C^*-algebras with the following recent theorem, due to N. T. Nemesh. (It is contained in his Ph.D. thesis).

Theorem 1. *Let A be a C^*-algebra, and I a closed left ideal of A. Then I is topologically projective (as a Banach A-module) if, and only if it has a right identity (or, equivalently, I has a closed left ideal in A as a complement).*

This theorem of Nemesh is rather difficult, and it is not known whether it has a quantum analogue.

Note that, defining both of already considered types of projectivity, we took into account norm, or completely bounded norm of the module in question only up to equivalence. Now we pass to the type that takes into account the precise value of these norms.

The closed unit ball of a normed space E will be denoted by B_E.

An operator between normed spaces E and F is called *strictly coisometric*, if it maps B_E onto B_F. An operator $\varphi : E \to F$ between quantum spaces is called *completely strictly coisometric* if, for any n, its amplification $\varphi_n : M_n(E) \to M_n(F)$ is strictly coisometric.

Definition 3. A quantum module P is called *metrically projective* if for every completely strictly coisometric morphism $\tau : Y \to X$ and an arbitrary morphism $\varphi : P \to X$ there exists a morphism $\psi : P \to Y$, making the diagram

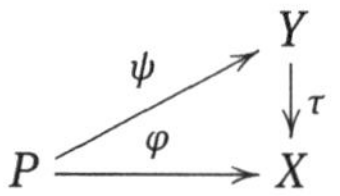

commutative and such that $\|\psi\|_{cb} = \|\varphi\|_{cb}$.

The classical version of this definition for Banach as well for non-complete (just normed) modules over a Banach (or normed) algebra are formulated in an obvious way.

This notion seems to be interesting even in the case of the simplest algebra $A := \mathfrak{B}C$. In the classical setting, that is just for Banach or normed spaces, we can thoroughly describe the projective objects. Namely, we have

Theorem 2 [8]. *(i) Every metrically projective Banach space coincides, up to an isometric isomorphism, to the space $l_1(\Lambda)$ for some index set Λ.*

(ii) Every metrically projective normed space coincides, up to an isometric isomorphism, to the normed subspace $l_1^0(\Lambda)$ of $l_1(\Lambda)$ for some index set Λ, consisting of all functions with finite support.

To run ahead, this theorem shows that for Banach, as well as for normed spaces the notion of projectivity coincides with that of freeness.

The first assertion resembles to the well known result of young Grothendieck [6], 1955, where some other type of projectivity, sometimes called extreme, was considered, and our proof heavily relies on some lemmas in his paper. The second part seems to be more original. It is based on some observations over extreme points of unit balls of normed dense subspaces of $l_1(\Lambda)$.

2 Category theory

Now, for a time, we pass from functional analysis to abstract nonsense (= category theory). We claim that there is a certain general-categorical scheme that contains, as particular cases, all mentioned versions of projectivity and a lot of other versions. The main definitions generalize those given by MacLane in his theory of relatively Abelian categories [13]. (Note that typical categories of functional analysis, used in this paper, are never Abelian and often even not additive). Apart from general overview, this scheme allows one to study projective objects by means of the so-called free objects, which will be soon defined.

Let $\mathcal{K}$ be an arbitrary category. A *rig* of $\mathcal{K}$ is a faithful covariant functor

$$\square : \mathcal{K} \to \mathcal{L},$$

where $\mathcal{L}$ is another category. A pair, consisting of a category and its rig, is called *rigged category*. If a rig is given, we shall call $\mathcal{K}$ main, and $\mathcal{L}$ auxiliary category.

Fix, for a time, a rigged category, say $(\mathcal{K}, \square : \mathcal{K} \to \mathcal{L})$. Let us call a morphism τ in $\mathcal{K}$ *admissible*, if $\square(\tau)$ is a retraction (that is, it has a right inverse) in $\mathcal{L}$. After this, we call an object P in $\mathcal{K}$ *$\square$-projective*, if, for every admissible morphism $\tau : Y \to X$ and an arbitrary morphism $\varphi : P \to X$ in $\mathcal{K}$, there exists a lifting (now in the obvious general-categorical sense) of φ across τ.

Let us denote the category of quantum Banach spaces (respectively,(just) Banach spaces) and completely bounded (respectively, (just) bounded) operators by **QBan** (respectively, **Ban**). If A is a quantum algebra, we denote the category of quantum left A-modules and their (completely bounded) morphisms by **QA – mod**. The subcategory of the latter with the same objects, but only completely contractive (as operators) morphisms we denote by $\mathbf{QA\text{-}mod_1}$.

Now we return to our three types of projectivity for quantum modules. First, one can easily show that

A quantum Banach A-module is relatively projective if, and only if it is $\square$-projective with respect to the rig

$$\square : \mathbf{QA\text{-}mod} \to \mathbf{QBan},$$

where $\square$ is the relevant forgetful functor. (It forgets about the outer multiplication).

The situation with the topological projectivity is somewhat more interesting. In fact, there are several rigs, providing this type of projectivity. The initial suggestion was to use the forgetful functor from **QA-mod** into the category **Bor** of bornological spaces – sets where the notion of a bounded set is axiomatically introduced. But this approach, as was noticed, does not give the best, the so-called freedom-loving category, in terms of our future definition. It seems that the rig, suggested by S. M. Shteiner in his Ph.D. thesis, is the most useful.

To introduce this rig, we begin with the definition of the so-called *hemi-linear space*. It is a set E, equipped with the operation of scalar multiplication (but not of the addition!), with the usual properties of such an operation, and the condition 'of common zero', that is $0x; 0 \in \mathfrak{B}C$ is the same element for all $x \in E$. The *hemi-norm* on such E is a function $x \mapsto \|x\|$, satisfying the usual properties of norm save, of course, the triangle inequality. A map, say $\varphi : E \to F$, between two hemi-normed spaces is called *bounded hemi-linear operator*, if it is homogeneous in the obvious sense, and there exists $C > 0$ such that for all $x \in E$ we have $\|\varphi(x)\| \leq C\|x\|$.

The category of hemi-normed spaces and bounded hemi-linear operators will be denoted by **HNor**.

Now, as a rig, we consider the functor

$$\boxdot : \mathbf{QA\text{-}mod} \to \mathbf{HNor},$$

taking a quantum module X to the hemi-normed space, appearing after taking the Banach l_∞-sum of the family of Banach spaces $\{M_n(X); n \in \mathfrak{B}N\}$ and then applying to it the natural forgetful functor from **Ban** into **HNor**. (Thus we 'forget' about the additive structure of the latter l_∞-sum and 'remember' only that it is a hemi-normed space.)

It is not difficult to verify that

A quantum A-module is topologically projective if, and only if it is $\boxdot$-projective

Finally, let us turn to the metric projectivity. Now we consider the rig

$$\odot : \mathbf{QA\text{-}mod_1} \to \mathbf{Set},$$

taking a quantum module X to the cartesian product $\mathsf{X}_{n=1}^{\infty} \mathsf{B}_{M_n(X)}$ of closed unit balls of the amplifications $M_n(X)$ of X.

Thus elements of the set $\odot(X)$ are sequences $(v_1, \ldots v_n \ldots); v_n \in M_n(X)$, where $\|v_n\| \leq 1$. The action of our functor on morphisms is defined in an obvious way.

It is easy to check that

A quantum module is metrically projective if, and only if it is $\odot$-projective.

(This is because the retractions in **Set** are just surjective maps, and therefore corresponding admissible morphisms are exactly those that are completely strictly coisomorphic as operators.)

The suggested frame-work gives the possibility to study projectivity by means of the so-called freeness. The following definition must be well known, perhaps under different names.

Let $\square : \mathcal{K} \to \mathcal{L}$ be a rig. Consider a pair, consisting of an object M in the auxiliary category $\mathcal{L}$ and an object $F(M)$ in the main category $\mathcal{K}$. A morphism $\mathsf{ua} : M \to \square(F(M))$

in $\mathcal{L}$ is called a *universal arrow for M* if, for every X in $\mathcal{K}$ and every morphism $\alpha : M \to \square(X)$ in $\mathcal{L}$, there exists a unique morphism $\beta : F(M) \to X$ in $\mathcal{K}$, making the diagram

$$\begin{array}{ccc} & & \square(F(M)) \\ & \overset{\text{ua}}{\nearrow} & \downarrow{\scriptstyle \square(\beta)} \\ M & \xrightarrow{\alpha} & \square(X) \end{array}$$

(in $\mathcal{L}$) commutative. If, for our pair, such a universal arrow exists, we call $F(M)$ a *free object in $\mathcal{K}$ with base M.*

It is a good thing to have a lot of free objects. A rigged category is called *freedom-loving*, if every object in $\mathcal{L}$ is a base of a free object. Such a category has a lot of advantages. Notably,

if a rigged category is freedom-loving, then an object P in $\mathcal{K}$ is projective if, and only if it is a retract of a free object (that is, there is free F in $\mathcal{K}$ and $\sigma : F \to P$ with a right inverse morphism).

Thus, if we are given a freedom-loving rigged category and we know its free objects, then we know a lot about its projective objects (and sometimes, like in the above-formulated Theorem 2 about normed spaces) we can describe in transparent terms all projective objects.

What are free objects in our main examples of rigged categories? We shall call a left quantum A-module *relatively, topologically or metrically free* if it is free with respect to the rigged categories (**QA-mod**, $\square$), (**QA-mod**, $\boxdot$) or (**QA-mod**$_1$, $\odot$), respectively.

Denote by $\overset{op}{\otimes}$ the operator-projective tensor product of (complete) quantum spaces. Recall that the 'raison d'etre' of this tensor product, discovered by Effros/Ruan and, independently, by Blecher/Paulsen, is that it linearizes completely bounded bilinear operators.

In what follows, the space $A \overset{op}{\otimes} E$, for a quantum space E, is considered as a left quantum A-module with the outer multiplication, well-defined by the equality $a \cdot (b \otimes x) := ab \otimes x;\ a, b \in A, x \in X$. Let e be the identity in A. A rather easy observation is

Proposition 3. *Let E be a quantum space. Then the module $A \overset{op}{\otimes} E$ is relatively free. The respective universal arrow is $ua : E \to A \overset{op}{\otimes} E : x \mapsto e \otimes x$. Every relatively free quantum A-module is completely topologically isomorphic to $A \overset{op}{\otimes} E$ for some E.*

I pass to free objects in two remaining rigs. It turns out, that there is an intimate connection between both types of freeness, observed by Shteiner. Namely, if some quantum module is metrically free, then the same module is topologically free, and topologically free quantum modules are exactly metrically free quantum modules up to a complete topological isomorphism. Moreover, for every hemi-normed space E there is a recipe, making the respective universal arrow in **HNor** for that E from a certain universal arrow in **Set** related to the rig (**QA-mod**$_1$, $\odot$). So, if we manage to find metrically free modules and respective universal arrows, then we know everything about topological freeness.

To do this, we shall use the special finite-dimensional quantum spaces $T_n; n \in \mathfrak{B}N$, distinguished in the known paper [1] by David Blecher (1992), where, among other things, somewhat another kind of projectivity was introduced and studied. (We regretfully leave this kind, of the so-called extreme projectivity, outside the scope of this paper.) This T_n is defined as the quantum dual to the space $\mathcal{B}(\mathfrak{B}C^n)$, the latter equipped with the standard quantum norm. (Thus, the underlying Banach space of T_n is just the space of $n \times n$ matrices with the trace-class norm.) The main virtue of the quantum space T_n, indicated by Blecher, is that, for every quantum space E, there exists a canonical complete isometric isomorphism $I_n : M_n(E) \to \mathcal{CB}(T_n, E)$ (cf. also [3]).

Another thing we shall essentially use, is the construction by Blecher of what he denotes as $\oplus_1$-sum of a given family of quantum spaces. It seems that the same construction, called operator l_1-sum, was independently introduced by Gilles Pisier in his book [15]. What matters for us is not the construction itself but the fact that $\oplus_1$-sum is exactly general-categorical coproduct in **QA-mod₁**.

Now, for a set M, denote by $G(M)$ the quantum space $\oplus_1\{T_{n,t}; n \in \mathfrak{B}N, t \in M\}$, where $T_{n,t}$ are copies of T_n, indexed by points of M. After this, consider the quantum A-module (our future free module) $F(M) := A \overset{op}{\otimes} G(M)$.

Our aim is to construct a special map (our future universal arrow) from M into the set $\odot F(M)$. Thus we must assign to every point $t \in M$ a sequence $((z_1)_t, \dots, (z_n)_t, \dots)$ where $(z_n)_t \in \mathsf{B}_{M_n(F(M))}$.

Consider the map $\Gamma_{n,t} : T_n \to F(M)$, that takes the matrix x in T_n to $e \otimes x_t$, where e is the identity in A and x_t is the copy of x in $T_{n,t}$, the latter space being identified with the relevant direct summand in $G(M)$. It is not difficult to see that $\Gamma_{n,t}$ is a completely contractive operator.

Finally, consider in $M_n(T_n) = M_n \otimes T_n$ the special element $\iota_n := \sum_{i,j=1}^n e_{ij} \otimes e_{ji}$, where $(e_{ij}); 1 \le i, j \le n$ are the respective elementary matrices, and set

$$(z_n)_t := [\Gamma_{n,t}]_n(\iota_n),$$

where $[\Gamma_{n,t}]_n : M_n(T_n) \to M_n(F(M))$ is the n-th amplification of $\Gamma_{n,t}$. One can show that $(z_n)_t$ is an element of norm 1 in $M_n(F(M))$. Then we have

Proposition 4 (cf. [8], [9]). *Let M be a set. Then the module*

$$F(M) := A \overset{op}{\otimes} (\oplus_1\{T_{n,t}; n \in \mathfrak{B}N, t \in M\})$$

is metrically free. The respective universal arrow is the map $ua : M \to \odot F(M)$, taking $t \in M$ to the sequence $((z_1)_t, \dots, (z_n)_t, \dots) \in \odot F(M)$. Every metrically free quantum A-module is completely isometrically isomorphic to $A \overset{op}{\otimes} (F(M))$ for some M.

Remark. If we are given a quantum module X and a map $\alpha : M \to \odot(X)$, then the morphism $\beta : F(M) \to X$, indicated in the definition of a universal arrow, acts as follows. Take $t \in M$. Let $\alpha(t)$ be the sequence $(\dots, x_n(t), \dots); x_n(t) \in \mathsf{B}_{M_n(X)}$. Consider

$\varphi_n(t) := I_n(x_n(t)) \in \mathcal{CB}(T_n, X)$ and (using the universal property of coproducts) the completely contractive operator $\varphi : G(M) = \oplus_1\{T_{n,t}; n \in \mathfrak{B}N, t \in M\} \to X$ with components $\varphi_n(t) : T_{n,t} \to X$. Then the desired β is well-defined on elementary tensors in $F(M) = A \overset{op}{\otimes} G(M)$ by taking $a \otimes u; a \in A, u \in G(M)$ to $a \cdot \varphi(u)$.

Recalling the interplay between the metric and the topological projectivity, we immediately obtain

Corollary. *A quantum A-module is topologically free if, and only if it is completely topologically isomorphic, for some set M, to a module of the form* $A \overset{op}{\otimes} (\oplus_1\{T_{n,t}; n \in \mathfrak{B}N, t \in M\})$.

Thus we see that all our three rigged categories are freedom-loving. Therefore we know, according to what was mentioned above, that their projective objects are retracts of free objects. On the other hand, retracts in our main categories can be described 'geometrically', as certain direct summands.

In what follows, we say that a quantum submodule of a quantum A-module Q is its *topological* (respectively, *metric*) *direct summand* if it is the range of a projection in Q which is a morphism (respectively, completely contractive morphism) of quantum modules.

Proposition 5. Let P be a quantum A-module. Then

(i,ii) P is relatively (respectively, topologically) projective if, and only if it is, up to a complete topological isomorphism, a topological direct summand of some quantum module of the form $A \overset{op}{\otimes} E$, where E is a quantum space (respectively, $A \overset{op}{\otimes} (\oplus_1\{T_{n,t}; n \in \mathfrak{B}N, t \in M\})$, where M is a set).

(iii) P is metrically projective if, and only if it is, up to a completely isometric isomorphism, a metric direct summand of some quantum module of the form $A \overset{op}{\otimes} (\oplus_1\{T_{n,t}; n \in \mathfrak{B}N, t \in M\})$, where M is a set.

Remark. Assertions (ii) and (iii) somewhat resemble to the description of operator spaces that are projective in the sense of Blecher (different from ours, see above). This description was given in terms of the so-called almost direct summands. Actually, these assertions were inspired by his mentioned result. On the other hand, his description also has an alternative proof in the general-categorical framework, as a particular case of the description of the so-called asymptotically projective objects in an asymptotic rigged category [9].

3 Relatively injective modules

All of considered three types of projectivity have natural injective analogues. Here we shall concentrate on the classical versions of injectivity; the quantum versions of relevant definitions are similar.

Definition 4. A Banach module J is called *relatively injective* (again, some people say traditionally injective, and some just injective) if for every morphism $i : X \to Y$ that has a left inverse bounded operator, and an arbitrary morphism $\varphi : X \to J$, there exists a morphism ψ, making the diagram

$$\begin{array}{ccc} X & \xrightarrow{\ i\ } & Y \\ \downarrow{\scriptstyle\varphi} & \swarrow{\scriptstyle\psi} & \\ J & & \end{array}$$

commutative.

Injective right modules and injective bimodules are defined in a similar way. Let us give some illustrations, showing the importance of this kind of injectivity.

Theorem 3. *(i) A C^*-algebra A is nuclear if, and only if its dual bimodule A^* is relatively injective.*

(ii) A von Neumann algebra A is hyperfinite (or amenable-after-Connes) if, and only if its predual bimodule A_* is relatively injective.

We recall these old results because recently they were strengthened in, let me say, quantitative sense by G. Racher [16], whose argument essentially differs from ours (and, we would say that, it is more elegant). Racher used a different type of injectivity, mentioned in some old paper of M. White [17]. Namely, let us call a (bi)module J *1-relatively injective*, if, for the same X, Y, i and φ as before, but with the additional assumption that i has a *contractive* left inverse operator, there exists a morphism ψ, making the diagram above commutative and such that $\|\psi\| \leq \|\varphi\|$. Clearly, the class of 1-relatively injective (bi)modules is less, and one can show that it is strictly less, than that of (just) relatively injective (bi)modules.

Theorem (Racher). *(i) If A is a nuclear C^*-algebra, then its dual bimodule A^* is 1-relatively injective*

(ii) If A is a hyperfinite von Neumann algebra, then its predual bimodule A_* is 1-relatively injective.

Topologically and metrically injective modules are defined in similar terms, involving the diagram above. The difference is that in the 'topological' case we demand from the morphism i that it must be (only) a topologically injective map, and in the 'metric' case that i is an isometry, and the norm of its 'extension' ψ is the same as $\|\varphi\|$.

It is very little known about topological injectivity, even in the context of just Banach spaces. At the same time metrically injective Banach spaces are thoroughly described (see, e.g., [14] and references therein), and there are considerable advances in the study of metrically injective quantum spaces (see, e.g., [4] and references in [5]). But we omit these topics: other people know them much better.

Now our point is that, as well as in the case of projectivity, all mentioned kinds of injectivity also can be included in the general-categorical framework. It is based on the notion of the so-called corigged category, the 'dual' counterpart of the rigged category. In this context the notion of the so-called cofree object, dual to that of a free object, comes to the forefront. We do not go into any details about it, we shall only mention that, under some general-categorical assumptions, we can describe all cofree objects in a given corigged category if we know free objects in some connected rigged category. This general observation, being applied to our concrete types of injectivity, enables one to characterize all relevant cofree modules. For example, topological as well as metric cofree quantum left A-modules are those of the form $\mathcal{CB}(A, \oplus_\infty\{\mathcal{B}(\mathfrak{B}C^n)_t; n \in \mathfrak{B}N, t \in M\})$ with the obviously defined outer multiplication. Here M is an arbitrary set, $\mathcal{B}(\mathfrak{B}C^n)_t$ are copies of the quantum space $\mathcal{B}(\mathfrak{B}C^n)$, indexed by points of M, and $\oplus_\infty$ is the symbol of the operator l_∞-sum, or, equivalently, the categorical product in $\mathbf{QA\text{-}mod_1}$.

Remark. The kind of injectivity, figuring in the above-mentioned paper of Racher, also can be presented as a partial case of some corigged category. Namely, the respective corig is the contravariant functor $\square_1 : (\mathbf{A} - \mathbf{mod} - \mathbf{A})_1 \to \mathbf{Ban_1}$, taking a Banach module X to the dual Banach space X^*. Here $(\mathbf{A} - \mathbf{mod} - \mathbf{A})_1$ is the category of Banach A-bimodules and their contractive morphisms, and $\mathbf{Ban_1}$ is the category of Banach spaces and contractive operators.

4 Flat modules

The third class of homologically best modules, closely connected with two others, consists of the so-called flat modules. Here we restrict ourselves to the classical case.

We begin with the following

Proposition 6. *If P is a relatively, topologically or metrically projective Banach A-module (left, right or two-sided), then its dual A-module A^* is injective in the same sense.*

But THE CONVERSE IS NOT TRUE, and this is one of the most important phenomena in homological algebra, pure and 'topological' alike. This makes substantial the following

Definition 5. A module F is called *relatively, topologically or metrically flat* if its dual module is relatively, topologically or injective, respectively.

For every type of flatness, there is an equivalent definition in terms of Banach module tensor product, but here we omit it (see, e.g., [7]).

It is worth to recall here that the amenability is not other but a particular case of flatness: a Banach algebra A is amenable if, and only if its unitization is a flat A-bimodule.

The difference between a rigid and demanding condition of projectivity and much more widespread and flexible condition of flatness has a lot of various manifestations, and in analysis, may be, not less than in algebra. For example, it transforms to the difference between finite-dimensional C^*-algebras and nuclear C^*-algebras, between compact groups and amenable groups, between paracompact locally compact spaces and arbitrary locally compact spaces, and so on.

We shall formulate in details only one of such manifestations.

Proposition 7. *Let I be a closed left ideal in a Banach algebra A, and $X := A/I$ the respective cyclic module. Then*

(i) X is topologically projective if, and only if I has a right identity.

(ii) X is topologically flat if, and only if I has a right bounded approximate identity.

Here the first assertion is immediate. As to the second one, the connection of flatness with approximate identities was indicated long ago by M. V. Sheinberg and myself (see, e.g., [7]) and then studied by M. White [17]. In the present form this assertion is contained in the Ph.D thesis of N. Nemesh.

We see, in particular, that if I has a right bounded approximate identity, X is certainly relatively flat. But the converse in general is false, and it is true only under some additional assumptions of a Banach-geometrical kind.

Curiously, this failure enables one to construct examples of non-complemented subspaces of Banach spaces. The following observation is rather old, but we are not sure that it was published.

Proposition 8. *Let I be a closed left ideal in an amenable Banach algebra A such that it does not have a right bounded approximate identity. (For example, a Malliavin ideal in Wiener algebra or, more general, in the group algebra $L_1(G)$ of a non-compact locally compact abelian group G). Then it is not complemented and even not weakly complemented as a Banach subspace.*

Bibliography

[1] Blecher, D.P. *The standard dual of an operator space*, Pacific J. of Math. v. 153, No. 1 (1992) 15–30.
[2] Blecher, D.P., Le Merdy, C. *On quotiens of function algebras and operator algebra structures on l^p*, J. Operator Theory. v. 34, No. 2 (1995) 315–346.
[3] Blecher, D.P., Paulsen, V.I. *Tensor products of operator spaces*, J. Funct. Anal., 99, No. 2 (1991) 262–292.
[4] Effros, E.G., Ruan, Z.-J. *Operator spaces*. Clarendon Press. Oxford. 2000.
[5] Frank, M., Paulsen, V.I. *Injective and projective Hilbert C^*-modules, and C^*-algebras of compact operators*, arXiv:math/0611349v2 [math.OA] (2008)
[6] Grothendieck, A. *Une caractérisation vectorielle-métrique des espaces L^1*, Canad. J. Math., v. 7 (1955) 552–561.
[7] Helemskii, A.Ya. *The Homology of Banach and Topological Algebras*. Kluwer, Dordrecht, 1989.
[8] Helemskii, A.Ya. *Metric freeness and projectivity for classical and quantum normed modules*, Mat. Sb., v. 204, No. 7 (2013) 1056–1083.

[9] Helemskii, A.Ya. *Projectivity for operator modules: approach based on freedom*. Rev. Roum. Math. Pures Appl, 53:2 (2014) 219–236.
[10] Helemskii, A.Ya. *Projective modules in the classical and quantum functional analysis*, Journal of Math. Sciences, v. 159 No.5, 2009, p.600–652.
[11] Helemskii, A.Ya. *Wedderburn-type theorems for operator algebras and modules: traditional fnd "quantized" homological approaches*. In: *Topological Homology*, Nova Sci. Publ., Huntington, NY, (2000) 57–92.
[12] Lykova, Z.A. *On homological characteristics of operators algebras*. Vest. Mosk. Univ. ser. mat. mech., 1986, 1, 8–13 (in Russian); Moscow Univ. Math. Bull., 1986, 41, no. 3, 10–15.
[13] Mac Lane, S. *Homology*. Springer-Verlag, Berlin, 1967.
[14] Pietsch, A. *History of Banach Spaces and Linear Operators*. Birkhäuser, Boston, 2007.
[15] Pisier, G. *Introduction to operator space theory*. Cam. Univ. Press. Cambridge, 2003.
[16] Racher, G. *On injective von Neumann algebras*, Proc. Amer. Math. Soc., v. 139, No. 7 (2011) 2529–2541.
[17] White, M.C. *Injective modules for uniform algebras*, Proc. London Math. Soc., v. 73, No. 3 (1996) 155–184.

Mohammad Sal Moslehian

Operator Grüss inequality

Abstract: Focusing on operator inequalities, we survey some recent significant developments of the Grüss inequality.

Keywords: Operator inequality, norm inequality, Grüss inequality

MSC: 47A63; 46L05; 47A30; 26D15

1 Introduction

In 1935, G. Grüss [12] provided an upper bound for $\int_a^b f(x)g(x)dx - \int_a^b f(x)dx \int_a^b g(x)dx$. It is known as the Grüss inequality, which is indeed a complement of Chebyshev's inequality. It says that if f and g are integrable real functions on $[a, b]$ and there exist real constants $\varphi, \phi, \gamma, \Gamma$ such that $\varphi \leq f(x) \leq \phi$ and $\gamma \leq g(x) \leq \Gamma$ hold for all $x \in [a, b]$, then

$$\frac{1}{b-a}\int_a^b f(x)g(x)dx - \frac{1}{(b-a)^2}\int_a^b f(x)dx \int_a^b g(x)dx \leq \frac{1}{4}(\Phi - \varphi)(\Gamma - \gamma)\,. \tag{1}$$

Several proofs are given in [10] for the cases when f and g are either absolutely monotone, completely monotone or anchored monotone. See also [7, 16] for several proofs of this inequality in the discrete form. It has been the subject of much investigation in which the conditions on the functions are varied to obtain different estimates; see [17] and references therein. This inequality has been investigated, applied and generalized by many mathematicians in different areas of mathematics, such as in inner product spaces [9], quadrature formulae [23], finite Fourier transforms [6], times scales [4], linear functionals [1], matrix traces [22], trace functional [5], inner product modules over H^*-algebras and C^*-algebras [2, 13], and so on. In addition, some complemented inequalities to the Grüss inequality can be found at [20].

2 Preliminaries

Let $\mathbf{B}(\mathcal{H})$ be the C^*-algebra of all bounded linear operators on a complex (separable) Hilbert space $(\mathcal{H}, \langle\cdot,\cdot\rangle)$ and I be its identity. Whenever $\dim \mathcal{H} = n$, we identify $\mathbf{B}(\mathcal{H})$ with the the full matrix algebra $\mathcal{M}_n$ of all $n \times n$ matrices with entries in the complex

Mohammad Sal Moslehian, Department of Pure Mathematics, Center of Excellence in Analysis on Algebraic Structures (CEAAS), Ferdowsi University of Mashhad, P.O. Box 1159, Mashhad 91775, Iran

https://doi.org/10.1515/9783110413557-010

field **C** and denote its identity by I_n A capital letter means an operator in $\mathbf{B}(\mathcal{H})$ and we identify a scalar with the identity operator I multiplied by this scalar.

For self-adjoint operators $A, B \in \mathbf{B}(\mathcal{H})$ the partially ordered relation $B \le A$ means that $\langle Bx, x\rangle \le \langle Ax, x\rangle$ for all $x \in \mathcal{H}$. In particular, if $0 \le A$, then A is called positive. If A is a positive invertible operator, then we write $0 < A$. A map $\Phi : \mathcal{A} \to \mathcal{B}$ between C^*-algebras is said to be positive if $\Phi(A) \ge 0$ whenever $A \ge 0$. It is called n-positive if $\Phi \otimes I_n : M_n(\mathcal{A}) \to M_n(\mathcal{B})$ is positive, where $M_n(\mathcal{A})$ is the C^*-algebra of $n \times n$ matrices with entries in $\mathcal{A}$. We say that Φ is completely positive if it is n-positive for all n. If Φ preserves the identity, then it is called unital. The reader is referred to [**?**] for undefined notations and terminologies.

3 Grüss type inequality concerning Gel'fand integral

Jocic, Krtinic and Moslehian [14] presented a Grüss inequality for inner product type integral transformers in norm ideals.

Let $(\Omega, \mathbf{M}, \mu)$ be a probability measurable space. By an $L^1(\Omega, \mathbf{M}, \mu)$-field of operators we mean a mapping $\mathbf{A} = (A_t)_{t\in\Omega}$ from Ω into $\mathbf{B}(\mathcal{H})$ such that the scalar functions on Ω given by $t \mapsto \operatorname{tr}(A_tY)$ $(Y \in \mathcal{C}_1(\mathcal{H}))$ are in $L^1(\Omega, \mu)$. Due to the fact that $\mathbf{B}(\mathcal{H})$ is the dual of $\mathcal{C}_1(\mathcal{H})$, for any $E \in \mathbf{M}$ there is a unique operator $\int_E A_t d\mu(t)$ (or $\int_E \mathbf{A} d\mu$) such that

$$\operatorname{tr}\left((\int_E \mathbf{A} d\mu) Y \right) = \int_E \operatorname{tr}(A_tY) d\mu(t)\,.$$

The operator $\int_E \mathbf{A} d\mu$ is called the Gel'fand integral of $\mathbf{A}$ over E; cf. [8]. Indeed, it is characterized by

$$\left\langle \left(\int_E A_t d\mu(t) \right) \xi, \xi \right\rangle = \int_E \langle A_t\xi, \xi\rangle d\mu(t) \quad (\xi \in \mathcal{H}) \tag{1}$$

In addition, $\int_E A_t^* d\mu(t) = \left(\int_E A_t d\mu(t)\right)^*$.

The following lemma is interesting on its own right.

Lemma 3.1. *Let $\mathbf{A} = (A_t)_{t\in\Omega}$ be an $L^1(\Omega, \mathbf{M}, \mu)$-field of operators, $E \in \mathbf{M}$ and $C_1, C_2 \in \mathbf{B}(\mathcal{H})$ such that $\int_E \left|A_t - \frac{C_1+C_2}{2}\right|^2 d\mu(t) \le \frac{1}{4}|C_1 - C_2|^2$. Then*

$$\int_E |\mathbf{A}|^2 d\mu - \left|\int_E \mathbf{A} d\mu\right|^2 \le \frac{1}{4}|C_1 - C_2|^2\,. \tag{2}$$

Proof. Set

$$G_1 = \left(C_1 - \int_E A_t d\mu(t) \right)^* \left(\int_E A_t d\mu(t) - C_2 \right)$$

and

$$G_2 = \int_E (C_1 - A_t)^* (A_t - C_2) d\mu(t)$$

Due to the fact that

$$\frac{1}{4}|C_1 - C_2|^2 - \left| T - \frac{C_1 + C_2}{2} \right|^2 = \mathrm{Re}\langle C_1 - T, T - C_2 \rangle$$

we observe that the condition

$$\int_E \left| A_t - \frac{C_1 + C_2}{2} \right|^2 d\mu(t) \le \frac{1}{4}|C_1 - C_2|^2$$

is equivalent to $\mathrm{Re}G_2 \ge 0$. We have

$$\begin{aligned} \int_E |A_t|^2 \, d\mu(t) - \left| \int_E A_t d\mu(t) \right|^2 &= G_1 - G_2 \\ &= \mathrm{Re}G_1 - \mathrm{Re}G_2 \\ &\le \mathrm{Re}G_1 \\ &\le \left| C_1 - \int_E A_t d\mu(t) + \int_E A_t d\mu(t) - C_2 \right|^2 \\ &\qquad (\text{by } 4\mathrm{Re}(T^*S) \le |T+S|^2 \ (T, S \in \mathbf{B}(\mathcal{H}))) \\ &= \frac{1}{4}|C_1 - C_2|^2 . \end{aligned}$$

□

We are ready to state our main result; see also [14] for more results.

Theorem 3.2. *Suppose that* $\mathbf{A} = (A_t)_{t\in\Omega}$ *and* $\mathbf{B} = (B_t)_{t\in\Omega}$ *are* $L^1(\Omega, \mathbf{M}, \mu)$*-fields of operators,* $E \in \mathbf{M}$ *and* $C_1, C_2, D_1, D_2 \in \mathbf{B}(\mathcal{H})$ *such that* $\int_E \left|A_t - \frac{C_1+C_2}{2}\right|^2 d\mu(t) \le \frac{1}{4}|C_1 - C_2|^2$ *and* $\int_E \left|B_t - \frac{D_1+D_2}{2}\right|^2 d\mu(t) \le \frac{1}{4}|D_1 - D_2|^2$*. Then*

$$\left\| \int_E \mathbf{B}\mathbf{A} d\mu - \int_E \mathbf{B} d\mu \int_E \mathbf{A} d\mu \right\| \le \frac{1}{4}\|C_1 - C_2\|\|D_1 - D_2\| . \tag{3}$$

Proof. First note that

$$\begin{aligned}\int_E B_tA_t d\mu(t) - \int_E B_t d\mu(t)\int_E A_t d\mu(t) &= \frac{1}{2}\Big(\int_E B_tA_t d\mu(t) + \int_E B_sA_s d\mu(s)\\ &\quad - \int_E\int_E B_tA_s d\mu(t)d\mu(s) - \int_E\int_E B_sA_t d\mu(t)d\mu(s)\Big)\\ &= \frac{1}{2}\int_E\int_E (B_t - B_s)(A_t - A_s) d\mu(t)d\mu(s). \end{aligned} \tag{4}$$

We have

$$\begin{aligned}
&\left\|\int_E B_tA_t d\mu(t) - \int_E B_t d\mu(t)\int_E A_t d\mu(t)\right\|\\
&= \left\|\frac{1}{2}\int_E\int_E (B_t - B_s)(A_t - A_s) d\mu(t)d\mu(s)\right\| && \text{(by (4))}\\
&= \frac{1}{2}\sup_{\|\xi\|=1}\left|\left\langle\left(\int_E\int_E (B_t - B_s)(A_t - A_s) d\mu(t)d\mu(s)\right)\xi, \xi\right\rangle\right|\\
&= \frac{1}{2}\sup_{\|\xi\|=1}\left|\int_E\left\langle\left(\int_E (B_t - B_s)(A_t - A_s) d\mu(t)\right)\xi, \xi\right\rangle d\mu(s)\right| && \text{(by (4))}\\
&= \frac{1}{2}\sup_{\|\xi\|=1}\left|\int_E\int_E \langle((B_t - B_s)(A_t - A_s))\xi, \xi\rangle d\mu(t)d\mu(s)\right| && \text{(by (4))}\\
&\le \frac{1}{2}\sup_{\|\xi\|=1}\int_E\int_E |\langle(A_t - A_s)\xi, (B_t^* - B_s^*)\xi\rangle| d\mu(t)d\mu(s)\\
&\le \frac{1}{2}\sup_{\|\xi\|=1}\int_E\int_E \|(A_t - A_s)\xi\|\,\|(B_t^* - B_s^*)\xi\| d\mu(t)d\mu(s)\\
& && \text{(by the Cauchy-Schwarz inequality)}\\
&\le \frac{1}{2}\sup_{\|\xi\|=1}\left[\left(\int_E\int_E \|(A_t - A_s)\xi\|^2 d\mu(t)d\mu(s)\right)^{\frac{1}{2}}\left(\int_E\int_E \|(B_t^* - B_s^*)\xi\|^2 d\mu(t)d\mu(s)\right)^{\frac{1}{2}}\right]\\
& && \text{(by the Cauchy-Schwarz inequality for double integrals)}\\
&\le \frac{1}{2}\sup_{\|\xi\|=1}\left(\int_E\int_E \langle(A_t - A_s)^*(A_t - A_s)\xi, \xi\rangle d\mu(t)d\mu(s)\right)^{\frac{1}{2}} \times
\end{aligned}$$

$$
\begin{aligned}
&\sup_{\|\xi\|=1}\left(\int_E\int_E \langle (B_t-B_s)(B_t-B_s)^*\xi,\xi\rangle\, d\mu(t)d\mu(s)\right)^{\frac{1}{2}}\\
&= \frac{1}{2}\sup_{\|\xi\|=1}\left\langle \int_E\int_E (A_t-A_s)^*(A_t-A_s)d\mu(t)d\mu(s)\xi,\xi\right\rangle^{\frac{1}{2}} \times\\
&\quad \sup_{\|\xi\|=1}\left\langle \int_E\int_E (B_t-B_s)(B_t-B_s)^* d\mu(t)d\mu(s)\xi,\xi\right\rangle^{\frac{1}{2}} \qquad \text{(by (1))}\\
&= \frac{1}{2}\left\|\int_E\int_E (A_t-A_s)^*(A_t-A_s)d\mu(t)d\mu(s)\right\|^{\frac{1}{2}}\left\|\int_E\int_E (B_t-B_s)(B_t-B_s)^* d\mu(t)d\mu(s)\right\|^{\frac{1}{2}}\\
&\le \left\|\int_E |A_t|^2\,d\mu(t)-\left|\int_E A_t d\mu(t)\right|^2\right\|^{\frac{1}{2}}\left\|\int_E |B_t^*|^2\,d\mu(t)-\left|\int_E B_t^* d\mu(t)\right|^2\right\|^{\frac{1}{2}} \qquad \text{(by(4))}\\
&\le \frac{1}{4}\|C_1-C_2\|\|D_1-D_2\| \qquad \text{(by (2))}
\end{aligned}
$$

□

Remark 3.3. *It is known that* $\frac{1}{4}$ *is sharp in* (5); *see [9].*

Corollary 3.4. *Let* $A_1,\dots,A_n$ *and* $B_1,\dots,B_n$ *be bounded linear operators acting on a Hilbert space* $\mathcal{H}$ *and* $C_1, C_2, D_1, D_2 \in \mathbf{B}(\mathcal{H})$ *such that* $\sum_{i=1}^n \left|A_i-\frac{C_1+C_2}{2}\right|^2 \le \frac{1}{4}|C_1-C_2|^2$ *and* $\sum_{i=1}^n \left|B_i-\frac{D_1+D_2}{2}\right|^2 \le \frac{1}{4}|D_1-D_2|^2$. *Then*

$$\left\|\sum_{i=1}^n B_iA_i-\sum_{i=1}^n B_i\sum_{i=1}^n A_i\right\| \le \frac{1}{4}\|C_1-C_2\|\|D_1-D_2\|\,. \tag{5}$$

Proof. Take $E=\Omega=\{1,\dots,n\}$ and μ to be the counting measure in Theorem 3.2. □

4 Grüss inequality for some types of positive linear maps

In this section $\mathcal{A}$ and $\mathcal{B}$ stand for unital C^*-algebras.

Let $\Delta(C,\|\cdot\|)=\inf_{\lambda\in\mathbf{C}}\|C-\lambda I\|$ be the $\|\cdot\|$-distance of C from the scalar operators. It is known that $\Delta(C,\|\cdot\|)\le\|C\|$ and $\Delta(C,\|\cdot\|)=c(C)$ for any normal operator C, where $c(C)$ denotes the radius of the smallest disk in the complex plane containing the spectrum $\sigma(C)$ of C.

Perić and Rajić [21] proved that if $\Phi : \mathcal{A} \to \mathbf{B}(\mathcal{H})$ is a unital completely bounded linear map between unital C^*-algebras, then

$$\|\Phi(AB) - \Phi(A)\Phi(B)\| \leq \|\Phi\|_{cb} \operatorname{diam}(W^1(A)) \operatorname{diam}(W^1(B))$$

for every $A, B \in \mathcal{A}$, where $W^1(\cdot) = \{\varphi(\cdot) : \varphi \text{ is a state of } \mathcal{A}\}$ denotes the generalized numerical range. Moslehian and Rajić [19] generalized this result to n-positive linear maps ($n \geq 3$) as follows.

Theorem 4.1. *[19, Theorem 2.6] Let $\Phi : \mathcal{A} \to \mathbf{B}(\mathcal{H})$ be a unital n-positive linear map between C^*-algebras for some $n \geq 3$. Then*

$$\|\Phi(AB) - \Phi(A)\Phi(B)\| \leq \Delta(A, \|\cdot\|)\, \Delta(B, \|\cdot\|) \tag{1}$$

for all operators $A, B \in \mathcal{A}$.

As a corollary one can get the next result.

Corollary 4.2. *[19, Corollary 2.8] Let $A, B \in \mathcal{M}_k$, where $k \geq 4$. Then*

$$\begin{aligned} &\|(k^2 - k - 1)tr(AB)I - kAB - (k-1)tr(A)tr(B)I + tr(B)A + tr(A)B\| \\ &\quad \leq \frac{(k^2 - k - 1)^2}{k-1} \Delta(A, \|\cdot\|)\, \Delta(B, \|\cdot\|). \end{aligned}$$

Assuming that a unitarily invariant norm $|||\cdot|||$ is given on a two-sided ideal of bounded linear operators acting on a separable Hilbert space, it induces some unitarily invariant norms $|||\cdot|||$ on matrix algebras $\mathcal{M}_n$ for all finite values of n via $|||A||| = |||A \oplus 0|||$. Matharu and Moslehian [15] showed that if $\mathcal{A}$ is a C^*-algebra of finite dimension k and $\Phi : \mathcal{A} \to \mathcal{M}_n$ is a unital completely positive map, then

$$|||\Phi(AB) - \Phi(A)\Phi(B)||| \leq |||I_n|||\, |||I_{kn}|||\Delta(A, \mathbf{C}I)\Delta(B, \mathbf{C}I)$$

for any $A, B \in \mathcal{A}$, where d_X denotes the diameter of the unitary orbit $\{UXU^* : U \text{ is unitary}\}$ of X and I_m stands for the identity of $\mathcal{M}_m$.

The next result reads as follows.

Corollary 4.3. *[15, Corollary 4.3] Let self-adjoint elements $A_1, \cdots, A_n, B_1, \cdots, B_n \in \mathcal{A}$ satisfy*

$$m_1 \leq A_j \leq M_1, \qquad m_2 \leq B_j \leq M_2 \qquad (j = 1, \cdots, n)$$

for some real numbers m_1, m_2, M_1, M_2. If $C_1, \cdots, C_n \in \mathcal{A}$ are such that $\sum_{j=1}^n C_j^* C_j = I$, then

$$\left| \sum_{j=1}^n C_j^* A_j B_j C_j - \sum_{j=1}^n C_j^* A_j C_j \sum_{j=1}^n C_j^* B_j C_j \right| \leq \frac{1}{4}(M_1 - m_1)(M_2 - m_2)I. \tag{2}$$

The inequality (2) is an extension of a the discrete case of the integral version (1) of the Grüss inequality. As a corollary we can say that if $m_1 \le a_j \le M_1, m_2 \le b_j \le M_2$ ($j = 1, \cdots, n$) are real numbers, then

$$\left| \frac{1}{n}\sum_{j=1}^{n} a_j b_j - \frac{1}{n}\sum_{j=1}^{n} a_j \frac{1}{n}\sum_{j=1}^{n} b_j \right| \le \frac{1}{4}(M_1 - m_1)(M_2 - m_2).$$

Biernacki, Pidek and Ryll-Nardjewski [3] presented a better estimate than the above inequality as follows: If $m_1 \le a_j \le M_1, m_2 \le b_j \le M_2$ ($j = 1, \cdots, n$) are real numbers, then

$$\left| \frac{1}{n}\sum_{j=1}^{n} a_j b_j - \frac{1}{n}\sum_{j=1}^{n} a_j \frac{1}{n}\sum_{j=1}^{n} b_j \right| \le \frac{1}{n}\left[\frac{n}{2}\right]\left(1 - \frac{1}{n}\left[\frac{n}{2}\right]\right)(M_1 - m_1)(M_2 - m_2).$$

In [18] the authors proved that if $A, B \in \mathbf{B}(\mathcal{H})$ are positive invertible operators and $\Phi : \mathbf{B}(\mathcal{H}) \to \mathbf{B}(\mathcal{K})$ is a positive linear map and $0 < m^2 A \le B \le M^2 A$ for some positive real numbers $m < M$, then the following Operator Grüss type inequality holds:

$$\Phi(A)\sharp\Phi(B) - \Phi(A\sharp B) \le \frac{\sqrt{M_1 M_2}\left(\sqrt{M_1 M_2} - \sqrt{m_1 m_2}\right)^2}{2\sqrt{m_1 m_2}} \min\left\{\frac{M_1}{m_1}, \frac{M_2}{m_2}\right\}.$$

5 A Grüss type inequality in Hilbert C^*-modules

In this section, we present a Grüss type inequality in the setting of semi-inner product C^*-modules. The notion of semi-inner product C^*-module is a natural generalization of that of semi-inner product space arising under replacement of the field of scalars $\mathbf{C}$ by a C^*-algebra. The next lemma is used to prove a Grüss type inequality in the semi-inner product C^*-module, see also [13].

Lemma 5.1. *[11, Lemma 5.1] Let $\mathcal{X}$ be a pre-inner product C^*-module over $\mathcal{A}$. Suppose that $x, h \in \mathcal{X}$ such that $\langle h, h\rangle$ is the unit element e of $\mathcal{A}$ and*

$$\langle Ay - x, x - ay\rangle \ge 0$$

holds for some positive invertible elements $a, A \in \mathcal{Z}(\mathcal{A})$. Then

$$0 \le \langle x, x\rangle - |\langle h, x\rangle|^2 \le \frac{1}{4}(A - a)^2.$$

Theorem 5.2. *[11, Theorem 5.2] Let $\mathcal{X}$ be a semi-inner product C^*-module over $\mathcal{A}$. Suppose that $x, y, h \in \mathcal{X}$ such that $\langle h, h\rangle$ is the unit element e of $\mathcal{A}$, $\langle y, y\rangle - |\langle h, y\rangle|^2$ is invertible and*

$$\langle Ah - x, x - ah\rangle \geq 0 \qquad and \qquad \langle Bh - y, y - bh\rangle \geq 0$$

hold for some positive invertible elements $a, A, b, B \in \mathcal{Z}(\mathcal{A})$. *Then*

$$|\langle x, y\rangle - \langle x, h\rangle\langle h, y\rangle| \leq \frac{1}{4}|A - a||B - b|.$$

Bibliography

[1] Andrica, D. and Badea, C. *Grüss' inequality for positive linear functionals*, Period. Math. Hungar. 19 (1988), no. 2, 155–167.

[2] Banić, S., Ilišević, D., and Varošanec, S. *Bessel- and Grüss-type inequalities in inner product modules*, Proc. Edinb. Math. Soc. (2) 50 (2007), no. 1, 23—36.

[3] Biernacki, M., Pidek, H., and Ryll-Nardzewski, C. *Mariae Curie-Sklodowska*, Ann. Univ. A4 (1950), 1–4.

[4] Bohner M. and Matthews, T. *The Grüss inequality on time scales*, Commun. Math. Anal. 3 (2007), no. 1, 1—8.

[5] Bourin, J.C. *Matrix versions of some classical inequalities*, Linear Algebra Appl. 416 (2006), no. 2-3, 890–907.

[6] Buse, C., Cerone, P., Dragomir, S.S., and Roumeliotis, J. *A refinement of Grüss type inequality for the Bochner integral of vector-valued functions in Hilbert spaces and applications*, J. Korean Math. Soc. 43 (2006), no. 5, 911–929.

[7] Cerone, P. and Dragomir, S.S. *Mathematical inequalities*, A perspective, CRC Press, Boca Raton, FL, 2011.

[8] Diestel, J. and Uhl Jr., J.J. *Vector measures. With a foreword by B. J. Pettis*, Mathematical Surveys, No. 15. American Mathematical Society, Providence, R.I., 1977.

[9] Dragomir, S.S. *Advances in inequalities of the Schwarz, Grüss and Bessel type in inner product spaces*, Nova Science Publishers, Inc., Hauppauge, NY, 2005.

[10] Fink, A.M. *A treatise on Grüss' inequality, Analytic and geometric inequalities and applications*, 93–113, Math. Appl., 478, Kluwer Acad. Publ., Dordrecht, 1999.

[11] Fujii, J.I., Fujii, M., Moslehian, M.S., Pecaric' J.E., and Seo, Y. *Reverses Cauchy-Schwarz type inequalities in pre-inner product* C^**-modules*, Hokkaido Math. J. , 40 (2011) , 1–17.

[12] Grüss, G. Über das Maximum des absoluten Betrages $\frac{1}{b-a}\int_a^b f(x)g(x)dx - \frac{1}{(b-a)^2}\int_a^b f(x)dx \int_a^b g(x)dx$, Math. Z. 39 (1935), 215–226.

[13] Ilišević, D., and Varošanec, S. *Grüss type inequalities in inner product modules*, Proc. Amer. Math. Soc. 133 (11) (2005), 3271—3280.

[14] Jocic, D., Krtinic, D., and Moslehian, M.S. *Landau and Grüss type inequalities for inner product type integral transformers in norm ideals*, Math. Inequal. Appl. 16 (2013), no. 1, 109-125.

[15] Matharu, J.S. and Moslehian, M.S. *Grüss inequality for some types of positive linear maps*, J. Operator Theory 73 (2015), no. 1, 265-278.

[16] Mercer, A.Mc.D. and Mercer, P.R. *New proofs of the Grüss inequality*, Aust. J. Math. Anal. Appl. 1 (2004), no. 2, Art. 12, 6 pp.

[17] Mitrinović, D.S., Pečarić, J.E., and Fink, A.M. *Classical and New Inequalities in Analysis*, Kluwer Academic, Dordrecht, 1993.

[18] Moslehian, M.S., Nakamoto, R., and Seo, Y. *A Diaz–Metcalf type inequality for positive linear maps and its applications*, Electron. J. Linear Algebra 22 (2011), 179–190.

[19] Moslehian, M.S. and Rajić, R. *A Grüss inequality for n-positive linear maps*, Linear Algebra Appl. 433 (2010), 1555–1560.
[20] Elezović, N., Marangunić, Lj., and Pečarić, J.E. *Unified treatment of complemented Schwarz and Grüss inequalities in inner product spaces*, Math. Inequal. Appl. 8 (2005), no.2, 223–231.
[21] Perić, I. and Rajić, R. *Grüss inequality for completely bounded maps*, Linear Algebra Appl. 390 (2004), 287–292.
[22] Renaud, P.F. *A matrix formulation of Grüss inequality*, Linear Algebra Appl. 335 (2001) 95–100.
[23] Ujević, N. *A generalization of the pre-Grüss inequality and applications to some quadrature formulae*, J. Inequal. Pure Appl. Math. 3 (2002), no. 1, Article 13, 9 pp.

Mustafa A. Muratov and Ben-Zion A. Rubshtein

Main embedding theorems for symmetric spaces of measurable functions

Abstract: Let m be the usual Lebesgue measure on $\mathbb{R}_+ = [0, +\infty)$. Dealing with symmetric (rearrangement invariant) spaces $\mathbf{E}$ on the standard measure space $(\mathbb{R}_+, m)$, we treat the following embeddings: $\mathbf{L}_1 \cap \mathbf{L}_\infty \subseteq \Lambda^0_{\widetilde{V}} \subseteq \mathbf{E}^0 \subseteq \mathbf{E} \subseteq \mathbf{E}^{11} \subseteq \mathbf{M}_{V_*} \subseteq \mathbf{L}_1 + \mathbf{L}_\infty$, where $\mathbf{E}^0 = cl_{\mathbf{E}}(\mathbf{L}_1 \cap \mathbf{L}_\infty)$ is the closure of $\mathbf{L}_1 \cap \mathbf{L}_\infty$ in $\mathbf{E}$, $\mathbf{E}^{11} = (\mathbf{E}^1)^1$ is the second associate space of $\mathbf{E}$, $V(x) = \|1_{[0,x]}\|_{\mathbf{E}}$ is the fundamental function of the symmetric space $\mathbf{E}$, $V_*(x) = \frac{x}{V(x)} 1_{(0,\infty)}(x)$ and $\widetilde{V}$ is the least concave majorant of V, $\Lambda_{\widetilde{V}}$ and $\mathbf{M}_{V_*}$ are the Lorentz and Marcinkiewicz spaces with the weights $\widetilde{V}$ and V_* respectively and $\Lambda^0_{\widetilde{V}} = cl_{\Lambda_{\widetilde{V}}}(\mathbf{L}_1 \cap \mathbf{L}_\infty)$. The embeddings and natural inequalities for corresponding norms are studied in detail.

Keywords: Symmetric spaces, Lorentz and Marcinkiewicz spaces, embedding theorems

MSC: 46E30, 46E35, 26D10, 26D15, 46B70

1 Three Embedding Theorems

1.1 Symmetric spaces

Let m be the usual Lebesgue measure on the set of nonnegative real numbers $\mathbb{R}_+ = [0, +\infty)$, and let $\mathbf{L}_0 = \mathbf{L}_0(\mathbb{R}_+, m)$ denote the set of all m-measurable functions $f\colon \mathbb{R}_+ \to \mathbb{R}$. Throughout the paper we deal with symmetric spaces $\mathbf{E}$ on the standard measure space $(\mathbb{R}_+, m)$.

A Banach space $\mathbf{E}$ of real measurable functions on $(\mathbb{R}_+, m)$ is called *symmetric (rearrangement invariant)* if

$$f \in \mathbf{L}_0\,,\ g \in \mathbf{E} \text{ and } f^* \le g^* \implies f \in \mathbf{E} \text{ and } \|f\|_{\mathbf{E}} \le \|g\|_{\mathbf{E}}.$$

Here, f^* is the decreasing right-continuous rearrangement of $|f|$ defined by

$$f^*(x) := \inf\{y \in [0, +\infty)\colon n_f(y) \le x\}\,,\ x \in [0, \infty)\,,$$

where

$$n_{|f|}(y) := m\,\{x \in [0, +\infty)\colon |f(x)| > y\}$$

Mustafa A. Muratov, Department of Mathematics and Computer Sciences, Crimean Federal University, Simferopol, 295007, Russian Federation; e-mail: mamuratov@gmail.com
Ben-Zion A. Rubshtein, Depatment of Mathematics, Ben-Gurion University of the Negev, Beer-Sheva, 84105, Israel

https://doi.org/10.1515/9783110413557-011

is the upper distribution function of $|f|$. We usually assume that $\lim_{x\to\infty} n_f(x) = 0$ to avoid the case $n_f(x) \equiv +\infty$.

It is easily seen that a Banach space $\mathbf{E} \subseteq \mathbf{L}_0$ is symmetric if and only if the following two conditions hold:

(1) $f \in \mathbf{L}_0$, $g \in \mathbf{E}$ and $|f| \leq |g|$ imply $f \in \mathbf{E}$ and $\|f\|_{\mathbf{E}} \leq \|g\|_{\mathbf{E}}$.
(2) $f \in \mathbf{L}_0$, $g \in \mathbf{E}$ and $f^* = g^*$ imply $f \in \mathbf{E}$ and $\|f\|_{\mathbf{E}} = \|g\|_{\mathbf{E}}$.

Condition (1) means that $\mathbf{E}$ is an ideal Banach lattice. Condition (2) is the *symmetry (or rearrangement invariance) property* of the norm $\|\cdot\|_{\mathbf{E}}$.

Thus a symmetric space is an ideal Banach lattice with a symmetric (rearrangement invariant) norm.

1.2 First embedding theorem

Recall that the fundamental function $\varphi_{\mathbf{E}}$ of a symmetric space $\mathbf{E}$ is defined as $\varphi_{\mathbf{E}}(x) = \|1_{[0,x]}\|_{\mathbf{E}}$, $x \geq 0$. In particular, $\varphi_{\mathbf{E}}(1) = \|1_{[0,1]}\|_{\mathbf{E}}$.

Theorem 1.1. *Let* $\mathbf{E}$ *be a symmetric space. Then there are continuous embeddings*

$$\mathbf{L}_1 \cap \mathbf{L}_\infty \subseteq \mathbf{E} \subseteq \mathbf{L}_1 + \mathbf{L}_\infty\,, \tag{1}$$

where

$$\varphi_{\mathbf{E}}(1)\|f\|_{\mathbf{L}_1\cap\mathbf{L}_\infty} \geq \|f\|_{\mathbf{E}}\,,\ f \in \mathbf{L}_1 \cap \mathbf{L}_\infty \ \textit{and}\ \|f\|_{\mathbf{E}} \geq \varphi_{\mathbf{E}}(1)\|f\|_{\mathbf{L}_1+\mathbf{L}_\infty}\,,\ f \in \mathbf{E} \tag{2}$$

Here, the space $\mathbf{L}_1 \cap \mathbf{L}_\infty$ is equipped with the norm $\|f\|_{\mathbf{L}_1\cap\mathbf{L}_\infty} = \max(\|f\|_{\mathbf{L}_1}, \|f\|_{\mathbf{L}_\infty})$, while the space

$$\mathbf{L}_1 + \mathbf{L}_\infty = \{f \in \mathbf{L}_0 : f = g + h\,,\ g \in \mathbf{L}_1\,,\ h \in \mathbf{L}_\infty\}$$

consists of all locally integrable functions on $(\mathbb{R}_+, m)$, and it is equipped with the norm

$$\|f\|_{\mathbf{L}_1+\mathbf{L}_\infty} = \inf\{\|g\|_{\mathbf{L}_1} + \|h\|_{\mathbf{L}_\infty} : f = g + h\,,\ g \in \mathbf{L}_1\,,\ h \in \mathbf{L}_\infty\}\,.$$

The spaces $\mathbf{L}_1 \cap \mathbf{L}_\infty$ and $\mathbf{L}_1 + \mathbf{L}_\infty$ are the smallest and largest symmetric spaces on $\mathbb{R}_+$.

The proof of the theorem is given in Section 2

1.3 Second embedding theorem

Let $\mathbf{E}^0 = cl_{\mathbf{E}}(\mathbf{L}_1 \cap \mathbf{L}_\infty)$ be the closure of $\mathbf{L}_1 \cap \mathbf{L}_\infty$ in a symmetric space $\mathbf{E}$, and let $\mathbf{E}^{11} = (\mathbf{E}^1)^1$ be the second associate space of $\mathbf{E}$, where

$$\mathbf{E}^1 := \left\{g \in \mathbf{L}_0 : \|g\|_{\mathbf{E}^1} := \sup_{\|f\|_{\mathbf{E}}\leq 1} \left|\int fg\,dm\right| < \infty,\ f \in \mathbf{E}\right\}\,.$$

The spaces $\mathbf{E}^0$ and $\mathbf{E}^{11}$ are respectively the minimal and maximal symmetric spaces corresponding to $\mathbf{E}$.

Theorem 1.2. *Let* $\mathbf{E}$ *be a symmetric space. Then there are continuous embeddings*

$$\mathbf{E}^0 \subseteq \mathbf{E} \subseteq \mathbf{E}^{11}\,, \tag{3}$$

where

$$\|f\|_{\mathbf{E}^0} = \|f\|_{\mathbf{E}} = \|f\|_{\mathbf{E}^{11}}\,,\ f \in \mathbf{E}^0 \text{ and } \|f\|_{\mathbf{E}} \geq \|f\|_{\mathbf{E}^{11}}\,,\ f \in \mathbf{E}\,. \tag{4}$$

Two particular cases, when $\mathbf{E} = \mathbf{E}^0$ ($\mathbf{E}$ is minimal), and when $\mathbf{E} = \mathbf{E}^{11}$ (E is maximal), are of a special interest here.

The proof of the theorem is given in Section 2

1.4 Third embedding theorem

Let W be an increasing function on $[0, +\infty)$ such that: $W(0) = 0$, W is concave on $(0, +\infty)$, and $W(x) > 0$ for some x. The Lorentz space Λ_W on $(\mathbb{R}_+, m)$ with a concave weight function W is defined as

$$\Lambda_W := \left\{ f \in \mathbf{L}_0 : \|f\|_{\Lambda_W} := \int_0^\infty f^*(x)\, dW(x) < \infty \right\}.$$

It is a symmetric space with the fundamental function $\varphi_{\Lambda_W} = W$.

Let V be a quasi-concave function on $\mathbb{R}_+$, i.e. $V(0) = 0$ and both the functions $V(x)$ and $V_*(x) := \dfrac{x}{V(x)} 1_{(0,\infty)}(x)$ are increasing. The Marcinkiewicz space $\mathbf{M}_V$ on $(\mathbb{R}_+, m)$ is defined as

$$\mathbf{M}_V := \{ f \in \mathbf{L}_0 : \|f\|_{\mathbf{M}_V} := \sup_{0<x<\infty} \frac{1}{V(x)} \int_0^x f^*(y)\, dy < \infty \}\,.$$

It is a symmetric space with the fundamental function $\varphi_{\mathbf{M}_V} = V_*$.

Theorem 1.3. *Let* $V(x) = \varphi_{\mathbf{E}}(x) = \|1_{[0,x]}\|_{\mathbf{E}}$ *be the fundamental function of a symmetric space* $\mathbf{E}$. *Let* $\widetilde{V}$ *denote the least concave majorant of* V, *and* $V_*(x) = \dfrac{x}{V(x)} 1_{(0,\infty)}(x)$. *Then there are continuous embeddings*

$$\Lambda^0_{\widetilde{V}} \subseteq \mathbf{E} \subseteq \mathbf{M}_{V_*}\,, \tag{5}$$

where

$$\|f\|_{\Lambda^0_{\widetilde{V}}} \geq \|f\|_{\mathbf{E}}\,,\ f \in \Lambda^0_{\widetilde{V}} \text{ and } \|f\|_{\mathbf{E}} \geq \|f\|_{\mathbf{M}_{V_*}}\,,\ f \in \mathbf{E}\,. \tag{6}$$

The space $\Lambda^0_{\widetilde{V}}$ is the minimal part of the Lorentz space $\Lambda_{\widetilde{V}}$. It is the smallest symmetric space on $\mathbb{R}_+$ whose fundamental function $\varphi_{\Lambda^0_{\widetilde{V}}} = \widetilde{V}$ is equivalent to V. The

Marcinkiewicz space $\mathbf{M}_{V_*}$ is the largest symmetric space on $\mathbb{R}_+$ satisfying $\varphi_{\mathbf{M}_{V_*}} = \varphi_{\mathbf{E}} = V$.

The inclusion $\Lambda_{\widetilde{V}} \subseteq \mathbf{E}$ claimed in [3, II.5.4, Th. 5.5] fails in general. Although, it is true, for example, if $V(+\infty) = \infty$ (the space $\Lambda^0_{\widetilde{V}}$ is minimal), or if the space $\mathbf{E}$ itself is maximal ($\mathbf{E} = \mathbf{E}^{11}$).

The proof of the theorem is given in Section 3

Combining embeddings (1), (3) and (5) we have for an arbitrary symmetric space $\mathbf{E}$ on $\mathbb{R}_+$:

$$\mathbf{L}_1 \cap \mathbf{L}_\infty \subseteq \Lambda^0_{\widetilde{V}} \subseteq \mathbf{E}^0 \subseteq \mathbf{E} \subseteq \mathbf{E}^{11} \subseteq \mathbf{M}_{V_*} \subseteq \mathbf{L}_1 + \mathbf{L}_\infty \,.$$

The inequalities (2), (4) and (6) determine natural relations between corresponding norms.

Various versions of the above embedding theorems were considered for example in: [3], [1], [4], [2].

In the present paper we restrict our attention to symmetric spaces on $\mathbb{R}_+$. In this case we present complete proofs of the theorems and some closely related results. Some exact references are used in several places in order to restrict the volume of the paper.

2 Embeddings $\mathbf{L}_1 \cap \mathbf{L}_\infty \subseteq \mathbf{E}^0 \subseteq \mathbf{E} \subseteq \mathbf{E}^{11} \subseteq \mathbf{L}_1 + \mathbf{L}_\infty$

2.1 Minimality and property (A)

A symmetric space $\mathbf{E}$ is called *minimal* if the set $\mathbf{F}_1$ of all simple (finite-valued) integrable functions is dense in $\mathbf{E}$. Since $\mathbf{F}_1$ is dense in $\mathbf{L}_1 \cap \mathbf{L}_\infty$ and $\mathbf{L}_1 \cap \mathbf{L}_\infty \subseteq \mathbf{E}$, the symmetric space $\mathbf{E}$ is minimal if and only if the closure $cl_{\mathbf{E}}(\mathbf{L}_1 \cap \mathbf{L}_\infty)$ coincides with $\mathbf{E}$.

In general case, the closure $\mathbf{E}^0 := cl_{\mathbf{E}}(\mathbf{L}_1 \cap \mathbf{L}_\infty) \subseteq \mathbf{E}$ is a symmetric space, which is referred as the *minimal part* of $\mathbf{E}$. So that $\mathbf{E}$ is a minimal symmetric spaces if and only if $\mathbf{E}^0 = \mathbf{E}$.

A symmetric space $\mathbf{E}$ is said to have property (A) (order continuous norm) if

(A) If $0 \leq f_n \in \mathbf{E}$ and $f_n \downarrow 0 \Longrightarrow \|f_n\|_{\mathbf{E}} \downarrow 0$.

Theorem 2.1. *Let* $\mathbf{E}$ *be a symmetric space. Then the following conditions are equivalent:*
(1) $\mathbf{E}$ *has property (A).*
(2) $\mathbf{E}$ *is minimal and* $\varphi_{\mathbf{E}}(0+) = 0$.
(3) $\mathbf{E}$ *is separable.*

Proof. See, [2, X.3, Th.3], [4, Prop.1.a.8, Th.1.b.16]. □

2.2 Minimality and property (C)

A symmetric space **E** is said to have property (C) (*order semi-continuous norm*) if
(C) $0 \le f_n \uparrow f \in \mathbf{E} \implies \sup_n \|f_n\|_{\mathbf{E}} = \|f\|_{\mathbf{E}}$.

Theorem 2.2. *Every minimal symmetric space has property (C).*

Proof. Case 1. $\varphi_{\mathbf{E}}(0+) = 0$.

If **E** is minimal and $\varphi_{\mathbf{E}}(0+) = 0$, then **E** has property (A) by Theorem 2.1. But (A), obviously, implies (C).

Case 2. $\varphi_{\mathbf{E}}(0+) > 0$.

Let $\varphi_{\mathbf{E}}(0+) = c > 0$ and $x_n \downarrow 0$. Then for every $f \in \mathbf{E}$, we have

$$\|f^*\|_{\mathbf{E}} \ge \|f^* \cdot 1_{[0,x_n]}\|_{\mathbf{E}} \ge f^*(x_n)\|1_{[0,x_n]}\|_{\mathbf{E}} = f^*(x_n)\varphi_{\mathbf{E}}(x_n) \ge c \cdot f^*(x_n)$$

Whence

$$\|f\|_{\mathbf{L}_\infty} = f^*(0) = \sup_n f^*(x_n) \le \frac{1}{c}\|f^*\|_{\mathbf{E}} < \infty ,$$

i.e. $f \in \mathbf{L}_\infty$. Thus $\mathbf{E} \subseteq \mathbf{L}_\infty$.

Now, we need the following simple lemma.

Lemma 2.3. *Let* **E** *be a minimal symmetric space such that* $\mathbf{E} \subset \mathbf{L}_\infty$ *(strict inclusion). Then for every* $f \in \mathbf{E}$,

$$\|f - f \cdot 1_{[0,n]}\|_{\mathbf{E}} \to 0 \textit{ as } n \to \infty . \tag{1}$$

Proof. Let $\mathbf{F}_0$ denote the set of all simple functions on $\mathbb{R}_+$ with bounded support. A simple function g belongs to $\mathbf{F}_0$ if and only if $g \cdot 1_{[a,\infty)} = 0$ for some finite $a > 0$.

Since $\mathbf{F}_0$ is dense in $\mathbf{L}_1 \cap \mathbf{L}_\infty$, and **E** is minimal

$$\mathbf{E} = \mathbf{E}^0 = cl_{\mathbf{E}}(\mathbf{L}_1 \cap \mathbf{L}_\infty) = cl_{\mathbf{E}}(\mathbf{F}_0) .$$

For every $\varepsilon > 0$ there exists $g \in \mathbf{F}_0$ such that $\|f - g\|_{\mathbf{E}} < \varepsilon$. Then we have

$$\|f - f \cdot 1_{[0,n]}\|_{\mathbf{E}} \le \|f - g\|_{\mathbf{E}} + \|g - g \cdot 1_{[0,n]}\|_{\mathbf{E}} + \|g \cdot 1_{[0,n]} - f \cdot 1_{[0,n]}\|_{\mathbf{E}} .$$

Moreover, $\|g - g \cdot 1_{[0,n]}\|_{\mathbf{E}} = 0$ for n large enough, and

$$\|g \cdot 1_{[0,n]} - f \cdot 1_{[0,n]}\|_{\mathbf{E}} \le \|f - g\|_{\mathbf{E}} < \varepsilon .$$

Therefore, $\|f - f \cdot 1_{[0,n]}\|_{\mathbf{E}} < 2\varepsilon$ for n large enough, and thus (1) holds. □

We now turn to the proof of Theorem 2.2. Let

$$0 \le f_n \uparrow f \in \mathbf{E} = \mathbf{E}^0 \subset \mathbf{L}_\infty .$$

Then by Lemma 2.3, we have

$$\sup_n \|f_n\|_{\mathbf{E}} = \sup_n \sup_k \|f_n \cdot 1_{[0,k]}\|_{\mathbf{E}} = \sup_k \|f \cdot 1_{[0,k]}\|_{\mathbf{E}} = \|f\|_{\mathbf{E}}; .$$

□

Remark 2.4. *The minimal spaces* $\mathbf{E}$ *were defined in [4, Def.2.a.1(ii)], by the equality* $\mathbf{E} = cl_{\mathbf{E}^{11}}(\mathbf{F}_1)$. *It is follows from Theorem 2.2 that*

$$cl_{\mathbf{E}^{11}}(\mathbf{F}_1) = cl_{\mathbf{E}}(\mathbf{F}_1) = \mathbf{E}^0.$$

Thus our definition of minimality ($\mathbf{E} = \mathbf{E}^0$), *is equivalent to the definition from [4].*

2.3 The natural embedding

Recall that the associate space $\mathbf{E}^1$ of a symmetric space $\mathbf{E}$ is defined as

$$\mathbf{E}^1 := \left\{ g \in \mathbf{L}_0 : \|g\|_{\mathbf{E}^1} := \sup_{\|f\|_{\mathbf{E}} \le 1} \left| \int fg \, dm \right| < \infty \right\} .$$

Proposition 2.5. *Let* $\mathbf{E}$ *be a symmetric space and let* $\mathbf{E}^0$ *denote its minimal part. Then*
(1) $\mathbf{E}^1$ *is a symmetric space.*
(2) $(\mathbf{E}^0)^1 = \mathbf{E}^1$.
(3) $\varphi_{\mathbf{E}^1}(x) = \varphi_*(x)$, *where* $\varphi_*(x) := \dfrac{x}{\varphi_{\mathbf{E}}(x)} 1_{(0,\infty)}, \; x \ge 0$.

Proof. See, [3, II.6.]. □

The associate space $\mathbf{E}^1$ can be identified with the subset $\{\upsilon_g : g \in \mathbf{E}^1\}$ of the dual space $\mathbf{E}^*$, where

$$\upsilon_g(f) := \int fg \, d\mu \, , \; f \in \mathbf{E} \, .$$

By the definition $\upsilon_g \in \mathbf{E}^*$ and $\|\upsilon_g\|_{\mathbf{E}^*} = \|g\|_{\mathbf{E}^1}$ for every $g \in \mathbf{E}^1$. Moreover

Proposition 2.6. *The natural embedding* $\upsilon : \mathbf{E}^1 \ni g \to \upsilon_g \in \mathbf{E}^*$ *is an isometric isomorphism from* $\mathbf{E}^1$ *onto a closed subspace* $\{\upsilon_g : g \in \mathbf{E}^1\}$ *of* $\mathbf{E}^*$.

Theorem 2.1 can be extended now by an equivalent condition (4).

Theorem 2.7. *The condition (1) - (3) in Theorem 2.1 are equivalent to the following condition*
(4) $\upsilon(\mathbf{E}^1) = \mathbf{E}^*$.

Proof. See, [2, X.4.]. □

Corollary 2.8. $(\mathbf{E}^0)^* = \upsilon(\mathbf{E}^1) \Longleftrightarrow \varphi_{\mathbf{E}}(0+) = 0$.

Proof. It follows from Theorems 2.1, 2.7 and Proposition 2.5. □

2.4 Embedding $\mathbf{E} \subseteq \mathbf{E}^{11}$ and property(C)

Since the associate space $\mathbf{E}^1$ itself is a symmetric space, we can consider its associate space $\mathbf{E}^{11} = (\mathbf{E}^1)^1$ The space $\mathbf{E}^{11}$ consists of all $h \in \mathbf{L}_0$ such that

$$\|h\|_{\mathbf{E}^{11}} = \sup \left\{ \left| \int_0^\infty hg\, dm \right| : g \in \mathbf{E}^1, \|g\|_{\mathbf{E}^1} \leq 1 \right\} < \infty .$$

If $f \in \mathbf{E}$ then for every $g \in \mathbf{E}^1$ we have

$$\left| \int_0^\infty fg\, dm \right| \leq \|f\|_{\mathbf{E}} \cdot \|g\|_{\mathbf{E}^1} .$$

This means that

$$\mathbf{E} \subseteq \mathbf{E}^{11} \text{ and } \|f\|_{\mathbf{E}^{11}} \leq \|f\|_{\mathbf{E}} ,\ f \in \mathbf{E} . \tag{2}$$

The inclusion $\mathbf{E} \subseteq \mathbf{E}^{11}$ may be strict.

Example 2.9. *Let* $\mathbf{R}_0 := \{f \in \mathbf{L}_1 + \mathbf{L}_\infty : f^*(\infty) = 0\}$. *Then*
1. $\mathbf{R}_0 = (\mathbf{L}_1 + \mathbf{L}_\infty)^0$;
2. $(\mathbf{L}_\infty)^0 = \mathbf{L}_\infty \cap \mathbf{R}_0 \subset \mathbf{L}_\infty = (\mathbf{L}_1)^1 = (\mathbf{L}_\infty)^{11}$.

Theorem 2.10. *Let* $\mathbf{E}$ *be a symmetric space. Then the following conditions are equivalent:*

(1) $\mathbf{E}$ *has property (C).*

(2) $\|f\|_{\mathbf{E}} = \sup \left\{ \left| \int_0^\infty fg\, dm \right| : g \in \mathbf{E}^1, \|g\|_{\mathbf{E}^1} \leq 1 \right\}$.

(3) $\|f\|_{\mathbf{E}^{11}} = \|f\|_{\mathbf{E}} ,\ f \in \mathbf{E}$.

Proof. See, [4, Prop.1.b.18], [2, X.4., Th.7] □

Condition (3) means that the embedding $\mathbf{E} \to \mathbf{E}^{11}$ is isometric. In contrast to the embeddings $\mathbf{E}^0 \to \mathbf{E}$ and $\mathbf{E}^0 \to \mathbf{E}^{11}$, the mapping $\mathbf{E} \to \mathbf{E}^{11}$ need not be isometric, in general. Moreover, it need not be an open mapping, i.e. the space $\mathbf{E}$ may be not closed in $\mathbf{E}^{11}$.

2.5 Maximality. Properties (B) and (BC)

A symmetric space $\mathbf{E}$ is called *maximal* if $\mathbf{E} = \mathbf{E}^{11}$.

A symmetric space $\mathbf{E}$ is said to have property (B) (*monotonically complete norm*) if

(B) If $0 \leq f_n \uparrow ,\ f_n \in \mathbf{E} ,\ \sup_n \|f_n\|_{\mathbf{E}} < \infty \implies f_n \uparrow f$ for some $f \in \mathbf{E}$.

Theorem 2.11. *Let* $\mathbf{E}$ *be a symmetric space. Then the following conditions are equivalent:*

(1) **E** *is maximal.*
(2) $\mathbf{E} = \mathbf{G}^1$ *for some symmetric space* $\mathbf{G}$.
(3) **E** *has property (B).*

Proof. 1) $\Longleftrightarrow$ 2). If $\mathbf{E} = \mathbf{G}^1$ for some symmetric space $\mathbf{G}$, then $\mathbf{E}^1 = \mathbf{G}^{11}$ and

$$\mathbf{E}^{11} = \mathbf{G}^{111} = \mathbf{G}^1 = \mathbf{E}.$$

If $\mathbf{E} = \mathbf{E}^{11}$, then $\mathbf{E} = \mathbf{G}^1$ for $\mathbf{G} = \mathbf{E}^1$.

2) $\Longrightarrow$ 3). Let $\mathbf{E} = \mathbf{G}^1$ and $\{f_n\} \subset \mathbf{E}$ such, that $0 \le f_n \uparrow$ and $\sup_n \|f_n\|_{\mathbf{E}} < \infty$.

Since both $\mathbf{E}$ and $\mathbf{G}^1$ are Banach spaces, the norms $||\cdot||_{\mathbf{E}}$ and $||\cdot||_{\mathbf{G}^1}$ are equivalent by the open mapping theorem. Hence $\sup_n ||f_n||_{\mathbf{G}^1} = c < \infty$.

Since

$$\|f_n\|_{\mathbf{G}^1} = \sup \left\{ \int_0^\infty f_n g dm : 0 \le g \in \mathbf{G},\ \|g\|_{\mathbf{G}} \le 1 \right\},$$

then for every $0 \le g \in \mathbf{G}$ with $\|g\|_{\mathbf{G}} \le 1$, the functions $f_n g$ are integrable and

$$\int_0^\infty f_n g dm \le \sup_n \|f\|_{\mathbf{G}^1} = c < \infty.$$

By the Fatou-Lebesgue theorem the function $\sup_n (f_n g) = (\sup_n f_n) g$ is integrable, and, hence it is finite almost everywhere. Let $f = \sup_n f_n$, then

$$\|f\|_{\mathbf{G}^1} = \sup \left\{ \int_0^\infty f g dm : 0 \le g \in \mathbf{G},\ \|g\|_{\mathbf{G}} \le 1 \right\} \le c < \infty,$$

i.e. $f \in \mathbf{G}^1 = \mathbf{E}$. Hence, the space $\mathbf{E} = \mathbf{G}^1$ has property (B).

3) $\Longrightarrow$ 1). Let $f \in \mathbf{E}^{11}$ and $f \ge 0$. Then there is a sequence $\{f_n\}$ such that $0 \le f_n \in \mathbf{F}_0$ and $f_n \uparrow f$. Since the embedding $\mathbf{E} \subseteq \mathbf{E}^{11}$ is continuous and $\mathbf{E}^0 = (\mathbf{E}^{11})^0$ as sets, the norms $\|\cdot\|_{\mathbf{E}^0}$ and $\|\cdot\|_{(\mathbf{E}^{11})^0}$ are equivalent by the open mapping theorem.

Hence for some $c > 0$,

$$\|f\|_{\mathbf{E}} \le c \cdot \|f\|_{\mathbf{E}^{11}},\ f \in \mathbf{E}^0 = (\mathbf{E}^{11})^0.$$

Since $\{f_n\} \subset \mathbf{F}_0 \subset \mathbf{E}^0$ and $f \in \mathbf{E}^{11}$, we have

$$\|f_n\|_{\mathbf{E}} \le c \cdot \|f_n\|_{\mathbf{E}^{11}} \le c \cdot \|f\|_{\mathbf{E}^{11}} < \infty$$

for all $n \ge 1$. Therefore, $\sup_n \|f_n\|_{\mathbf{E}} < \infty$ and property (B) implies $f \in \mathbf{E}$.

We have shown that $\mathbf{E}^{11} \subseteq \mathbf{E}$. So $\mathbf{E}^{11} = \mathbf{E}$, i.e. $\mathbf{E}$ is maximal. □

Combining Theorems 2.10 and 2.11 we have

Theorem 2.12. *Let* $\mathbf{E}$ *be a symmetric space. Then the following conditions are equivalent:*

(1) $\mathbf{E} = \mathbf{E}^{11}$ *and* $\|\cdot\|_{\mathbf{E}} = \|\cdot\|_{\mathbf{E}^{11}}$.

(2) $\mathbf{E} = \mathbf{G}^1$ *and* $\|\cdot\|_{\mathbf{E}} = \|\cdot\|_{\mathbf{G}^1}$ *for some symmetric space* $\mathbf{G}$.

(3) $\mathbf{E}$ *has both property (B) and (C).*

Remark 2.13. Fatou's property. *The equivalent conditions in Theorem 2.12 can be reformulated as follows:*

$$(F): \quad \begin{array}{l} \{f_n\} \subset \mathbf{E},\ f_n \to f \text{ almost everywhere and } \sup\limits_n \|f_n\|_{\mathbf{E}} < \infty \quad \Longrightarrow f \in \mathbf{E} \\ \text{and } \|f\|_{\mathbf{E}} \le \underline{\lim\limits_n} \|f_n\|_{\mathbf{E}}. \end{array}$$

In the case $\mathbf{E} = \mathbf{L}_1$ *property (F) is just the statement of Fatou's theorem. Therefore, (F) is known as Fatou's property. (See, [4, Th.1.b.18, Rem.2])*

2.6 Proof of Theorem 1.1 end Theorem 1.2

We begin with the proof of Theorem 1.2. Taking into account (2), it remains to prove

Proposition 2.14. *The equalities* $\|f\|_{\mathbf{E}^0} = \|f\|_{\mathbf{E}} = \|f\|_{\mathbf{E}^{11}}$ *hold for all* $f \in \mathbf{E}^0$.

Proof. The symmetric space $\mathbf{E}^0$ is minimal, and $(\mathbf{E}^0)^1 = \mathbf{E}^1$ by Proposition 2.5. Whence $(E^0)^{11} = E^{11}$.

Since $\mathbf{E}^0$ is minimal, it has Property (C) (Theorem 2.2). Therefore, the embedding $\mathbf{E}^0 \subseteq (\mathbf{E}^0)^{11} = \mathbf{E}^{11}$ is isometric by Theorem 2.10. □

Turning now to Theorem 1.1 we refer the reader to [3, Theorem II.4.1] for the proof of the embeddings (1) and the inequality

$$\|f\|_{\mathbf{E}} \ge \varphi_{\mathbf{E}}(1)\|f\|_{\mathbf{L}_1+\mathbf{L}_\infty}\ ,\ f \in \mathbf{E}\ , \tag{3}$$

It remains to prove the inequality

$$\varphi_{\mathbf{E}}(1)\|f\|_{\mathbf{L}_1\cap\mathbf{L}_\infty} \ge \|f\|_{\mathbf{E}}\ ,\ f \in \mathbf{L}_1 \cap \mathbf{L}_\infty\ , \tag{4}$$

instead a more weak inequality

$$2\varphi_{\mathbf{E}}(1)\|f\|_{\mathbf{L}_1\cap\mathbf{L}_\infty} \ge \|f\|_{\mathbf{E}}\ ,\ f \in \mathbf{L}_1 \cap \mathbf{L}_\infty\ ,$$

that was proved in [3, II.4., Th.4.1].

If $f \in \mathbf{L}_1 \cap \mathbf{L}_\infty \subseteq \mathbf{E}$, we have $\|f\|_{\mathbf{E}^0} = \|f\|_{\mathbf{E}} = \|f\|_{\mathbf{E}^{11}}$ by Proposition 2.14. Therefore,

$$\|f\|_{\mathbf{E}} = \|f\|_{\mathbf{E}^{11}} = \sup\left\{\left|\int fg\,dm\right| : \|g\|_{\mathbf{E}^1} \le 1\right\}\ .$$

Replacing $\mathbf{E}$ by $\mathbf{E}^1$ in (1) and (3), we have

$$\mathbf{E}^1 \subseteq \mathbf{L}_1 + \mathbf{L}_\infty \text{ and } \|g\|_{\mathbf{E}^1} \geq \varphi_{\mathbf{E}^1}(1)\|g\|_{\mathbf{L}_1+\mathbf{L}_\infty}, \; g \in \mathbf{E}^1 .$$

Hence, $\|g\|_{\mathbf{E}^1} \leq 1$ implies $\varphi_{\mathbf{E}^1}(1)\|g\|_{\mathbf{L}_1+\mathbf{L}_\infty} \leq 1$, and

$$\|f\|_{\mathbf{E}} \leq \sup \left\{ \left| \int fg \, dm \right| : \varphi_{\mathbf{E}^1}(1)\|g\|_{\mathbf{L}_1+\mathbf{L}_\infty} \leq 1 \right\} .$$

Putting $h = \varphi_{\mathbf{E}^1}(1)g$, we have

$$\|f\|_{\mathbf{E}} \leq \sup \left\{ \left| \int f \frac{h}{\varphi_{\mathbf{E}^1}(1)} \, dm \right| : \|h\|_{\mathbf{L}_1+\mathbf{L}_\infty} \leq 1 \right\} = .$$

$$= \frac{1}{\varphi_{\mathbf{E}^1}(1)} \sup \left\{ \left| \int fh \, dm \right| : \varphi_{\mathbf{E}^1}\|h\|_{\mathbf{L}_1+\mathbf{L}_\infty} \leq 1 \right\} =$$

$$= \frac{1}{\varphi_{\mathbf{E}^1}(1)} \|f\|_{\mathbf{L}_1 \cap \mathbf{L}_\infty} = \varphi_{\mathbf{E}}(1)\|f\|_{\mathbf{L}_1 \cap \mathbf{L}_\infty} .$$

The latter equality follows from $\varphi_{E^1} = \varphi_*(x) = \dfrac{x}{\varphi_{\mathbf{E}}(x)} 1_{(0,\infty)}$, $x \geq 0$ with $x = 1$ (Proposition 2.5).

3 Embeddings $\Lambda^0_{\widetilde{V}} \subseteq \mathbf{E}^0 \subseteq \mathbf{E} \subseteq \mathbf{E}^{11} \subseteq \mathbf{M}_{V_*}$

3.1 Lorentz spaces

The Lorentz space $(\Lambda_W, \|\cdot\|_{\Lambda_W})$ is defined by

$$\Lambda_W := \left\{ f \in \mathbf{L}_0 : \|f\|_{\Lambda_W} := \int_0^\infty f^*(x) \, dW(x) < \infty \right\} ,$$

where the weight $W : \mathbb{R}_+ \to \mathbb{R}_+$ is assumed to be increasing and concave on $(0, \infty)$ with $W(0) = 0$ and $W(x) > 0$ for $x > 0$. Here,

$$\int_0^\infty f^*(x) \, dW(x) = f^*(0)W(0+) + \int_0^\infty f^*(x)W'(x)dx$$

is the improper Riemann-Stieltjes integral of decreasing function f^* with respect to the increasing function W on $\mathbb{R}_+$.

Recall some basic properties of the spaces.

Proposition 3.1. *(1)* $(\Lambda_W, \|\cdot\|_{\Lambda_W})$ *is a symmetric space.*
(2) The fundamental function φ_{Λ_W} *coincides with* W.
(3) $(\Lambda_W, \|\cdot\|_{\Lambda_W})$ *is maximal* ($(\Lambda_W)^{11} = \Lambda_W$) *and* $\|\cdot\|_{(\Lambda_W)^{11}} = \|\cdot\|_{\Lambda_W}$.

Proof. See, [3, II.5.1]. □

Recall that the symmetric space $\mathbf{R}_0$ is defined as

$$\mathbf{R}_0 := \{f \in \mathbf{L}_0 : f^*(+\infty) = \lim_{x\to+\infty} f^*(x) = 0\} .$$

The space coincides with the closure $cl_{\mathbf{L}_1+\mathbf{L}_\infty}(\mathbf{L}_1 \cap \mathbf{L}_\infty)$ of $\mathbf{L}_1 \cap \mathbf{L}_\infty$ in $\mathbf{L}_1 + \mathbf{L}_\infty$, i.e. $\mathbf{R}_0 = (\mathbf{L}_1 + \mathbf{L}_\infty)^0$ is the minimal part of $\mathbf{L}_1 + \mathbf{L}_\infty$.

Proposition 3.2. *(1) The minimal part of Λ_W has the form $\Lambda_W^0 = \Lambda_W \cap \mathbf{R}_0$.*
(2) Λ_W is minimal $\Longleftrightarrow W(+\infty) = +\infty$.
(3) Λ_W is separable $\Longleftrightarrow W(+\infty) = +\infty$ and $W(0+) = 0$.

Proof. See, [3, II.5.1]. □

3.2 Four Types of Lorentz Spaces

Consider the four main types of Lorentz spaces.

Case (1). $W(0+) = 0$, $W(\infty) = \infty$.
Since $W(\infty) = \infty$, by Proposition 3.2, $\boldsymbol{\Lambda}_W$ is minimal with

$$\varphi_{\Lambda_W}(0+) = W(0+) = 0.$$

This implies the separability of $\boldsymbol{\Lambda}_W$ (Theorem 2.1). By Theorem 2.1 and Proposition 3.1, the assumption of Case (**1**) are equivalent to property (A) and $\upsilon(\boldsymbol{\Lambda}_W^1) = \boldsymbol{\Lambda}_W^*$. Thus, $\boldsymbol{\Lambda}_W \subseteq \mathbf{R}_0$, $\boldsymbol{\Lambda}_W \nsubseteq \mathbf{L}_\infty$ and $\boldsymbol{\Lambda}_W$ is separable.

Example. $\boldsymbol{\Lambda}_W = \mathbf{L}_1$ for $W(x) = x$, $x \ge 0$.

Case (2). $W(0+) > 0$, $W(\infty) = \infty$.
As in case (1), the condition $W(\infty) = \infty$ ensures the minimality of the Lorentz space $\boldsymbol{\Lambda}_W$. However, $\boldsymbol{\Lambda}_W$ is not separable, since the fundamental function $\varphi_{\boldsymbol{\Lambda}_W} = W$ satisfies $\varphi_{\boldsymbol{\Lambda}_W}(0+) = W(0+) > 0$ (Theorem 2.1).
The condition $W(0+) > 0$ implies:

$$\|f\|_{\boldsymbol{\Lambda}_W} \ge f^*(0)W(0+) = W(0+)\|f\|_{\mathbf{L}_\infty}$$

for all $f \in \boldsymbol{\Lambda}_W$. Hence, $\boldsymbol{\Lambda}_W \subseteq \mathbf{L}_\infty$.
On the other hand,

$$\|1_{(0,\infty)}\|_{\boldsymbol{\Lambda}_W} = \int_0^\infty dW = W(\infty) = \infty,$$

i.e $1_{(0,\infty)} \notin \boldsymbol{\Lambda}_W$ and therefore $\boldsymbol{\Lambda}_W \neq \mathbf{L}_\infty$. Whence,

$$\boldsymbol{\Lambda}_W \subseteq \mathbf{L}_\infty \cap \mathbf{R}_0 = \mathbf{L}^0_\infty .$$

Thus $\boldsymbol{\Lambda}_W \subseteq \mathbf{L}_\infty \cap \mathbf{R}_0$, $\boldsymbol{\Lambda}_W$ is minimal and is not separable.

Example. $\boldsymbol{\Lambda}_W = \mathbf{L}_1 \cap \mathbf{L}_\infty$ for $W(x) = (x+1) \cdot 1_{(0,\infty)}(x)$, $x \geq 0$.

Case (3). $W(0+) = 0$, $W(\infty) < \infty$.
The space $\boldsymbol{\Lambda}_W$ is not minimal, since $W(\infty) < \infty$ (Proposition 3.2). Moreover,

$$\|1_{(0,\infty)}\|_{\boldsymbol{\Lambda}_W} = W(0+) + \int_0^\infty dW = W(\infty) < \infty,$$

i.e. $1_{(0,\infty)} \in \boldsymbol{\Lambda}_W$, and hence $\mathbf{L}_\infty \subseteq \boldsymbol{\Lambda}_W$.
The condition $W(0+) = 0$ implies $\boldsymbol{\Lambda}_W \neq \mathbf{L}_\infty$.
Thus $\boldsymbol{\Lambda}_W \supset \mathbf{L}_\infty$ and $\boldsymbol{\Lambda}_W$ is not minimal.

Example. $\boldsymbol{\Lambda}_W = \mathbf{L}_1 + \mathbf{L}_\infty$ for $W(x) = \min\{x, 1\}$, $x \geq 0$. In this case,

$$\|f\|_{\boldsymbol{\Lambda}_W} = \int_0^\infty f^* \, dW = \int_0^\infty f^* W' \, dm = \int_0^1 f^* \, dm = \|f\|_{\mathbf{L}_1 + \mathbf{L}_\infty} .$$

Case (4). $W(0+) > 0$, $W(\infty) < \infty$.
Those conditions imply, respectively, $\boldsymbol{\Lambda}_W \subseteq \mathbf{L}_\infty$ and $\boldsymbol{\Lambda}_W \supseteq \mathbf{L}_\infty$, i.e. $\boldsymbol{\Lambda}_W = \mathbf{L}_\infty$. The inequalities

$$0 < W(0+) \leq W(x) \leq W(\infty) < \infty$$

imply

$$W(0+)\|f\|_{\mathbf{L}_\infty} \leq \|f\|_{\boldsymbol{\Lambda}_W} \leq W(\infty)\|f\|_{\mathbf{L}_\infty},$$

where $\|f\|_{\mathbf{L}_\infty} = f^*(0)$. Thus $\boldsymbol{\Lambda}_W = \mathbf{L}_\infty$.

Example. $\boldsymbol{\Lambda}_W = \mathbf{L}_\infty$ with $W(x) = 1_{(0,\infty)}(x)$, $x \geq 0$. Clearly, the space $\boldsymbol{\Lambda}_W = \mathbf{L}_\infty$ is not minimal, $\mathbf{L}^0_\infty = \mathbf{L}_0 \cap \mathbf{R}_0$.

Example 3.3. ***Spaces* $\mathbf{L}_p$, $1 < p < \infty$.** *The fundamental function of $\mathbf{L}_p$ space is $\varphi_{\mathbf{L}_p} = x^{\frac{1}{p}}$ for $x \geq 0$.*

We set $W_p = \varphi_{\mathbf{L}_p}$, $1 < p < \infty$, and consider the family of Lorentz spaces $\boldsymbol{\Lambda}_{W_p}$ where $1 < p < \infty$.

The functions W_p are concave, and $W_p(0+) = 0$, $W_p(\infty) = \infty$. Thus, the Lorentz spaces $\boldsymbol{\Lambda}_{W_p}$ as well as the spaces $\mathbf{L}_p$ are separable. It can be show that $\boldsymbol{\Lambda}_{W_p} \subset \mathbf{L}_p$, $1 < p < \infty$, and all the inclusions are strict.

3.3 Marcinkiewicz spaces

The Marcinkiewicz space $(\mathbf{M}_V, \|\cdot\|_{\mathbf{M}_V})$ can be defined as

$$\mathbf{M}_V := \{f \in \mathbf{L}_0 : V_* \cdot f^{**} \in \mathbf{L}_\infty\} ,$$

with

$$\|f\|_{\mathbf{M}_V} := \|V_* \cdot f^{**}\|_{\mathbf{L}_\infty} , \; f \in \mathbf{M}_V ,$$

where

$$V_*(0) = 0 \, , \; V_*(x) = \frac{x}{V(x)} 1_{(0,\infty)}(x) \text{ and } f^{**}(x) = \frac{1}{x}\int_0^x f^*(y)\,dy \, , \; x > 0$$

In this definition the weight V is not necessarily concave. It is assumed to be *quasi-concave* on $\mathbb{R}_+$, i.e. $V(0) = 0$ and both the functions $V(x)$ and $V_*(x)$ are increasing.

Proposition 3.4. *(1)* $(\mathbf{M}_V, \|\cdot\|_{\mathbf{M}_V})$ *is a symmetric space.*
(2) The fundamental function $\varphi_{\mathbf{M}_V}$ *coincides with* V_*.
(3) $\mathbf{M}_V$ *is maximal:* $(\mathbf{M}_V)^{11} = \mathbf{M}_V$

Proof. See, [3, II.5.2]. □

To describe the duality between Lorentz and Marcinkiewicz spaces, we recall that for every quasi-concave function on $\mathbb{R}_+$, there exists the least concave majorant $\widetilde{V}$ of V, and it satisfies the inequality

$$\frac{1}{2}\widetilde{V} \le V \le \widetilde{V}. \tag{1}$$

See, [3, II.1].

This yields $\mathbf{M}_{\widetilde{V}} = \mathbf{M}_V$, i.e. the Marcinkiewicz spaces $\mathbf{M}_{\widetilde{V}}$ and $\mathbf{M}_V$ coincides as sets, while the norms $\|\cdot\|_{\mathbf{M}_{\widetilde{V}}}$ and $\|\cdot\|_{\mathbf{M}_V}$ are equivalent,

$$\frac{1}{2}\|\cdot\|_{\mathbf{M}_{\widetilde{V}}} \le \|\cdot\|_{\mathbf{M}_V} \le \|\cdot\|_{\mathbf{M}_{\widetilde{V}}}.$$

Proposition 3.5. *(1) For every concave weight* W, *we have*

$$(\Lambda^1_W, \|\cdot\|_{\Lambda^1_W}) = (\mathbf{M}_W, \|\cdot\|_{\mathbf{M}_W}) \text{ and } (\mathbf{M}^1_W, \|\cdot\|_{\mathbf{M}^1_W}) = (\Lambda_W, \|\cdot\|_{\Lambda_W})$$

(2) For every quasi-concave weight V , *we have* $\mathbf{M}^1_V = \Lambda_{\widetilde{V}}$ *as sets and the norms* $\|\cdot\|_{\mathbf{M}^1_V}$ *and* $\|\cdot\|_{\Lambda_{\widetilde{V}}}$ *are equivalent.*

3.4 Embeddings $\Lambda^0_{\widetilde{V}} \subseteq \mathbf{E}$

Let V be a quasiconcave function on $\mathbb{R}^+$, $\widetilde{V}$ denotes the least concave majorant of V, and the quasi-concave function V_* is defined by $V_*(x) = \frac{x}{V(x)}$, $x > 0$ and $V_*(0) = 0$.

The Lorentz space $\boldsymbol{\Lambda}_{\widetilde{V}}$ as well as its minimal part $\boldsymbol{\Lambda}_{\widetilde{V}}^0 = cl_{\boldsymbol{\Lambda}_{\widetilde{V}}}(\mathbf{F}_1)$ have the fundamental function $\varphi_{\boldsymbol{\Lambda}_{\widetilde{V}}} = \widetilde{V}$, where $\widetilde{V}$ and V are equivalent.

First, we prove the embedding $\boldsymbol{\Lambda}_{\widetilde{V}}^0 \subseteq \mathbf{E}$ for every symmetric space $\mathbf{E}$ having the fundamental function $\varphi_{\mathbf{E}} = V$.

Proposition 3.6. *Let* $\mathbf{E}$ *be a symmetric space with the fundamental function* $\varphi_{\mathbf{E}} = V$. *Then* $\boldsymbol{\Lambda}_{\widetilde{V}}^0 \subseteq \mathbf{E}$ *and* $\|f\|_{\boldsymbol{\Lambda}_{\widetilde{V}}} \geq \|f\|_{\mathbf{E}}$ *for all* $f \in \boldsymbol{\Lambda}_{\widetilde{V}}^0$.

Proof. We prove that

$$\|f\|_{\boldsymbol{\Lambda}_{\widetilde{V}}} \geq \|f\|_{\mathbf{E}} \tag{2}$$

for all $f \in \mathbf{F}_1 \subseteq \boldsymbol{\Lambda}_{\widetilde{V}}^0$.

Suppose that $f = 1_A$ and $0 \leq mA < \infty$.

In this case, $f = 1_A \in \boldsymbol{\Lambda}_{\widetilde{V}}$, $f^* = 1_{[0,x]} \in \boldsymbol{\Lambda}_{\widetilde{V}}$ for $x = mA$, and

$$\|1_A\|_{\boldsymbol{\Lambda}_{\widetilde{V}}} = \|1_{[0,x]}\|_{\boldsymbol{\Lambda}_{\widetilde{V}}} = \widetilde{V}(x) \geq V(x) = \|1_{[0,x]}\|_{\mathbf{E}} = \|1_A\|_{\mathbf{E}}.$$

Let $f \in \mathbf{F}_1$, then $f^* \in \mathbf{F}_0 \subseteq \boldsymbol{\Lambda}_{\widetilde{V}}^0$ and f^* can be written in the form

$$f^* = \sum_{i=1}^{m} c_i \cdot 1_{[0,b_i]},\ c_i > 0,\ 0 \leq b_1 < b_2 < \cdots < b_m.$$

For such functions we have

$$\|f\|_{\boldsymbol{\Lambda}_{\widetilde{V}}} = \int_0^\infty f^*\, d\widetilde{V} = \sum_{i=1}^{m} c_i \widetilde{V}(b_i) \geq \sum_{i=1}^{m} c_i V(b_i) = \sum_{i=1}^{m} c_i \|1_{[0,b_i]}\|_{\mathbf{E}}$$

$$\geq \left\| \sum_{i=1}^{m} c_i \cdot 1_{[0,b_i]} \right\|_{\mathbf{E}} = \|f\|_{\mathbf{E}}.$$

Thus, the inequality (2) holds for all $f \in \mathbf{F}_1$.

By definition $\boldsymbol{\Lambda}_{\widetilde{V}}^0 = cl_{\boldsymbol{\Lambda}_{\widetilde{V}}}(\mathbf{F}_1)$. Therefore for any $f \in \boldsymbol{\Lambda}_{\widetilde{V}}^0$ one can choose a sequence $f_n \in \mathbf{F}_1$ such that $\|f_n - f\|_{\boldsymbol{\Lambda}_{\widetilde{V}}} \to 0$ as $n \to \infty$.

On the other hand, the inequality (2) provides that f_n is a Cauchy sequence in $\mathbf{X}$ and hence $\|f_n - f_0\|_{\mathbf{E}} \to 0$ for some $f_0 \in \mathbf{E}$. Since

$$\|f_n - f\|_{\boldsymbol{\Lambda}_{\widetilde{V}}} \to 0 \Longrightarrow \|f_n - f\|_{\mathbf{L}_1 + \mathbf{L}_\infty} \to 0$$

and

$$\|f_n - f_0\|_{\mathbf{E}} \to 0 \Longrightarrow \|f_n - f_0\|_{\mathbf{L}_1 + \mathbf{L}_\infty} \to 0$$

we have $f = f_0 \in \mathbf{E}$. Thus $\boldsymbol{\Lambda}_{\widetilde{V}}^0 \subseteq \mathbf{E}$ and (2) is valid for all $f \in \boldsymbol{\Lambda}_{\widetilde{V}}^0$. □

3.5 Embeddings $\mathbf{E} \subseteq \mathbf{M}_{V_*}$

Further we show that the Marcinkiewicz space $\mathbf{M}_{V_*}$ is the largest symmetric space $\mathbf{E}$ having the fundamental function $\varphi_E = V$

Proposition 3.7. *Let* $\mathbf{E}$ *be a symmetric space with the fundamental function* $\varphi_{\mathbf{E}} = V$. *Then* $\mathbf{E} \subseteq \mathbf{M}_{V_*}$ *and*

$$\|f\|_{\mathbf{M}_{V_*}} \leq \|f\|_{\mathbf{E}} \tag{3}$$

for all $f \in \mathbf{E}$.

Proof. For any symmetric space $\mathbf{E}$ we have

$$\|f^* \cdot 1_{[0,x]}\|_{\mathbf{L}_1} \leq \frac{x}{\varphi_{\mathbf{E}}(x)} \|f^* \cdot 1_{[0,x]}\|_{\mathbf{E}}, \quad x > 0$$

for all $f \in \mathbf{E}$ (see, [3, II.4.1]). This inequality with $\varphi_{\mathbf{E}} = V$ is rewritten as

$$\frac{\|f^* \cdot 1_{[0,x]}\|_{\mathbf{L}_1}}{\|f^* \cdot 1_{[0,x]}\|_{\mathbf{E}}} \leq \frac{x}{V(x)} = V_*(x), \; x > 0, \; f \in \mathbf{E}.$$

Thus,

$$\|f\|_{\mathbf{E}} = \|f^*\|_{\mathbf{E}} \geq \|f^* \cdot 1_{[0,x]}\|_{\mathbf{E}} \geq \frac{1}{V_*(x)} \|f^* \cdot 1_{[0,x]}\|_{\mathbf{L}_1} = \frac{1}{V_*(x)} \int_0^x f^* \, dm$$

and

$$\|f\|_{\mathbf{M}_{V_*}} = \sup_{0<x<\infty} \left\{ \frac{1}{V_*(x)} \int_0^x f^* \, dm \right\} \leq \|f\|_{\mathbf{E}}.$$

□

3.6 Theorem 1.3 and its consequences

It is worthwhile combining Propositions 3.6 and 3.7 with Theorem 1.2 in the following embedding theorem.

Theorem 3.8. *Let* $\mathbf{E}$ *be a symmetric space with the fundamental function* $\varphi_{\mathbf{E}} = V$. *Let* $\widetilde{V}$ *be the least concave majorant of the quasi-concave function* V *and* $V_*(x) = \dfrac{x}{V(x)}$, $x > 0$. *Then*

$$\Lambda^0_{\widetilde{V}} \subseteq \mathbf{E}^0 \subseteq \mathbf{E} \subseteq \mathbf{E}^{11} \subseteq \mathbf{M}_{V_*}, \tag{4}$$

where

$$\|f\|_{\Lambda_{\widetilde{V}}} = \|f\|_{\Lambda^0_{\widetilde{V}}} \geq \|f\|_{\mathbf{E}^0} = \|f\|_{\mathbf{E}}, \; f \in \Lambda^0_{\widetilde{V}}$$

and

$$\|f\|_{\mathbf{E}} \geq \|f\|_{\mathbf{X}^{11}} \geq \|f\|_{\mathbf{M}_{V_*}}, \; f \in \mathbf{E}.$$

Proof. From Proposition 3.6 we have $\boldsymbol{\Lambda}^0_{\widetilde{V}} \subseteq \mathbf{E}$, whence

$$\boldsymbol{\Lambda}^0_{\widetilde{V}} \subseteq cl_{\mathbf{X}}(\boldsymbol{\Lambda}^0_{\widetilde{V}}) = cl_{\mathbf{X}}(\mathbf{F}_1) = \mathbf{X}^0.$$

Using Proposition 3.7 we have $\mathbf{E} \subseteq \mathbf{M}_{V_*}$, whence by duality relations between Lorentz and Marcinkiewicz spaces (Section 3.3), we have also

$$\mathbf{E} \subseteq \mathbf{M}_{V_*} \Longrightarrow \mathbf{E}^1 \supseteq \mathbf{M}^1_{V_*} = \boldsymbol{\Lambda}_{\widetilde{V_*}} \Longrightarrow \mathbf{E}^{11} \subseteq \mathbf{M}^{11}_{V_*} = \mathbf{M}_{V_*}.$$

The embedding $\mathbf{E} \subseteq \mathbf{E}^{11}$ was proved in Section 2.3. All corresponding inequalities for norms also follow. □

It should be emphasized that all the spaces $\mathbf{E}^0, \mathbf{E}, \mathbf{E}^{11}$ and $\mathbf{M}_{V_*}$ have the same quasi-concave fundamental function V, while the spaces $\boldsymbol{\Lambda}^0_{\widetilde{V}}$ as well as $\boldsymbol{\Lambda}_{\widetilde{V}}$ have the fundamental function $\widetilde{V}$. The concave function $\widetilde{V}$ is equivalent but not necessarily equal to V. Namely, $\frac{1}{2}\widetilde{V} \leq V \leq \widetilde{V}$ (see, (1)).

Remark 3.9. 1. *The Lorentz space $\boldsymbol{\Lambda}_{\widetilde{V}}$ is minimal if and only if $\widetilde{V}(\infty) = \infty$ (Section 3.1) or, what is equivalent, that $V(\infty) = \infty$. Obviously, we have $\boldsymbol{\Lambda}^0_{\widetilde{V}} = \boldsymbol{\Lambda}_{\widetilde{V}} \subseteq \mathbf{E}$ in this case.*

Putting $\mathbf{E} = \mathbf{L}^0_\infty = \mathbf{L}_\infty \cap \mathbf{R}_0$ we obtain $V(x) = \varphi_{\mathbf{E}}(x) = 1_{(0,\infty)}$, $x \geq 0$ and $\boldsymbol{\Lambda}_V = \boldsymbol{\Lambda}_{\widetilde{V}} = \mathbf{L}_\infty$. Thus $\boldsymbol{\Lambda}_{\widetilde{V}}$ is not minimal, $\boldsymbol{\Lambda}^0_{\widetilde{V}} = \mathbf{L}^0_\infty = \mathbf{E}$. We see that $\boldsymbol{\Lambda}_{\widetilde{V}} \not\subseteq \mathbf{E}$ in this example.

2. *Consider the case, when the space $\mathbf{E}$ is maximal, i.e. $\mathbf{E} = \mathbf{E}^{11}$ as sets. Then*

$$\boldsymbol{\Lambda}^0_{\widetilde{V}} \subseteq \mathbf{X} \Longrightarrow (\boldsymbol{\Lambda}^0_{\widetilde{V}})^1 = \boldsymbol{\Lambda}^1_{\widetilde{V}} \supseteq \mathbf{X}^1 \Longrightarrow \boldsymbol{\Lambda}_{\widetilde{V}} = \boldsymbol{\Lambda}^{11}_{\widetilde{V}} \subseteq \mathbf{E}^{11} = \mathbf{E}.$$

Thus $\boldsymbol{\Lambda}_{\widetilde{V}} \subseteq \mathbf{E}$ in this case even if $\boldsymbol{\Lambda}_{\widetilde{V}}$ is not minimal, i.e. the inclusion $\boldsymbol{\Lambda}^0_{\widetilde{V}} \subset \boldsymbol{\Lambda}_V$ is strict.

3. *The equality $\mathbf{E} = \mathbf{X}^{11}$ does not imply in general the equality of the norms $\|\cdot\|_{\mathbf{E}}$ and $\|\cdot\|_{\mathbf{E}^{11}}$. For example, let $\mathbf{E} = \mathbf{L}_\infty$ is equipped with the norm*

$$\|f\|_{\mathbf{E}} = f^*(0) + f^*(\infty), \quad f \in \mathbf{E} = \mathbf{L}_\infty.$$

We have in this case

$$\|f\|_{\mathbf{E}} = \|f\|_{\mathbf{L}_\infty} = f^*(0), \quad f \in \mathbf{L}^0_\infty$$

since the seminorm $f \mapsto f^*(\infty)$ vanishes identically on $\mathbf{L}^0_\infty$. Thus we have

$$V(x) = \varphi_{\mathbf{E}}(x) = 1_{(0,\infty)}(x), \quad x \in \mathbb{R}^+$$

and $\boldsymbol{\Lambda}_V = \mathbf{L}_\infty = \mathbf{E}$, while $\boldsymbol{\Lambda}^0_V = \mathbf{L}^0_\infty \neq \boldsymbol{\Lambda}_V = \mathbf{L}_\infty$. However, be putting $f = 1_{[0,\infty)}$ we obtain

$$\|1_{[0,\infty)}\|_{\mathbf{E}} = 2 > \|1_{[0,\infty)}\|_{\mathbf{L}_\infty} = 1.$$

Corollary 3.10. *Let $\mathbf{E}$ be a maximal symmetric space with the fundamental function $\varphi_{\mathbf{E}}$. Then*

(1) $\varphi_{\mathbf{E}}(0+) > 0 \iff \mathbf{E} \subseteq \mathbf{L}_\infty$;

(2) $\varphi_{\mathbf{E}}(\infty) < \infty \iff \mathbf{E} \supseteq \mathbf{L}_\infty$;

(3) $\varphi_{\mathbf{E}}(0+) > 0$ *and* $\varphi_{\mathbf{E}}(\infty) < \infty \iff \mathbf{E} = \mathbf{L}_\infty$.

Proof. Let $V = \varphi_{\mathbf{E}}$. Since $\mathbf{E}$ is maximal we have $\boldsymbol{\Lambda}_{\widetilde{V}} \subseteq \mathbf{E}$.

(1). If $\mathbf{E} \subseteq \mathbf{L}_\infty$, then $\boldsymbol{\Lambda}_{\widetilde{V}} \subseteq \mathbf{L}_\infty$ and $V(0+) > 0$.

Conversely, the condition $V(0+) > 0$ implies that $\mathbf{M}_{V_*} \subseteq \mathbf{L}_\infty$. Indeed,

$$f \in \mathbf{M}_{V_*} \iff \|Vf^{**}\|_{\mathbf{L}_\infty} < \infty \implies V(0+)f^{**}(0+) < \infty$$

$$\implies f^{**}(0+) = f^*(0) < \infty \iff f \in \mathbf{L}_\infty.$$

Whence

$$\mathbf{E} \subseteq \mathbf{M}_{V_*} \subseteq \mathbf{L}_\infty.$$

(2). $V(\infty) < \infty$ implies $1_{[0,\infty)} \in \boldsymbol{\Lambda}_{\widetilde{V}}$, since

$$\|1_{[0,\infty)}\|_{\boldsymbol{\Lambda}_{\widetilde{V}}} = \int_0^\infty d\widetilde{V} = \widetilde{V}(\infty) = V(\infty) < \infty.$$

Whence

$$1_{[0,\infty)} \in \boldsymbol{\Lambda}_{\widetilde{V}} \subseteq \mathbf{E} \text{ and } \mathbf{E} \supseteq \mathbf{L}_\infty.$$

Conversely,

$$\mathbf{E} \supseteq \mathbf{L}_\infty \implies \mathbf{M}_{V_*} \supseteq \mathbf{L}_\infty \iff \mathbf{M}_{V_*} \ni 1_{[0,\infty)}.$$

This means that

$$\|1_{[0,\infty)}\|_{\mathbf{M}_{V_*}} = \|V \cdot 1^{**}_{[0,\infty)}\|_{\mathbf{L}_\infty} = \|V\|_{\mathbf{L}_\infty} = V(\infty) < \infty.$$

(3). Follows from (1) and (2).

□

In conclusion, we consider the classic spaces $\mathbf{E} = \mathbf{L}_p$.

Example 3.11. *Let* $1 < p < \infty$ *and* $V(x) = \varphi_{\mathbf{L}_p}(x) = x^{\frac{1}{p}}$, $x \geq 0$. *Then* $\widetilde{V} = V$ *and* $V_*(x) = \varphi_{\mathbf{L}_q}(x) = x^{\frac{1}{q}}$ *with* $\frac{1}{p} + \frac{1}{q} = 1$. *Thus we have embeddings* $\boldsymbol{\Lambda}_V \subseteq \mathbf{L}_p \subseteq \mathbf{M}_{V_*}$, *where*

$$\boldsymbol{\Lambda}_V = \left\{ f \in \mathbf{L}_0 : \|f\|_{\boldsymbol{\Lambda}_V} = \int_0^\infty f^*(x)\, d(x^{\frac{1}{p}}) < \infty \right\}$$

and

$$\mathbf{M}_{V_*} = \left\{ f \in \mathbf{L}_0 : \|f\|_{\mathbf{M}_{V_*}} = \sup_{0<x<\infty} x^{\frac{1}{p}} f^{**}(x) < \infty \right\}.$$

The spaces $\boldsymbol{\Lambda}_V$ *and* $\mathbf{L}_p$ *are minimal and separable in contrast to the space* $\mathbf{M}_{V_*}$, *which is not minimal.*

The minimal part $\mathbf{M}^0_{V_*} = cl_{\mathbf{M}_{V_*}}(\mathbf{L}_1 \cap \mathbf{L}_\infty)$ *of* $\mathbf{M}_{V_*}$ *has the form*

$$\mathbf{M}^0_{V_*} = \left\{ f \in \mathbf{M}_{V_*} : \lim_{x \to 0+} x^{\frac{1}{p}} f^{**}(x) = \lim_{x \to +\infty} x^{\frac{1}{p}} f^{**}(x) = 0 \right\}.$$

(See, [3, II.2.3]). The minimal space $\mathbf{M}^0_{V_*}$ *is separable as well as* Λ_V *and* $\mathbf{L}_p$.

It is easy to verify that all the inclusions

$$\Lambda_V \subset \mathbf{L}_p \subset \mathbf{M}^0_{V_*} \subset \mathbf{M}_{V_*}$$

are strict.

In the case $p = 1$*, we have:* $V(x) = x$, $V_*(x) = 1_{(0,\infty)}(x)$ *and*

$$(\Lambda_V, \|\cdot\|_{\Lambda_V}) = (\mathbf{L}_1, \|\cdot\|_{\mathbf{L}_1}) = (\mathbf{M}_{V_*}, \|\cdot\|_{\mathbf{M}_{V_*}}).$$

In the case $p = \infty$*, we have:* $V(x) = 1_{(0,\infty)}(x)$, $V_*(x) = x$*, and*

$$(\Lambda_V, \|\cdot\|_{\Lambda_V}) = (\mathbf{L}_\infty, \|\cdot\|_{\mathbf{L}_\infty}) = (\mathbf{M}_{V_*}, \|\cdot\|_{\mathbf{M}_{V_*}}).$$

Bibliography

[1] Bennet,C., Sharpley, R. *Intropolation of operators*. London, Academic Press, 1988, 469 p.

[2] Kantorovich, L., Akilov, G. *Functional Analysis in Normed Spaces*. Translated from the Russian by D.E. Brown. Edited by A.P.Robertson. International series of monographs in pure and applied mathematics, vol. 46, Pergamon Press, New York, USA, 1964 [In Russian: Functional Analysis, Nauka, Moskow, 1977].

[3] Krein, S.G., Petunin, Yu.I., Semenov, E.M. *Interpolation of liniar operators*. Trans. Math. Mon., 54, AMS, Perovidance, 1982 [In Russian: Interpolation of liniar operators, Nauka, Moskow, 1978].

[4] Lindenstrauss, J., Tzafriri, L. *Classical Banach Spaces II. Function Spaces*. Springer, 1979.

Igor V. Nikolaev

Mapping class groups are linear

Abstract: We construct a faithful representation of the mapping class group of genus $g \geq 2$ into the matrix group $GL_{6g-6}(\mathbf{Z})$.

Keywords: AF-algebras, mapping class groups

MSC: 14H55; 46L85

1 Introduction

The mapping class group $Mod\,(X)$ is a group of isotopy classes of the orientation-preserving diffeomorphisms of a two-sided closed surface X of genus $g \geq 1$; such a group was introduced by M. Dehn [Dehn 1938] [4]. The groups $Mod\,(X)$ are known to be prominent in algebraic geometry [Hain & Looijenga 1997] [7], topology [Thurston 1982] [13] and dynamics [Thurston 1988] [14]; we refer the reader to an excellent monograph by [Farb & Margalit 2011] [6] for a detailed account. When X is a torus, the $Mod\,(X)$ is isomorphic to the group $SL_2(\mathbf{Z})$. (The $SL_2(\mathbf{Z})$ is called a modular group, hence our notation for the mapping class groups.) A little is known about the representations of $Mod\,(X)$ beyond the case $g = 1$. Recall, that the group is called *linear*, if there exists a faithful representation into the matrix group $GL_m(R)$, where R is a commutative ring. The braid groups are known to be linear [Bigelow 2001] [2]. Using a modification of the argument for the braid groups, it is possible to prove, that $Mod\,(X)$ is linear in the case $g = 2$ [Bigelow & Budney 2001] [3]. Whether the mapping class group is linear for $g \geq 3$, is an open problem, known as the *Harvey conjecture* [Harvey 1979] [8, p. 267].

Denote by $T_S(g)$ the Teichmüller space of genus $g \geq 1$ with a distinguished point S. Let $q \in H^0(S, \Omega^{\otimes 2})$ be a holomorphic quadratic differential on the Riemann surface S, such that all zeroes of q (if any) are simple. By $\widetilde{S}$ we mean a double cover of S ramified over the zeroes of q and by $H_1^{odd}(\widetilde{S})$ the odd part of the integral homology of $\widetilde{S}$ relative to the zeroes. Note that $H_1^{odd}(\widetilde{S}) \cong \mathbf{Z}^n$, where $n = 6g-6$ if $g \geq 2$ and $n = 2$ if $g = 1$. The result of [Hubbard & Masur 1979] [9] implies, that $T_S(g) - \{S\} \cong Hom\,(H_1^{odd}(\widetilde{S}); \mathbf{R}) - \{0\}$, where 0 is the zero homomorphism. Finally, denote by $\lambda = (\lambda_1, \dots, \lambda_n)$ the image of a basis of $H_1^{odd}(\widetilde{S})$ in the real line $\mathbf{R}$, such that $\lambda_1 \neq 0$. Note that such an option always exists, since the zero homomorphism is excluded. We let $\theta = (\theta_1, \dots, \theta_{n-1})$, where $\theta_i = \lambda_{i-1}/\lambda_1$. Recall that, up to a scalar multiple, the vector $(1, \theta) \in \mathbf{R}^n$ is the limit of a generically convergent Jacobi-Perron continued fraction [Bernstein 1971] [1]:

$$\begin{pmatrix} 1 \\ \theta \end{pmatrix} = \lim_{k\to\infty} \begin{pmatrix} 0 & 1 \\ I & b_1 \end{pmatrix} \cdots \begin{pmatrix} 0 & 1 \\ I & b_k \end{pmatrix} \begin{pmatrix} 0 \\ \mathbf{I} \end{pmatrix},$$

Igor V. Nikolaev, Department of Mathematics and Computer Science, St. John's University, 8000 Utopia Parkway, New York, NY 11439, U.S.A. e-mail: igor.v.nikolaev@gmail.com
https://doi.org/10.1515/9783110413557-012

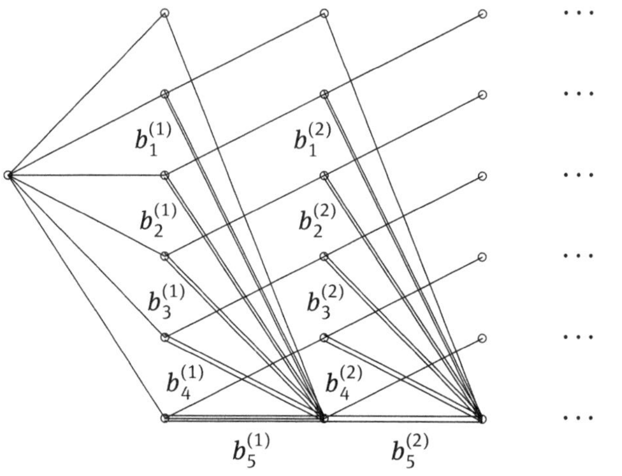

Fig. 1: Toric AF-algebra $\mathbf{A}_\theta$ of genus $g = 2$.

where $b_i = (b_1^{(i)}, \ldots, b_{n-1}^{(i)})^T$ is a vector of the non-negative integers, I the unit matrix and $\mathbf{I} = (0, \ldots, 0, 1)^T$. We introduce an AF-algebra, $\mathbf{A}_\theta$, via the Bratteli diagram [Effros 1981] [5], shown in Fig.1. (The numbers $b_j^{(i)}$ of the diagram indicate the multiplicity of edges of the graph.) Let us call $\mathbf{A}_\theta$ a *toric AF-algebra*.

Let $R, R' \in T_S(g)$ be a pair of isomorphic Riemann surfaces; let $\mathbf{A}_\theta$ and $\mathbf{A}_{\theta'}$ be the corresponding toric AF-algebras. Recall, that the stable isomorphism between the C^*-algebras is a fundamental equivalence in noncommutative geometry; the C^*-algebras $\mathbf{A}$ and $\mathbf{A}'$ are said to be *stably isomorphic*, whenever $\mathbf{A} \otimes \mathcal{K}$ is isomorphic to $\mathbf{A}' \otimes \mathcal{K}$, where $\mathcal{K}$ is the C^*-algebra of compact operators. Denote by V the maximal subset of $T_S(g)$, such that for each Riemann surface $R \in V$, there exists a convergent Jacobi-Perron continued fraction. Let F be the map which sends the Riemann surfaces into the toric AF-algebras according to the formula $R \mapsto \mathbf{A}_\theta$; we shall call F the *Teichmüller functor*. Let W be the image of V under F; the following lemma relates the algebras $\mathbf{A}_\theta$ and $\mathbf{A}_{\theta'}$.

Lemma 1. **([11])** *The set V is a generic subset of $T_S(g)$ and the map F has the following properties: (i) $V \cong W \times (0, \infty)$ is a trivial fiber bundle, whose projection map $p : V \to W$ coincides with F; (ii) $F : V \to W$ is a covariant functor, which maps isomorphic Riemann surfaces $R, R' \in V$ to stably isomorphic toric AF-algebras $\mathbf{A}_\theta, \mathbf{A}_{\theta'} \in W$.*

Recall that $Mod\,(X)$ acts on $T_S(g)$ by isomorphisms of the Riemann surfaces; the action is properly discontinuous and free for a finite index subgroup of $Mod\,(X)$ [Hain & Looijenga 1997] [7]. Lemma 1 extends the action to the toric AF-algebras, where $Mod\,(X)$

acts by the stable isomorphisms. This fact is remarkable: it is known, that the stable isomorphism group of a non-stationary toric AF-algebra admits a faithful representation into the matrix group $GL_n(\mathbf{Z})$ [Effros 1981] [5]. Roughly speaking, our observation implies the Harvey conjecture.

Theorem 1. *For every surface X of genus $g \geq 2$, there exists a faithful representation $\rho : Mod\,(X) \to GL_{6g-6}(\mathbf{Z})$.*

The structure of the article is as follows. We briefly review the AF-algebras in Section 2 and prove theorem 1 in Section 3. A numerical example illustrating theorem 1 is constructed in Section 4.

2 *AF*-algebras

We refer the reader to [Murphy 1990, Ch. 6] [10] for the basics of AF-algebras; the subject is briefly reviewed below. Recall that the C^*-algebra A is an algebra over the complex numbers $\mathbf{C}$ with a norm $a \mapsto ||a||$ and an involution $a \mapsto a^*, a \in A$, such that A is complete with the respect to the norm, and such that $||ab|| \leq ||a||\,||b||$ and $||a^*a|| = ||a||^2$ for every $a, b \in A$. If A is commutative, then the Gelfand theorem says that A is isometrically $*$-isomorphic to the C^*-algebra $C_0(X)$ of the continuous complex-valued functions on a locally compact Hausdorff space X. For otherwise, the algebra A represents a noncommutative topological space. The C^*-algebras A and A' are said to be stably isomorphic (Morita equivalent), if and only if, $A \otimes \mathcal{K} \cong A' \otimes \mathcal{K}$, where $\mathcal{K}$ is the C^*-algebra of all compact operators on a Hilbert space $\mathcal{H}$. Roughly speaking, such an isomorphism means that A and A' are homeomorphic as the noncommutative spaces.

An *AF-algebra* (approximately finite-dimensional C^*-algebra) is defined to be the norm closure of the union of an ascending sequence of the finite dimensional C^*-algebras M_n's, where M_n is the C^*-algebra of the $n \times n$ matrices with the entries in $\mathbf{C}$. Here the index $n = (n_1, \ldots, n_k)$ represents a semi-simple matrix algebra $M_n = M_{n_1} \oplus \cdots \oplus M_{n_k}$. The ascending sequence mentioned above can be written as

$$M_1 \xrightarrow{\varphi_1} M_2 \xrightarrow{\varphi_2} \ldots,$$

where M_i are the finite dimensional C^*-algebras and φ_i the homomorphisms between such algebras. The set-theoretic limit $A = \lim M_n$ has a natural algebraic structure given by the formula $a_m + b_k \to a + b$; here $a_m \to a, b_k \to b$ for the sequences $a_m \in M_m, b_k \in M_k$. The homomorphisms φ_i can be arranged into a graph as follows. Let $M_i = M_{i_1} \oplus \cdots \oplus M_{i_k}$ and $M_{i'} = M_{i'_1} \oplus \cdots \oplus M_{i'_k}$ be the semi-simple C^*-algebras and $\varphi_i : M_i \to M_{i'}$ the homomorphism. One has the two sets of vertices $V_{i_1}, \ldots, V_{i_k}$ and $V_{i'_1}, \ldots, V_{i'_k}$ joined by the a_{rs} edges, whenever the summand M_{i_r} contains a_{rs} copies of the summand $M_{i'_s}$ under the embedding φ_i. As i varies, one obtains an infinite graph called a *Bratteli diagram* of the AF-algebra. The Bratteli diagram defines a unique AF-algebra. If the homomorphisms $\varphi_1 = \varphi_2 = \cdots = Const$ in the definition of the AF-algebra A, the AF-algebra A is called *stationary*. The Bratteli diagram of a stationary AF-algebra looks like a periodic graph with the incidence matrix $A = (a_{rs})$ repeated over and over again. Since matrix A is a non-negative integer matrix, one can take a power of A to obtain a strictly positive integer matrix – which we always assume to be the case. The stationary AF-algebra has a non-trivial group of the automorphisms [Effros 1981, Ch.6] [5].

3 Proof of theorem 1

Proof. Let W be the set of toric AF-algebras of genus $g \geq 2$. Let G be a finitely generated group and $G \times W \to W$ be an action of G on W by the stable isomorphisms of toric AF-algebras; in other words, $\gamma(\mathbf{A}_\theta) \otimes \mathcal{K} \cong \mathbf{A}_\theta \otimes \mathcal{K}$ for all $\gamma \in G$ and all $\mathbf{A}_\theta \in W$. The following preparatory lemma will be important.

Lemma 2. *For each $\mathbf{A}_\theta \in W$, there exists a representation $\rho_{\mathbf{A}_\theta} : G \to GL_{6g-6}(\mathbf{Z})$.*

Proof. The proof of lemma is based on the following well known criterion of the stable isomorphism for the AF-algebras: a pair of such algebras $\mathbf{A}_\theta$, $\mathbf{A}_{\theta'}$ are stably isomorphic if and only if their Bratteli diagrams coincide, except (possibly) a finite part of the diagram, see [Effros 1981] [5, Theorem 2.3]. (Note, that the order isomorphism between the dimension groups mentioned in the original text, can be reformulated in the language of the Bratteli diagrams as stated.)

Let G be a finitely generated group on the generators $\{\gamma_1, \ldots, \gamma_m\}$ and $\mathbf{A}_\theta \in W$. Since G acts on the toric AF-algebra $\mathbf{A}_\theta$ by stable isomorphisms, the toric AF-algebras $\mathbf{A}_{\theta_1} := \gamma_1(\mathbf{A}_\theta), \ldots, \mathbf{A}_{\theta_m} := \gamma_m(\mathbf{A}_\theta)$ are stably isomorphic to $\mathbf{A}_\theta$; moreover, by transitivity, they are also pairwise stably isomorphic. Therefore, the Bratteli diagrams of $\mathbf{A}_{\theta_1}, \ldots, \mathbf{A}_{\theta_m}$ coincide everywhere except, possibly, some finite parts. We shall denote by $\mathbf{A}_{\theta_{\max}} \in W$ a toric AF-algebra, whose Bratteli diagram is the maximal common part of the Bratteli diagrams of $\mathbf{A}_{\theta_i}$ for $1 \leq i \leq m$; such a choice is unique and defined correctly because the set $\{\mathbf{A}_{\theta_i}\}$ is a finite set. By the definition of a toric AF-algebra, the vectors $\theta_i = (1, \theta_1^{(i)}, \ldots, \theta_{6g-7}^{(i)})$ are related to the vector $\theta_{\max} = (1, \theta_1^{(\max)}, \ldots, \theta_{6g-7}^{(\max)})$ by the formula:

$$\begin{pmatrix} 1 \\ \theta_1^{(i)} \\ \vdots \\ \theta_{6g-7}^{(i)} \end{pmatrix} = \underbrace{\begin{pmatrix} 0 & 0 & \ldots & 0 & 1 \\ 1 & 0 & \ldots & 0 & b_1^{(1)(i)} \\ \vdots & \vdots & & \vdots & \vdots \\ 0 & 0 & \ldots & 1 & b_{6g-7}^{(1)(i)} \end{pmatrix} \cdots \begin{pmatrix} 0 & 0 & \ldots & 0 & 1 \\ 1 & 0 & \ldots & 0 & b_1^{(k)(i)} \\ \vdots & \vdots & & \vdots & \vdots \\ 0 & 0 & \ldots & 1 & b_{6g-7}^{(k)(i)} \end{pmatrix}}_{A_i} \begin{pmatrix} 1 \\ \theta_1^{(\max)} \\ \vdots \\ \theta_{6g-7}^{(\max)} \end{pmatrix}$$

The above expression can be written in the matrix form $\theta_i = A_i\theta_{\max}$, where $A_i \in GL_{6g-6}(\mathbf{Z})$. Thus, one gets a matrix representation of the generator γ_i, given by the formula $\rho_{\mathbf{A}_\theta}(\gamma_i) := A_i$. The map $\rho_{\mathbf{A}_\theta} : G \to GL_{6g-6}(\mathbf{Z})$ extends to the rest of the group G via its values on the generators; namely, for every $g \in G$ one sets $\rho_{\mathbf{A}_\theta}(g) = A_1^{k_1} \dots A_m^{k_m}$, whenever $g = \gamma_1^{k_1} \dots \gamma_m^{k_m}$. Clearly, $\rho_{\mathbf{A}_\theta}(1) := I$, where I is the identity matrix. It is verified inductively, that the map $\rho_{\mathbf{A}_\theta} : G \to GL_{6g-6}(\mathbf{Z})$ is a homomorphism, since $\rho_{\mathbf{A}_\theta}(g_1g_2) = \rho_{\mathbf{A}_\theta}(g_1)\rho_{\mathbf{A}_\theta}(g_2)$ for $\forall g_1, g_2 \in G$. Lemma 2 follows.

□

Let $W_{aper} \subset W$ be a set consisting of the toric AF-algebras, whose Bratteli diagrams do not contain periodic (infinitely repeated) blocks; these are known as non-stationary toric AF-algebras and they are generic in the set W endowed with the natural topology. We call the action of G on the toric AF-algebra $\mathbf{A}_\theta \in W$ *free*, if $\gamma(\mathbf{A}_\theta) \neq \mathbf{A}_\theta$ for all $\gamma \in G$ except $\gamma = Id$.

Lemma 3. *Let $\mathbf{A}_\theta \in W_{aper}$ and G be free on the $\mathbf{A}_\theta$. Then $\rho_{\mathbf{A}_\theta}$ is a faithful representation.*

Proof. Since the action of G is free, to prove that $\rho_{\mathbf{A}_\theta}$ is faithful, it remains to show, that in the formula $\theta_i = A_i\theta_{\max}$, it holds $A_i = I$, if and only if, $\theta_i = \theta_{\max}$, where I is the unit matrix. Indeed, it is immediate that $A_i = I$ implies $\theta_i = \theta_{\max}$. Suppose now that $\theta_i = \theta_{\max}$ and, let to the contrary, $A_i \neq I$. One gets $\theta_i = A_i\theta_{\max} = \theta_{\max}$. Such an equation has a non-trivial solution, if and only if, the vector $\theta_{\max}$ has a periodic Jacobi-Perron fraction; the period of such a fraction is given by the matrix A_i. This is impossible, since it has been assumed, that $\mathbf{A}_{\theta_{\max}} \in W_{aper}$. The contradiction finishes the proof of lemma 3. □

Let $G = Mod\,(X)$, where X is a surface of genus $g \geq 2$. The group G is finitely generated [Dehn 1938] [4]; it acts on the Teichmüller space $T(g)$ by isomorphisms of the Riemann surfaces. Moreover, the action of G is free on a generic set, $U \subset T(g)$, consisting of the Riemann surfaces with the trivial group of the automorphisms. Let $F : V \to W$ be the Teichmüller functor between the Riemann surfaces and toric AF-algebras (lemma 1); the following is true.

Lemma 4. *The pre-image $F^{-1}(W_{aper})$ is a generic set in the space $T(g)$.*

Proof. Note, that the set of stationary toric AF-algebras is a countable set. The functor F is a surjective map, which is continuous with respect to the natural topology on the sets V and W. Therefore, the pre-image of the complement of a countable set is a generic set. □

To finish the proof, consider the set $U \cap F^{-1}(W_{aper})$; this set is a non-empty set, since it is the intersection of two generic subsets of $T(g)$. Let R be a point (a Riemann surface) in the above set. In view of lemma 1, group G acts on the toric AF-algebra $\mathbf{A}_\theta = F(R)$ by the stable isomorphisms. By the construction, the action is free and $\mathbf{A}_\theta \in W_{aper}$. In view of lemma 3, one gets a faithful representation $\rho = \rho_{\mathbf{A}_\theta}$ of the group $G = Mod\,(X)$ into the matrix group $GL_{6g-6}(\mathbf{Z})$. Theorem 1 follows.

□

4 Numerical example

We shall consider a numerical example illustrating theorem 1. Let $\gamma_1, \ldots, \gamma_{3g-1}$ be the Lickorish generators of $Mod\,(X)$. Following Penner ([Penner 1984] [12, p. 46]), consider a matrix representation $\rho : Mod\,(X) \to GL_{6g-6}(\mathbf{Z})$ given by the formulas:

$$\rho(\gamma_1) = \begin{pmatrix} 1 & & & & \pm 1 & & & \\ & 1 & & & & & & \\ & & \ddots & & & & & \\ & & & 1 & & & & \\ & & & & 1 & & & \\ & & & & & 1 & & \\ & & & & & & \ddots & \\ & & & & & & & 1 \end{pmatrix}$$

$$\rho(\gamma_2) = \begin{pmatrix} 1 & & & & & & & \\ & 1 & & & & \pm 1 & & \\ & & \ddots & & & & & \\ & & & 1 & & & & \\ & & & & 1 & & & \\ & & & & & 1 & & \\ & & & & & & \ddots & \\ & & & & & & & 1 \end{pmatrix}$$

$$\vdots$$

$$\rho(\gamma_{3g-3}) = \begin{pmatrix} 1 & & & & & & & \\ & 1 & & & & & & \\ & & \ddots & & & & & \\ & & & 1 & & & & \pm 1 \\ & & & & 1 & & & \\ & & & & & 1 & & \\ & & & & & & \ddots & \\ & & & & & & & 1 \end{pmatrix}$$

and

$$\rho(\gamma_{3g-2}) = E_1 \dots E_{k_1} \begin{pmatrix} 1 & & & & & & & & \\ & 1 & & & & & t_{ij} & & \\ & & \ddots & & & & & & \\ & & & 1 & & & & & \\ & & & & 1 & & & & \\ & & & & & 1 & & & \\ & & & & & & \ddots & & \\ & & & & & & & & 1 \end{pmatrix} E_{k_1}^{-1} \dots E_1^{-1}$$

$$\rho(\gamma_{3g-1}) = E_1 \dots E_{k_2} \begin{pmatrix} 1 & & & & & & & & \\ & 1 & & & & & t_{ij} & & \\ & & \ddots & & & & & & \\ & & & 1 & & & & & \\ & & & & 1 & & & & \\ & & & & & 1 & & & \\ & & & & & & \ddots & & \\ & & & & & & & & 1 \end{pmatrix} E_{k_2}^{-1} \dots E_1^{-1},$$

where $t_{ij} \in \mathbf{Z}$ and E_i are the matrix form of elementary transformations listed in [Penner 1984] [12, pp. 52-53]. Then $\rho(\gamma_i)$ define a faithful representation of $Mod\ (X)$ by the integral linear transformations of a cone in $\mathbf{R}^{6g-6}$, see [Penner 1984] [12, p. 45]; in view of theorem 1, the transformations extend to entire $\mathbf{R}^{6g-6}$ and define a faithful representation $\rho : Mod\ (X) \to GL_{6g-6}(\mathbf{Z})$.

Bibliography

[1] Bernstein, L. *The Jacobi-Perron Algorithm, its Theory and Applications*, Lect. Notes in Math. 207, Springer 1971.

[2] Bigelow, S. *Braid groups are linear*, J. Amer. Math. Soc. 14 (2001), 471–486.

[3] Bigelow, S. and Budney, R. *The mapping class group of a genus two surface is linear*, Algebr. Geom. Topol. 1 (2001), 699–708.

[4] Dehn, M. *Die Gruppe der Abbildungsklassen. (Das arithmetische Feld auf Flachen)*, Acta Math. 69 (1938), 135–206.

[5] Effros, E.G. *Dimensions and C^*-Algebras*, in: Conf. Board of the Math. Sciences, Regional conference series in Math., No.46, AMS, 1981.

[6] Farb, B. and Margalit, D. *A Primer on Mapping Class Groups*, Princeton Mathematical Series, Vol. 49, 2011.

[7] Hain, R. and Looijenga, E. *Mapping class groups and moduli spaces of curves*, Proc. Symp. Pure Math., Vol. 62.2 (1997), 97–142.

[8] Harvey, W.J. *Geometric structure of surface mapping class groups*, Homological Group Theory, LMS Lect. Notes. 36 (1979), 255–269.

[9] Hubbard, J. and Masur, H. *Quadratic differentials and foliations*, Acta Math. 142 (1979), 221–274.
[10] Murphy, G.J. *C^*-Algebras and Operator Theory*, Academic Press, 1990.
[11] Nikolaev, I. *Noncommutative geometry of algebraic curves*, Proc. Amer. Math. Soc. 137 (2009), 3283–3290.
[12] Penner, R.C. *The action of the mapping class group on curves in surfaces*, L'Enseignement Mathématique 30 (1984), 39–55.
[13] Thurston, W.P. *Three dimensional manifolds, Kleinian groups and hyperbolic geometry*, Bull. Amer. Math. Soc. 6 (1982), 357–381.
[14] Thurston, W.P. *On the geometry and dynamics of diffeomorphisms of surfaces*, Bull. Amer. Math. Soc. 19 (1988), 417–431.

Mohamed Oudadess

Subnormable *A*-convex algebras

Abstract: We consider particular A-convex algebras, named separately A-convex ones. They are not uniformly A-convex. However, they are subnormable. We give an example of a uniformly convex algebra which is not A-convex. We also consider a comparison with a class of algebras introduced by Chilana and Sharma.

Keywords: Algebra, A-convex, subnormable, separately A-convex, uniformly A-convex.

MSC: Primary 05C38, 15A15; Secondary 05A15, 15A18

1 Introduction

Locally A-convex algebras have been introduced by Cochran, Keown and Williams [2]. Uniformly A-convex ones have been noticed by Cochran, alone [3]. They have been studied in several papers ([4], [9] to [15]). Here, we introduce particular A-convex algebras which are not uniformly A-convex; the so called separately A-convex ones (s-A-$c.a.$). However, they are subnormable in the unitary case (Proposition 1). But the norm, the topology of which is stronger than the initial topology, is not necessarily an algebra norm (Example 2); thus, pointing out a difference with the uniformly A-convex case. On another hand, we show that an A-convex algebra is uniformly convex if, and only if, it is A-subnormable (Proposition 5).

A comparison with *the locally boundedly multiplicatively convex algebras,* of Chilana and Sharma, is provided (Section 4). It is shown, in particular, that a $l.b.m.c.a.$ is separately A-convex if, and only if, it admits an absorbent bounded disc (Proposition 8).

A generalization of uniformly A-convex algebras is the one of uniformly convex algebras ([9], [8]). We give an example of a uniformly convex algebra which is not A-convex (Example 11); and consider the normalization of the unit (Proposition 13).

2 Preliminaries

Let (E, τ) be a locally convex algebra ($l.c.a.$), with a separately continuous multiplication, whose topology τ is given by a family $(p_\lambda)_\lambda$ of seminorms. The algebra (E, τ) is said to be locally A-convex (l-A-$c.a.$; [2], [3]) if, for every x and every λ, there is

Mohamed Oudadess, c/o A. El Kinani, Ecole Normale Supérieure, B.P. 5118, Takaddoum,10105 Rabat. e-mail: oudadesshha@gmail.com

https://doi.org/10.1515/9783110413557-013

$M(x,\lambda) > 0$ such that

$$\max\left[p_\lambda(xy), p_\lambda(yx)\right] \le M(x,\lambda)p_\lambda(y); \forall y \in E.$$

In the case of a single space norm, $(E, \|.\|)$ is called an A-normed algebra. If $M(x,\lambda) = M(x)$ depends only on x, we say that (E,τ) is a locally uniformly A-convex algebra ($l.u$-A-$c.a.$; [3]. If it happens that, for every λ,

$$p_\lambda(xy) \le p_\lambda(x)p_\lambda(y); \forall x, y \in E,$$

then (E,τ) is named a locally m-convex algebra ($l.m.c.a.$; [8], see also [7]). Recall also that a $l.c.a.$ has a continuous multiplication if, for evry λ, there is λ'such that

$$p_\lambda(xy) \le p_{\lambda'}(x)p_{\lambda'}(y); \forall x, y \in E.$$

Definition 2.1. *We will say that a $l.c.a.$ (E,τ) is subnormable (resp. A-subnormable, resp. m-subnormable) if there is a vector space norm (resp. an A-norm, resp. an algebra norm) the topology of which is stronger than τ.*

A $l.c.a.$ is said to be locally uniformly convex ($l.u.c.a.$, [11] and [10]) if,

$$(\forall x)(\forall\lambda)(\exists M(x) > 0)(\exists\lambda') : p_\lambda(xy) \le M(x)p_{\lambda'}(y), \forall y$$

with $M(x)$ depending only on x (not on λ) and λ' depending only on λ (not on x).

3 Separately A-convex algebras

The idea is to replace $M(x,\lambda)$, of the general case, by $M(\lambda)M(x)$

Definition 3.1. *A locally convex algebra $(E,(p_\lambda)_\lambda)$ is said to be separately A-convex (s-A-convex; s-A-$c.a.$) if*

$$\forall x, \forall\lambda, \exists M(\lambda) > 0, \exists M(x) > 0 \colon \max\left[p_\lambda(xy), p_\lambda(yx)\right] \le M(\lambda)M(x)p_\lambda(y); \forall y \in E.$$

Remark 3.2. *Clearly, one can assume that $p_\lambda(e) = 1$, for every λ, where e is the unit of E.*

Remark 3.3. *An equivalent formulation of the definition above is the existence of a 0-neighborhoods base $\mathcal{V}$ such that*

$$\forall U \in \mathcal{V}, \forall x \in E, \exists M(U) > 0, \exists M(x) > 0 : xU \subset M(U)M(x)U, xU \subset M(U)M(x)U.$$

Now, here is a first interesting property.

Proposition 3.4. *A unital s-A-convex algebra is subnormable.*

Proof. Put $p'_\lambda(x) = \sup\{p_\lambda(xu) : p_\lambda(u) \le 1\}$. The topology τ' given by $(p'_\lambda)_\lambda$ is m-convex and stronger than that defined by $(p_\lambda)_\lambda$. Moreover, one has

$$p'_\lambda(xy) \le M(\lambda)M(x)p'_\lambda(y); \forall y \in E.$$

Then, taking $y = e$, one gets

$$\frac{1}{M(\lambda)}p'_\lambda(x) \le M(x); \forall y \in E.$$

Put

$$\|x\| = \sup\left\{\frac{1}{M(\lambda)}p'_\lambda(x) : \lambda\right\}.$$

The map $x \longrightarrow \|x\|$ is a vector space norm the topology of which is stronger than τ', hence than τ. □

Proposition 3.5. *The norm $\|.\|$ is the coarsest among vector space norms the topologies of which are stronger than τ.*

Proof. Let $(p_\lambda)_\lambda$ be a family of seminorms defining the topology τ of E, and a norm $\|.\|'$ the topology of which is stronger than τ. Without loss of generality, we may suppose that, for every λ, there is an α_λ such that

$$p'_\lambda(xy) \le \alpha_\lambda M(x)p'_\lambda(y); \forall y \in E$$

and

$$p'_\lambda(x) \le \alpha_\lambda \|x\|' ; \forall \in E.$$

Now

$$\|x\| = \sup_\lambda \frac{1}{\alpha_\lambda} p'_\lambda(x).$$

Then, one has

$$\|x\| = \sup \frac{1}{\alpha_\lambda} p'_\lambda(x) \le \|x\|'.$$

The map $x \longrightarrow \|x\|$ is a vector space norm. □

Remark 3.6. *The adjunction of a unit does not provide an algebra of the same type. Using the previous result and standard calculations (see [9], for example), one shows the following.*

Proposition 3.7. *Let (E, τ) be a s-A-c.a. and (E_1, τ_1) its unitization. Then (E_1, τ_1) is of the same type if, and only if, (E, τ) is subnormable.*

Proof. Barrelledness has here a particular consequence. □

Proposition 3.8. *Let (E, τ) be a unital s-A-c.a., which is also separated.*

(a) If (E, τ) is barrelled, then it is an A-normed algebra.
(b) If (E, τ) is barrelled and complete, then it is a Banach algebra.

Proof. The first assertion is due to the fact that $B_1 = \{x : \|x\| \leq 1\}$ is a τ-barrel. For the second, it is known that a complete A-normed algebra is Banach. □

Example 3.9. *Every l.u-A-c.a is a s-A-c.a.*

The norm $\|.\|$ is not necessarily an algebra norm as the following example shows.

Example 3.10. *Consider the algebra $C[X]$ of complex polynomials and $(z_m)_m$ a sequence of complex numbers such that $|z_m| \to +\infty$. Endow $C[X]$ with the topology τ given by the seminorms $P \longmapsto |P|_m = |P(z_m)|$. Then $(C[X], \tau)$ is a unital commutative and metrizable l.m.c.a.. Since it has a denumerable algebraic basis, it is subnormable by a lemma in [6], p. 1039. Denote by $\|.\|$ the vector space norm stronger than τ. Thus*

$$\forall m, \exists k_m > 1 : |P|_m \leq k_m \|P\| ; \forall P.$$

Then

$$|PQ|_m = |P|_m |Q|_m \leq \|P\| k_m |Q|_m ; \forall Q.$$

But the topology τ can also be defined by the family of seminorms $(|.|_m)_m \cup (\alpha |.|_m)_{m,\alpha \geqslant 1}$ which make of E a $l.u.c.a.$. It can not be a $l.u$-A-$c.a.$ for otherwise it could be endowed with a stronger algebra norm $\|.\|_0$ than τ. But then the characters $P \longmapsto P(z_m)$ should be continuous for $\|.\|_0$. Whence

$$|P(z_m)| \leq \|P\|_0 ; \forall m$$

which contradicts $|P(z_m)| \to +\infty$.

Here also, not every element is bounded as it is the case in $l.u$-A-$c.a.$'s . Indeed

$$\sup_m \left(\limsup_l |P^l|_m^{\frac{1}{l}} \right) = \sup_m |P(z_m)| = +\infty,$$

for every non constant P.

This example shows also that a unital $l.u.c.a.$ (which is always subnormable) is not necessarily m-subnormable. Actually it is not even A-subnormable for otherwise it should be m-subnormable.

Finally $(C[X], \tau)$ can not be a Q-algebra, otherwise there should exist $|.|_1, ..., |.|_n$ and $k > 0$ such that $\rho(P) \le k \sup\{|.|_i, 1 \le i \le n\}$, for every P. But then $C[X]$ should be finite dimensional ([9]); a contradiction.

Example 3.11. *The standard cartesian product of the algebras in the preceding examples is s-A-convex.*

Example 3.12. *Let E be a l.u-A-c.a. which is not a l.m.c.a. e.g.,* $(\mathcal{C}_b(\mathbb{R}), \beta)$ *in [2]. Then* $\mathcal{C}_b(\mathbb{R}) \times C[X]$ *is s-A-c.a which is not u-A-convex nor m-convex.*

Example 3.13. *Let X be a completely regular space. A Nachbin family is a set V of functions (over X) which are positive and upper semi-continuous (u.s.c.). It is said to be directed if*

$$\forall (v_1, v_2) \in V \times V, \exists v \in V, \exists \alpha > 0: v_1 \le \alpha v, v_2 \le \alpha v.$$

Put

$C(X) = \{f : X \longrightarrow \mathbb{C} / f \text{ is continuous}\}$,
$C_b(X) = \{f \in C(X) / f \text{ is bounded}\}$,
$CV_b(X) = \{f \in C(X) / vf \text{ is bounded}, \forall v, v \in V\}$,
$|f|_v = \sup\{v(x) f(x) : x \in X\}$.

And consider the topology τ_V *given by the family* $(|.|_v)_v$ *of seminorms. Now, put* $E = C_b(X) \cap CV_b(X)$*. One has*

$$|fg|_v = \sup\{v(x)|f(x)||g(x)| : x \in X\} \le \|f\| \, |g|_v .$$

So $(E, (|.|_v)_v)$ *is a l.u-A-c.a., hence a s-A-c.a. It is not always unital.*

Example 3.14. *Other examples can be obtained, using stability properties.*

Example 3.15. *Let* $C_b(\mathbb{R})$ *be the algebra of complex continuous functions on the real field* $\mathbb{R}$ *with the usual pointwise operations. Consider the weights given by* $w_n(x) = e^{n|x|}$, *with* $n \in \mathbb{N}$*. Put*

$$E = \{f \in C(\mathbb{R}) : \sup|f(x)| w_n(x)\} < +\infty, \forall n.$$

Let $|.|_n$ *be the seminorm defined by*

$$|f|_n = \sup\{|f(x)| w_n(x) : t \in \mathbb{R}\}.$$

Then $(E, (|.|_n)_n)$ *is a Fréchet l.u − A − c.a. (hence a s-A-c.a.) which is not unital. Its unitization* E_1 *can not be a s-A-c.a. Indeed there should be ((Proposition 1) a norm, on* E_1*, the unit ball of which is a* τ*-barrel. So* $(E, (|.|_n)_n)$ *should be normable; which is not the case.*

It is worthwhile to notice the following application.

Example 3.16. *Let X be a σ-compact locally compact space and $\mathcal{K}(X)$ be the algebra of complex continuous functions on the real field $\mathbb{X}$ with compact supports. Endowed with the measure topology τ_L. It becomes a l.u-A-c.a. If its unitization $\mathcal{K}_1(X)$ is a s-A-c.a., then X is compact. Indeed there should be ((Proposition 1) a norm $\|.\|$, the topology of which is stronger than τ_L. Hence the unit ball B_1 of $\|.\|$ is bounded. So, there must exist a compact subset K, of X, such that $B_1 \subset \mathcal{K}_K(X)$; which is true only if $X = K$.*

We have seen that a unital s-A-c.a is subnormable. The converse is true, with an additional condition.

Proposition 3.17. *If $(E, (p_\lambda)_\lambda)$ is a l-A-c.a. which is A-subnormable, then it is s-A-convex.*

Proof. Let $\|.\|$ be an A-norm the topology of which is stronger than the topology of E. Take the algebra norm $\|.\|_o$ associated to it; that is

$$\|x\|_o = \sup\{\|xu\|_o : \|u\|_o \leq 1\}.$$

It is also stronger than the m-convex topology τ', of Proposition 1, for the latter is the coarsest m-convex topology stronger than τ. Then

$$p_\lambda(xy) \leq p_{\lambda'}(x)p_\lambda(y) \leq k_\lambda \|x\| p_\lambda(y) \leq M(\lambda)M(x)p'_\lambda(y); \forall y \in E; y \in E.$$

□

In the m-convex case, one can say more.

Proposition 3.18. *Let $(E, (p_\lambda)_\lambda)$ be a unital l.m.c.a. It is a s-A-c.a. if, and only if, it is subnormable.*

Proof. Subnormability is necessary. For the converse, let $\|.\|$ be a norm stronger than τ. One then has

$$\forall\lambda, \exists k_\lambda > 0: p_\lambda(xy) \leq p_{\lambda'}(x)p_\lambda(y) \leq k_\lambda \|x\| p_\lambda(y) \forall y \in E; y \in E.$$

□

4 Comparison with locally boundedly *m*-convex algebras

Chilana and Sharma have introduced the algebras in the title.

Definition 4.1. *([1]) Let $(E, (p_\lambda)_\lambda)$ be a l-A-c.a. It is said to be a locally boundedly m-convex algebra (l.b.m.c.a.) if*

$$\forall\lambda, \forall B \in \mathbb{B}\tau,\ \exists k_{\lambda,B} > 0 : p_\lambda(xy) \leq k_{\lambda,B}\,|x|_B\, p_\lambda(y) \leq k_\lambda\,\|x\|\, p_\lambda(y), \forall y \in E; y \in E,$$

where $|x|_B = \inf\{\alpha > 0 : x \in \alpha B\}$.

Remark 4.2. *In the definition of Chilana and Sharma, $|x|_B$ may be infinite. Now, if there is a bounded set B_0 which is absorbent then $|x|_{B_0}$ is finite for every x. So that, one has*

$$\forall\lambda, \exists k_{\lambda,B_0} > 0 : p_\lambda(xy) \leq k_{\lambda,B_0}\,|x|_{B_0}\, p_\lambda(y).$$

Hence $(E, (p_\lambda)_\lambda)$ becomes a *s-A-c.a.* So, we do have the following characterization.

Proposition 4.3. *Let $(E, (p_\lambda)_\lambda)$ be a l.b.m.c.a. It is a s-A-c.a. if, and only if, it admits an absorbent bounded set.*

Example 4.4. *Let $\mathcal{C}(\mathbb{R})$ be the algebra of complex continuous functions on the real field $\mathbb{R}$ with the usual pointwise operations. Endow it with the topology defined by the seminorms $|.|_n$, $n \in \mathbb{N}^*$, given by*

$$|f|_n = \int_{-n}^{n} |f(t)|\, dt.$$

It is a *l-A-c.a.*, for

$$|fg|_n \leq [\sup\{|f(t)| : |t| \leq n\}]\, |g|_n\,.$$

Arguing as for $(\mathcal{C}_b(\mathbb{R}), \beta)$ in [2], one shows that it is not a *l.m.c.a.* It is not a *l.u-A-c.a.*, since it admits non bounded elements. Actually, it can not be a *s-A-c.a.* Indeed, if it were, then the unit ball of the norm in Proposition 1 would be an absorbent bounded subset. So *E* will be a *l.b.m.c.a* (Proposition 7). But then, it should be a *l.m.c.a.* ([1], Theorem 2.12, p. 143); which is not the case.

Chilana and Sharma have shown that the algebra $(\mathcal{C}_b(\mathbb{R}), \beta)$ is a *l.b.m.c.a.* One can give a more general statement.

Proposition 4.5. *A unital l.u-A-c.a. $(E, (p_\lambda)_\lambda)$ is a l.b.m.c.a.*

Proof. The algebra $(E, (p_\lambda)_\lambda)$ is *m*-subnormable. The claim follows then, by Proposition 7. □

In ([1], Corollary 2.15, p. 144), conditions are given to ensure that a *l-A-c.a* $(E, (p_\lambda)_\lambda)$ is a *l.b.m.c.a.* More precisely, multiplication has to be hypocontinuous and there must exist an idempotent bounded bornovorous subset. It has been shown that multiplication is hypocontinuous in any Mackey complete (*M*-complete) *l-A-c.a.*([15]). So, one has the following.

Proposition 4.6. *Let $(E,(p_\lambda)_\lambda)$ be a unital M-complete l-A-c.a. If it admits an idempotent bounded bornovorous subset, then it is a l.b.m.c.a.*

Avoiding completeness, one can give a condition ensuring the hypocontinuity of multiplication; using the very definition of a s-A-$c.a$. Indeed, one has

$$p'_\lambda(xy) \leq M(\lambda)M(x)p'_\lambda(y); \forall y \in E.$$

Put

$$\|\mathbf{x}\|' = \inf\left\{M(x) > 0 : p'_\lambda(xy) \leq M(\lambda)\, M(x)p'_\lambda(y); \forall y \in E.\right\}$$

One still has

$$p'_\lambda(xy) \leq M(\lambda)\, m(x)p'_\lambda(y); \forall y \in E.$$

The map$\|\mathbf{x}\|' : E \longrightarrow \mathbb{R}^*_+$ is a vector space norm.

Proposition 4.7. *Let $(E,(p_\lambda)_\lambda)$ be a unital l-A-c.a. If the norm $\|\mathbf{x}\|'$ is bounded on bounded nets tending to zero, then*

(a) Multiplication is hypocontinuous.
(b) The bounded completion of $(E,(p_\lambda)_\lambda)$ is an algebra.

Remark 4.8. *Some examples in [1] can now be viewed in a less technical manner, using the previous propositions.*

Remark 4.9. *Given a unital l-A-c.a (E,τ) and $M(\tau)$ the associated m-convex topology, it has been shown that $\mathbb{B}\tau = \mathbb{B}M(\tau)$, if (E,τ) is M-complete. One can give another condition, using barrelledness. Recall (cf. [1]) that a l.c.a. is said to be almost barrelled (resp. almost m- barrelled) if every barrel (resp. m- barrel) is bornovorous.*

Proposition 4.10. *Let $(E,(p_\lambda)_\lambda)$ be a unital l-A-c.a.*
(a) *If it is almost m- barrelled, then $\mathbb{B}\tau = \mathbb{B}M(\tau)$.*
(b) *If it is a s-A-c.a. and almost m- barrelled, then $\mathbb{B}\tau = \mathbb{B}M(\tau) = \mathbb{B}\tau_{\|.\|}$.*

Proof. The first claim is true, for $B_\lambda = \{x : q_\lambda(x) \leq 1\}$ is a m-τ-barrel . The second, because $B_{\|.\|} = \{x : \|x\| \leq 1\}$ m-τ-barrel. □

Example 4.11. *The algebra $(\mathcal{C}_b(\mathbb{R}), \beta)$ is a unital complete l.u-A-c.a. which is not m-convex. It is not barrelled, for it should be a l.m.c.a.[8]. It is almost barrelled, for in a complete space every barrel is bornivorous.*

5 Locally uniformly convex algebras

A unital $l.u.c.a.$ is subnormable ([10] and [9]). But it is not necessarily separately A-convex (not even A-convex), as Example 1 shows. Now, examine the normalization of the unit.

Proposition 5.1. *Let $(E, (p_\lambda)_\lambda)$ be a $l.u.c.a.$ with a unit e. The following are equivalent.*

(a) $p_\lambda(e) = 1$, for every λ.
(b) $p'_\lambda(e)\,[p_\lambda(e)]^{-1} \le m$, for every λ.

Proof. The first assertion obviously implies the second. Conversely, one has

$$\begin{aligned}[p_\lambda(e)]^{-1}\, p_\lambda(xy) &\le M(x)\,[p_\lambda(e)]^{-1}\, p'_\lambda(y)\\ &\le M(x)p'_\lambda(e)\,[p_\lambda(e)]^{-1}\, p'_\lambda(e)p'_\lambda(y)\\ &\le mM(x)\left[p'_\lambda(e)\right]^{-1} p'_\lambda(e).\end{aligned}$$

One can then replace p_λ by $[p_\lambda(e)]^{-1}\, p_\lambda$. □

The boundedness of $\left(p'_\lambda(e)\right)_\lambda$ leads to an interesting property, concerning m-subnormability.

Proposition 5.2. *Let $(E, (p_\lambda)_\lambda)$ be a $l.u.c.a.$ with a unit e. If there is an $\alpha > 0$ such that $\{p_\lambda(e) : \lambda\} \le \alpha$, then $(E, (p_\lambda)_\lambda)$ is m-subnormable.*

Proof. One has, by definition

$$p'_\lambda(xy) \le M(x)p'_\lambda(y), \text{ for every } \lambda.$$

So,

$$p_\lambda(x) \le M(x)p'_\lambda(y) \le \alpha M(x), \text{ for every } \lambda.$$

Put $\|x\| = \sup\{p_\lambda(x) : \lambda\}$. The map $x \longrightarrow \|x\|$ is an A-norm. Indeed,

$$\|xy\| \le \sup\left\{p'_\lambda(xy) : \lambda\right\} \le M(x)\,\|y\|\,, \text{ for every } y.$$

Take the algebra norm $\|.\|_o$ associated to it, that is $\sup\{\|xu\| : \|u\| \le 1\}$. □

Remark 5.3. *We have seen that the algebra $(C\,[X]\,, \tau)$ of Example 2 can not be m-subnormable. One notices that $(|1|_m)_m$ is not bounded. Indeed*

$$|1|_m = |1(z_m)| = |z_m|.$$

Remark 5.4. *Any s-A-convex algebra is a l.u.c.a. Indeed,*

$$\begin{aligned} [p_\lambda(xy), p_\lambda(yx)] &\leq M(\lambda)M(x)p_\lambda(y); \forall y \in E \\ &\leq M(x)p'_\lambda(y); \forall y \in E, \end{aligned}$$

where $p'_\lambda = M(\lambda)p_\lambda$.

We now give an example of a *l.u.c.a.* which is not *A*-convex.

Example 5.5. *Let* $(E, \|.\|)$ *be a commutative unital* C^**-algebra. Endow it with the weak topology* σ *given by co* $(M(E))$*, the convex hull of the carrier space. It is Hausdorff. If*

$$f = \sum \alpha_i \chi_i, \quad \text{with} \quad 0 \leq \alpha_i \leq 1 \quad \text{and} \quad \sum \alpha_i = 1,$$

then

$$|xy|_f = |f(xy)| \leq \|x\| \sum \alpha_i |\chi_i(y)| \leq \|x\| \sum |\chi_i(y)|, \text{for every } y.$$

But,

$$y \longrightarrow \sum |\chi_i(y)|$$

is a continuous seminorm. So, (E, σ) *is a l.u.c.a. It can not be A-convex. Indeed, the vector space spanned by co* $(M(E))$ *is the whole topological dual of* $(E, \|.\|)$*; for this claim cf. ([5], p. 21) and [16]. Thus, it is total. One can then apply a result of S. Warner [17].*

Acknowledgment: This paper is in final form and no version of it will be submitted for publication elsewhere.

Bibliography

[1] Chilana, A. K., Sharma, S. *The locally boundedly Multiplicatively Convex Algebras*, Math. Nachr. 77 (1977), 139–161.

[2] Cochran, A. C., Keown, R., Williams, C. R. *On a class of topological algebras*, Pacific J. Math. 34 (1970), 17–25.

[3] Cochran, A. C. *Representation of A*-convex algebras, Proc. Amer. Math. Soc. 41 (1973), 473–479.

[4] Cochran, A. C. *Weak A-convex algebras*, Proc. Amer. Math. Soc. 26 (1970), 73–77.

[5] Dixmier, J. *Les* C^**-algebres et leurs representations*, Guauthier-Villars Paris (1969).

[6] Esterle, J. *Sur la non normabilité de certaines algèbres d'opérateurs*, C. R. Acad. Sc. Paris, t 278, Série A (1974), 1037–1040.

[7] Mallios, A. *Topological algebras. Selected topics*, North-Holland, Amsterdam, 1986.

[8] Michael, E. A. *Locally multiplicatively convex topological algebras,* Mem. Amer. Math. Soc. 11, (1952).

[9] Oudadess, M. *Remarks on locally A*-convex algebras, Bull. Greek Math. Soc. 56 (2009), 47–55

[10] Oudadess, M. *Locally Uniformly Convex algebras,* Bull. Greek Math. Soc. (2009).

[11] Oudadess, M. *Bounded structures in locally A*-convex algebras, Math. Studies 4, Estonian Math. Soc., Tartu (2008).

[12] Oudadess, M. *Théorèmes de structures et propriétés fondamentales des algèbres localement uniformément A*-convexes, C. R. Acad. Sc. Paris, t 296, Série J (1983), 851–853.

[13] Oudadess, M. *Une norme d'algèbre de Banach dans les algèbres uniformément A*-convexes, Africa Math., Vol. IX (1987), 15–22.

[14] Oudadess, M. *Théorème du type Gelfand-Naimark dans les algèbres uniformément A-convexes*, Ann. Sc. Math. Québec, Vol 9(1), (1985), 73–82.

[15] Oudadess, M. *Discontinuity of the product in multiplier algebras*, Publications Mathématiques 34 (1990), 397–401.

[16] Rudin, W. *Functional Analysis,* Mc Graw-Hill, 1973.

[17] Warner, S. *Weakly topologized algebras*, Proc. Amer. Math. Soc. 8 (1957), 314–16.

[18] Warner, S. *Weak locally multiplicatively convex algebras*, Pacific J. Math. 5 (1955), 1025–1032.

Mohamed Oudadess

Commutative BP^*-algebras and Gelfand-Naimark's theorem

Abstract: We reexamine a Gelfand-Naimark theorem type, of Husain and Warsi, on commutative BP^*-algebras. There is a gap in the proof. We fix hypotheses and give an alternative one.

Keywords: BP^*-algebra, Gelfand-Naimark, algebras of continuous functions.

MSC: 46H05, 46H20

1 Introduction

In 1976, Husain and Warsi have introduced BP^*-algebras ([8], see also [9]; and studied their representations. No concrete examples are given. In 1977, they considered positive functionals; and gave examples of BP^*-algebras [9]. In the first paper, it is given a Gelfand-Naimark's theorem type. The proof relies, in particular, on the non voidness of the carrier space. But that claim is not valid [11]. The aim of this manuscript is to provide an alternative proof. The classical examples are examined (Section 3). All of them can be seen as algebras of continuous functions on a compact space. It turns out that, with appropriate hypotheses, this is the case in a general context. It is shwon that a unital, commutative, hermitian and semi-simple BP^*-algebra is an algebra of continuous functions (Proposition 4.2).

2 Preliminaries

Let (E, τ) be a locally convex algebra ($l.c.a.$), with a separately continuous multiplication, whose topology τ is given by a family $(p_\lambda)_{\lambda\in\Lambda}$ of seminorms. The algebra (E, τ) is said to be locally A-convex (l-A-$c.a.$; [2], [3]) if, for every x and every λ, there is $M(x, \lambda) > 0$ such that

$$\max\,[p_\lambda(xy), p_\lambda(yx)] \le M(x, \lambda)p_\lambda(y); \forall y \in E.$$

In the case of a single space norm, $(E, \|.\|)$ is called an A-normed algebra. If $M(x, \lambda) = M(x)$ depends only on x, we say that (E, τ) is a locally uniformly A-convex algebra ($l.u$-A-$c.a.$; [3]). If it happens that, for every λ,

$$p_\lambda(xy) \le p_\lambda(x)p_\lambda(y); \forall x, y \in E,$$

Mohamed Oudadess, Ecole Normale Supérieure, B.P. 5118, Takaddoum, 10105 Rabat, Morocco;
Email: oudadessm@yahoo.fr, oudadesshha@gmail.com
https://doi.org/10.1515/9783110413557-014

then (E, τ) is named a locally m -convex algebra ($l.m.c.a.$; cf. [10] or [5]). Recall also that a $l.c.a.$ has a continuous multiplication if, for every λ, there is λ' such that

$$p_\lambda(xy) \le p_{\lambda'}(x)p_{\lambda'}(y); \forall x, y \in E.$$

If $(E, (p_\lambda)_\lambda)$ is a unital $l.$-A-$c.a.$, then it can be endowed with a stronger m -convex topology $M(\tau)$, where τ is the topology on E [12]. It is determined by the family $(q_\lambda)_{\lambda\in\Lambda}$ of seminorms given by

$$q_\lambda(x) = \sup\{p_\lambda(xu) : p_\lambda(u) \le 1\}.$$

If $(E, (p_\lambda)_\lambda)$ is a $l.u$-A-$c.a.$, then there is yet [13] an algebra norm $\|.\|_0$ which induces a topology $\tau_{\|.\|_0}$ stronger than $M(\tau)$. It is given by

$$\|x\|_0 = \sup\{q_\lambda(x) : \lambda\}.$$

The bounded structure (bornology) of a locally convex space ($l.c.s.$) (E, τ) is the collection $\mathbb{B}\tau$ of all the subsets B of E which are bounded in the sense of von Neumann-Kolmogorov, that is B is absorbed by every neighborhood of the origin. If $\tau_{\|.\|}$ is the topology induced by a norm $\|.\|$, we write $\mathbb{B}\tau_{\|.\|}$.

If the topology of a $l.c.s.$ (E, τ) is given by a family $(p_\lambda)_{\lambda\in\Lambda}$ of seminorms, with Λ a directed set, we will often, for simplicity, write only $(p_\lambda)_\lambda$, especially when there is no risk of confusion.

An element x of E is said to be bounded [1] (i-bounded in the sense of S. Warner ([15]) if there is $\alpha > 0$ such that $\{(\alpha x)^n : n = 1, 2, \dots\}$ is bounded. A subset B of E is said to be m -bounded [1] or regular (i-bounded in the sense of S. Warner ([15]) if it is absorbed by an idempotent bounded subset of E. A $l.c.a.$ (E, τ) is said to be pseudo-Banach if it is pseudo-complete and every $B \in B\tau$ is regular ([1], [6]). It is then a bornological inductive limit of Banach algebras, as it has been explicitly stated in [6].

3 Definition and examples

Let (E, τ) be a commutative locally convex algebra ($l.c.a.$), endowed with an involution (*-$l.c.a.$) $x \longrightarrow x^*$; the product is only separtely continuous. Designate by $\mathbb{B}\tau$ the von Neumann-Kolmogorof bounded structure of (E, τ). Put

$$\mathcal{B} = \left\{B \in \mathbb{B}\tau / B \text{ is a closed bounded disc: } B = B^* \text{ and } B^2 \subset B\right\}.$$

Remark 3.1. *Husain and Warsi considered a sub-collection $\mathcal{B}_0$, of $\mathcal{B}$, such that every element B in $\mathcal{B}$ is contained in an element of $\mathcal{B}_0$; that is a bornological base of $\mathcal{B}$.*

Definition 3.2. *([7], [8]). A *-l.c.a. (E,τ) is said to be a BP^*-algebra if every element is bounded and, for every B in $\mathcal{B}$, the subalgebra $E(B)$ generated by B, endowed with the gauge of B, is an involutive Banach algebra.*

Example 3.3. *Let Ω be the first non countable ordinal and endow the set $[0,\Omega[$ with the order topology. Consider $C([0,\Omega[)$ the complex algebra of continuous functions, on $[0,\Omega[$, endowed with the topology of uniform convergence on compacta and the involution $f\mapsto\overline{f}$. It is known that every f in $C([0,\Omega[)$ is bounded. Thus $C([0,\Omega[)$ is a BP^*-algebra.*

Example 3.4. *Let $C_b(\mathbb{R})$ be the algebra of complex continuous bounded functions on the real field $\mathbb{R}$ with the usual pointwise operations and the complex conjugation as an involution. Denote by $C_0^+(\mathbb{R})$ the srictly positive elements of $C_b(\mathbb{R})$. Consider the family $\{p_\varphi : \varphi\in C_0^+(\mathbb{R})\}$ of seminorms given by*

$$p_\varphi(f)=\sup\{f(x)\varphi(x):x\in\mathbb{R}\};f\in C_b(\mathbb{R}).$$

*They determine a locally convex topology β. The space $(C_b(\mathbb{R}),\beta)$ is a complete locally convex *-algebra. It is not a l.m.c.a. ([3]), nor a Q-algebra. But it is a l.u-A-c.a. It is a BP^*-algebra.*

Example 3.5. *Let $\mathcal{C}[0,1]$ be the algebra of complex continuous functions on the interval $[0,1]$. Endow it with the semi norm given by*

$$p(f)=\sup\{|f(x)|\,\varphi(x):x\in[0,1]\},$$

where

$$\begin{aligned}\varphi(x)&=1-x,\ \frac{1}{2}\le x\le 1\\ \varphi(x)&=x,\ 0\le x\le\frac{1}{2}.\end{aligned}$$

Example 3.6. *Let $L^\infty[0,1]$ be the algebra of complex functions on the interval $[0,1]$, which are measurable and essentially bounded. Endow it with the topology given by the family $\left(\|.\|_p\right)_p$, $p\ge 1$, of seminorms, with*

$$\|f\|_p=\left(\int_0^1|f(t)|^p\right)^{\frac{1}{p}}.$$

Example 3.7. *We now give an example of a BP^*-algebra which is not A-convex. Let $(E,\|.\|)$ be an infinite dimensional commutative semisimple hermitian Banach algebra e.g., a C^*-algebra. Endow it with the weak topology $\sigma(E,E'$. Then $(E,\sigma(E,E'))$ is a locally convex *-algebra. By ([17], Theorem 2), it is not with continuous product. So it can not be A -convex [16]. Now $\mathbb{B}\sigma=\mathbb{B}\tau_{\|.\|}$. Hence $(E,\sigma(E,E'))$ is a BP^* -algebra.*

It is whorthwhile to determine a large class of locally convex algebras which are BP^*-algebras. Here is a chracterization which shows that there are plenty of them.

Proposition 3.8. *Let (E, τ) be a unital commutative and pseudo-complete l.c.a., with a continuous involution. It is a BP^*-algebra if, and only if, every element is bounded.*

Proof. Necessity is by definition. For the converse, every element is aborbed by an idempotent bounded disc B_1; hence by the self-adjoint bounded disc $B_2 = \Gamma(B_1 \cup B_1^*)$, the absolutely convex hull of $B_1 \cup B_1^*$. Now, E being commutative, the idempotent hull of B_2 is nothing else than $B_2 \cup B_2^* \cup (B_2 B_2^*)$: hence it is bounded. So, one can consider

$$\mathcal{B} = \left\{B \in \mathbb{B}\tau / B \text{ is a closed bounded disc: } B = B^* \text{ and } B^2 \subset B\right\}.$$

□

Remark 3.9. *Actually, we do not need pseudo-completeness. The algebra has, in some sense, to be only *-pseudo-complete as it is asked for by Husain and Warsi; by the very definition of a BP^*-algebra.*

4 Examination of the examples

We will notice that, as claimed in [8], the algebras involved are of continuous functions on a compact space.

4.1 The algebra $C([0, \Omega[)$

The space $[0, \Omega[$ is pseudo-compact. Hence

$$C([0, \Omega[) = C_b([0, \Omega[) = \{f : [0, \Omega[\longrightarrow \mathbb{C} / f \text{ is continuous and bounded}\}.$$

But $[0, \Omega[$ is also completely regular. So $C_b([0, \Omega[)$ is in bijection with $C(\beta[0, \Omega[)$, where $\beta[0, \Omega[$ is the Stone-Cech compactification of $[0, \Omega[$. Actually, one can say more. The algebra $C_b([0, \Omega[)$ endowed with the supremum norm and the involution $f \longrightarrow \overline{f}$ is a C^*-algebra.

4.2 The algebra $C_b(\mathbb{R})$

The functions are bouned by hypothesis, and $\mathbb{R}$ is completely regular. The situation is as in 3.1.

4.3 The algebra $L^\infty [0, 1]$

This algebra, endowed with the essential supremum norm $\|.\|_{s.e}$ and the involution $f \longrightarrow \overline{f}$ is a C^*-algebra. So the Gelfand transform $f \longrightarrow \widehat{f}$ is an isometry of $L^\infty [0, 1]$ on the algebra $\mathcal{C}(\Delta)$ of complex continuous functions on the carrier space Δ of $L^\infty [0, 1]$. W. Rudin gives a direct and detailed treatement ([14], p. 273).

4.4 The algebra $(E, \sigma(E, E'))$

This example can be well examined in the bornological context. It is not only a BP^*-algebra. Indeed $(E, \sigma(E, E'))$ is clearly Pseudo-Banach. If E is unital, then the space $\mathcal{M}$ of its non zero characters is non void; and it is compact. If moreover E is semisimple, then the Gelfand transform $\mathcal{G} : E \longrightarrow \mathcal{C}(\mathcal{M})$ is one-to-one ([1] and [6]).

5 General case

In the examples looked at, the algebras are hermitian and semisimple. Moreover in a commutative C^*-algebra [4], one has $\|x\| = \rho(x)$ and $\|x\|^2 = \|x^*x\| = \rho(x^*x)$. This suggests the examination of the spectral radius in BP^*-algebras which are candidates to be algebras of continuous functions. Fortunately a result in the frame of Banach algebras, also extended to locally m-convex ones, remains valid. The proof goes along the lines of that given in ([5], Theorem 22.1, p. 281); in the frame of m-convex algebras, except for a single argument which can be conducted, here, in a different manner. For the convenience of the reader, we give the whole proof.

Proposition 5.1. *Let (E, τ) be a unital commutative BP^*-algebra. The following assertions are equivalent.*

(i) (E, τ) is hermitian.
(ii) $\rho(x) \le p(x)$, where $p(x) = [\rho(x^*x)]^{\frac{1}{2}}$.
(iii) $\rho(x^*x) = [\rho(x)]^2$.

Proof. (i)$\Longrightarrow$(ii) Suppose that $\rho(y) > p(y)$, for some y. Then, there is $\lambda \in Sp(y)$ such that $|\lambda| > p(y)$. Putting $z = \lambda^{-1}x$, one has

$$\rho[e - (e - z^*z)] < 1 \text{ with } (e - z^*z) \in H(E).$$

But E is the algebraic inductive limit of the Banach algebras $E(B)$, with B in $\mathcal{B}$. Thus, for every x,

$$\rho(x) = \inf\{\rho_B(x), B \in \mathcal{B}(x)\}, \text{ where } \mathcal{B}(x) = \{B \in \mathcal{B} : x \in E(B)\}.$$

Hence, there is a B_0 such that $\rho_{B_0}[e-(e-z^*z)] < 1$. Whence the existence of an $u \in H(E(B_0)) \subset H(E)$ such that

$$u^2 = e - z^*z \text{ and } \rho_{B_0}[e-u] < 1.$$

But then u is in $G(E(B_0)) \subset G(E)$. On the other hand,

$$(e+z^*)(e-z) = u^2 = e-(z-z^*) = iu\left[ie - iu^{-1}(z-z^*)u^{-1}\right]u,$$

where $v = -iu^{-1}(z-z^*)u^{-1} \in H(E)$; and so $Sp(v) \subset \mathbb{R}$. Hence $ie - v \in G(E)$. So $(e+z^*)(e-z) \in G(E)$. Thus $e-z$ has a left inverse. One shows also that it has a right inverse ([5], p. 282). Whence $\lambda \notin Sp(u)$; a contradiction.

(ii)$\Longrightarrow$(iii) We have $\rho(x^*x) \le \rho(x^*)\rho(x) \le [\rho(x)]^2$, hence $p(x) \le \rho(x)$; and, by hypothesis, $\rho(x) \le p(x)$.

(iii)$\Longrightarrow$(i) Let $x \in H(E)$ and $\alpha + i\beta \in Sp(x)$, with α, β in $\mathbb{R}$ and $\beta \neq 0$. Then

$$y = \beta^{-1}(x - \alpha e) \in H(E) and i \in Sp(y).$$

Putting $z_n = y + ine$, for every $n \in \mathbb{N}^*$, one has

$$z_n^* z_n = y^2 + n^2 e \text{ and } (n+1)i \in Sp(z_n).$$

Therefore

$$|(n+1)i|^2 \le [\rho(z_n)]^2 \le \rho\left(y^2 + n^2 e\right) \le \rho\left(y^2\right) + n^2.$$

Whence

$$2n + 1 \le \rho\left(y^2\right) < +\infty, \forall n;$$

a contradiction. □

Proposition 5.2. *A unital, commutative, hermitian and semi-simple BP^*-algebra is an algebra of continuous functions.*

Proof. By Proposition 4.1, the specral radius ρ is a C^*-seminorm. In fact, it is a norm by semisimplicity. So (E, ρ) is a C^*-normed algebra. Its completion $\widehat{E}$ is isomorphic to $\mathcal{C}(\mathcal{M})$, by Gelfand-Naimark's theorem. □

Bibliography

[1] Allan, G. R., Dales, H. G., Mc Clure, J. P. *Pseudo-Banach algebras*, *Studia Math.* XL (1971), 55–56.

[2] Cochran, A. C., Keown, R., Williams, C. R. *On a class of topological algebras*, *Pacific J. Math.* 34 (1970), 17–25.

[3] Cochran, A., C. *Representation of A-convex algebras, Proc. Amer. Math. Soc.* 41 (1973), 473–479.
[4] Dixmier, J., *Les C^*-algebres et leurs representations.* Guauthier-Villars Paris, 1969.
[5] Fragoulopoulou, M., *Topological Algebras with Involution.* North-Holland, Math. Studies 200, 2005.
[6] Hogbé-Nlend, H. *Les fondements de la théorie spectrale des algèbres bornologiques, Bol. Soc. Brasil Mat.* 3 (1972), 19–56.
[7] Husain, T., Warsi, S. A. *Positive functionals on BP^*-algebras, Per. Math. Hungarica* 8 (1), (1977), 15–28.
[8] Husain, T., Warsi, S. A. *Representations of BP^*-algebras, Math. Japonica* 21 (1976), 237–247.
[9] Husain, T. *Multiplicative functionals on topological algebras.* Pitman Res. Notes Math. Ser. 85, Longman, 1983.
[10] Mallios, A. *Topological algebras. Selected topics.* North-Holland, Amsterdam, 1986.
[11] Oudadess, M. *Sur le spectre ponctuel dans les algèbres localement A-convexes, Bull. Soc. Roy. Sc. Liège* (2), (1985), pp. 65–68.
[12] Oudadess, M. *Théorèmes de structures et propriétés fondamentales des algèbres localement uniformément A-convexes, C. R. Acad. Sc. Paris* 296, Série J (1983), 851–853.
[13] Oudadess, M. *Une norme d'algèbre de Banach dans les algèbres uniformément A-convexes, Africa Math.*, Vol. IX (1987), 15–22.
[14] Rudin, W. *Functional Analysis.* Mc Graw-Hill, 1973.
[15] Warner, S. *Inductive limits of normed algebras, Trans. Amer. Math. Soc.* 82 (1956), 190–216.
[16] Warner, S. *Weakly topologized algebras, Proc. Amer. Math. Soc.* 8 (1957), 314–316.
[17] Warner, S. *Weak locally multiplicatively convex algebras, Pacific J. Math.* 5 (1955), 1025–1032.

Wilson Toko, Yevhen Zelenyuk, and Yuliya Zelenyuk

Discrete nonclosed subsets in maximally nondiscrete topological groups

Abstract: A topological group is maximally nondiscrete if its topology is maximal among all nondiscrete Hausdorff group topologies. We show that every infinite Abelian group G admits a maximally nondiscrete group topology $\mathcal{T}$ such that $(G, \mathcal{T})$ contains a discrete nonclosed subset which is a convergent sequence in some weaker group topology.

Keywords: topological groups; ultrafilters

MSC: Primary 22A05, 54G05; Secondary 03E35, 54H11

1 Introduction

A topological group is *maximally nondiscrete* if its topology is maximal among all nondiscrete Hausdorff group topologies. In [4] a question was raised whether every maximally nondiscrete topological group has the property that all its discrete subsets are closed (Question 5.2.3). It was answered in the negative in [5] by showing that $\bigoplus_\omega \mathbb{Z}_2$ admits a maximally nondiscrete group topology in which it contains a discrete subset with a unique accumulation point (see also [8, Corollary 5.25]).

In this paper we show that

Theorem 1.1. *Every infinite Abelian group G admits a maximally nondiscrete group topology $\mathcal{T}$ such that $(G, \mathcal{T})$ contains a discrete nonclosed subset which is a convergent sequence in some weaker group topology.*

Notice that such a subset has a unique accumulation point and converges to that point in that weaker topology.

As is well known, every infinite Abelian group contains as a subgroup either a direct sum of countably many nontrivial finite groups or the the group of integers or a quasicyclic group. Therefore, in order to prove Theorem 1.1, it suffices to prove it in these three cases. (Indeed, every group topology defined on a subgroup H of an

Wilson Toko, School of Mathematics, University of the Witwatersrand, Private Bag 3, Wits 2050, South Africa; E-mail: wilson.toko@wits.ac.za
Yevhen Zelenyuk, School of Mathematics, University of the Witwatersrand, Private Bag 3, Wits 2050, South Africa; E-mail: yevhen.zelenyuk@wits.ac.za
Yuliya Zelenyuk, School of Mathematics, University of the Witwatersrand, Private Bag 3, Wits 2050, South Africa; E-mail: yuliya.zelenyuk@wits.ac.za

https://doi.org/10.1515/9783110413557-015

Abelian group G extends to a group topology on G by declaring H to be open.) We do this in sections 2, 3, 4, respectively. Then in section 5 we show that

Theorem 1.2. *For every Abelian group G and for every nondiscrete metrizable group topology $\mathcal{T}_0$ on G, there is a maximally nondiscrete group topology $\mathcal{T}$ on G finer than $\mathcal{T}_0$ such that $(G, \mathcal{T})$ contains a discrete nonclosed subset which is a convergent sequence in $\mathcal{T}_0$.*

Although Theorem 1.1 is a consequence of Theorem 1.2, the proof of the former in the case of integers and quasicyclic groups captures more topologies than that of the latter.

The proof involves the notion of a rapid ultrafilter. This idea came from [7], where it was shown that, assuming there is no rapid ultrafilter, every nondiscrete group topology on $\bigoplus_\omega \mathbb{Z}_2$ finer than the direct sum topology has a discrete subset with a unique accumulation point. (The *direct sum topology* is one induced by the product topology.) It had been known previously that, assuming there is no P-point, every nondiscrete group topology on $\bigoplus_\omega \mathbb{Z}_2$ finer than the direct sum topology and satisfying one more condition has a discrete subset with a unique accumulation point (see [8, Theorem 5.19]), and this was used in [5].

Recall that

(a) a nonprincipal ultrafilter p on ω is a *P-point* if whenever $\{A_n : n < \omega\}$ is a partition of ω with $A_n \notin p$, there is $A \in p$ such that $|A \cap A_n| < \omega$ for each n, and

(b) p is *rapid* if whenever $\{A_n : n < \omega\}$ is a partition of ω with finite A_n, there is $A \in p$ such that $|A \cap A_n| \leq n$ for each n.

Martin's Axiom implies the existence of P-points and rapid ultrafilters. However, it is consistent with ZFC, the system of usual axioms of set theory, that there is no P-point [6, VI, §4], and it is consitent with ZFC that there is no rapid ultrafiter [3].

2 Direct sums

For every $n < \omega$, let G_n be a nontrivial finite group written additively, and let $G = \bigoplus_{n<\omega} G_n$. For every nonzero $x \in G$, let

$$\phi(x) = \max \operatorname{supp}(x) \text{ and } \theta(x) = \min \operatorname{supp}(x),$$

where as usual $\operatorname{supp}(x) = \{n < \omega : x(n) \neq 0\}$. Let $\mathcal{T}_0$ denote the direct sum topology on G.

Proposition 2.1. *Let $\mathcal{T}$ be a nondiscrete group topology on G finer than $\mathcal{T}_0$. Suppose $(G, \mathcal{T})$ contains no discrete nonclosed subset which is a convergent sequence in $\mathcal{T}_0$. Then for every nonprincipal ultrafilter p on G converging to 0 in $\mathcal{T}$, the ultrafilter $\phi(p)$ on ω is rapid.*

Here, $\phi(p)$ is the ultrafilter on ω with a base consisting of subsets $\phi(A)$, where $A \in p$.

In the proof of Proposition 2.1 we use the following characterization of rapid ultrafiters [3]:

an ultrafilter p on ω is rapid if and only if there is a function $h : \omega \to \omega$ with the property that, whenever $f : \omega \to \omega$, there exists $A \in p$ such that for every $n < \omega$, $|A \cap f(n)| \leq h(n)$, where $A \cap f(n) = A \cap \{0, 1, \ldots, f(n) - 1\}$.

Proof. Let $f : \omega \to \omega$ be given. Without loss of generality one may suppose that f is increasing. Define $U \subseteq G$ by

$$U = \{x : f(\theta(x)) \leq \phi(x)\} \cup \{0\}.$$

It then follows that U is a neighborhood of 0 in $\mathcal{T}$.

Indeed, assume the contrary. Then 0 is an accumulation point of

$$G \setminus U = \{x : f(\theta(x)) > \phi(x)\}.$$

But for every $n < \omega$, there are only finitely many $x \in G \setminus U$ with $\theta(x) = n$, because $\phi(x) < f(\theta(x))$. Consequently, $G \setminus U$ is a sequence converging to 0 in $\mathcal{T}_0$, a contradiction.

Pick a neighborhood V of 0 in $\mathcal{T}$ such that $V - V \subseteq U$. We claim that for every $n < \omega$,

$$|\phi(V) \cap f(n)| \leq \prod_{i<n} |G_i|.$$

To see this, assume the contrary. Then there exist distinct nonzero $x, y \in V$ such that $\phi(x), \phi(y) < f(n)$ and $h(x)(i) = h(y)(i)$ for each $i < n$. Let $z = x - y$. Then $0 \neq z \in U$, $\theta(z) \geq n$, and $\phi(z) < f(n)$. Hence, $f(n) \leq f(\theta(z)) \leq \phi(z) < f(n)$, a contradiction. □

From Proposition 2.1 and [3] we obtain that

Corollary 2.2. *It is consistent with ZFC that for every nondiscrete group topology $\mathcal{T}$ on G finer than $\mathcal{T}_0$, $(G, \mathcal{T})$ contains a discrete nonclosed subset which is a convergent sequence in $\mathcal{T}_0$.*

Now we are going to prove Theorem 1.1 for our G using Proposition 2.1. To this end, we need an ultrafilter on ω which is not rapid.

Lemma 2.3. *There is a nonprincipal ultrafilter on ω which is neither rapid nor a P-point.*

Of course, this is a known fact, but we prove it for convenience of the reader.

Proof. Pick a partition $\{A_{i,n} : i, n < \omega\}$ of ω such that $n^2 \leq |A_{i,n}| < \omega$ for all $i, n < \omega$. Let $\mathcal{A}$ be the family of all subsets $A \subseteq \omega$ for which there is a sequence $(m_i)_{i<\omega}$ in ω such that

(i) for every $i < \omega$ and $n < m_i$, $A \cap A_{i,n} = \emptyset$, and
(ii) for every $i < \omega$ and $n \geq m_i$, $|A \cap A_{i,n}| \geq |A_{i,n}| - n$.

The family $\mathcal{A}$ has the finite intersection property, so there is an ultrafilter p on ω extending $\mathcal{A}$, and since $\bigcap \mathcal{A} = \emptyset$, p is nonprincipal. By (i), the partition $\{\bigcup_{n<\omega} A_{i,n} : i < \omega\}$ witnesses that p is not a P-point, and by (ii), the partition $\{A_{i,n} : i, n < \omega\}$ witnesses that p is not rapid. □

Now to prove Theorem 1.2 for our G, pick a nonprincipal ultrafilter u on ω which is not rapid. Let $\mathcal{T}^u$ be the group topology on G with a neighborhood base at 0 consisting of subgroups $H_A = \{x \in G : \operatorname{supp}(x) \subseteq A\}$, where $A \in u$, and let $\mathcal{T}$ be any nondiscrete group topology on G finer than $\mathcal{T}^u$. For every nonprincipal ultrafilter p on G converging to 0 in $\mathcal{T}$, one has $\phi(p) = u$. Hence by Proposition 2.1, $(G, \mathcal{T})$ contains a discrete nonclosed subset which is a convergent sequence in $\mathcal{T}_0$.

3 Integers

Every integer x can be written uniquely as

$$x = \sum_{i=0}^{n} x_i(-2)^i,$$

where $n < \omega$, $x_i \in \{0, 1\}$ for each $i \leq n$, and $x_n \neq 0$ unless $n = 0$, and so x is represented by $x_n \dots x_0$ in negabinary. The digits x_i can be found by repeated division by -2. For example, for $x = -3$,

$$\begin{aligned} -3 &= 2 \cdot (-2) + 1, \\ 2 &= (-1) \cdot (-2) + 0, \\ -1 &= 1 \cdot (-2) + 1, \\ 1 &= 0 \cdot (-2) + 1, \end{aligned}$$

consequently, $-3 = 1 + (-2)^3 + (-2)^4$, and so -3 is represented by 11001.

Arithmetic operations over negabinary numbers are similar to those in other bases. In particular, to subtract two numbers, inductively, subtract the digits plus the carry and from the resulting number write down the digit and get the next carry, using the following decompositions (there are only three possibilities for the resulting number to be distinct from 0, 1):

$$\begin{aligned} -2 &= 1 \cdot (-2) + 0, \\ -1 &= 1 \cdot (-2) + 1, \\ 2 &= (-1) \cdot (-2) + 0. \end{aligned}$$

For example, to subtract 1010 ($-2-8=-10$) and 101 ($1+4=5$),

$$\begin{array}{r} 1\,0\,1\,0 \\ 1\,0\,1 \\ 1\,1\,0\,0\,0\,1 \end{array}$$

so the result is 110001 ($1+16-32=-15$).

For every integer x, extend the negabinary expansion $x = \sum_{i=0}^{n} x_i(-2)^i$ to $x = \sum_{i<\omega} x_i(-2)^i$ by putting $x_i = 0$ for all $i > n$. Let

$$\sigma(x) = \{i < \omega : x_i \neq 0\},\ \phi(x) = \max\sigma(x),\ \text{ and } \theta(x) = \min\sigma(x).$$

(The functions ϕ and θ are not defined at $x = 0$.)

Lemma 3.1. *For any distinct nonzero $x, y \in \mathbb{Z}$,*
(a) $\theta(x-y) = \min\{i : x_i \neq y_i\}$, *in particular,*
$\theta(x-y) \geq \min\{\theta(x), \theta(y)\}$,
(b) $\phi(x-y) \leq \max\{i : x_i \neq y_i\} + 2$, *in particular,*
$\phi(x-y) \leq \max\{\phi(x), \phi(y)\} + 2$, *and*
(c) $\sigma(x-y) \subseteq \bigcup_{j=0}^{2}((\sigma(x) \cup \sigma(y)) + j)$.

Proof. Let $z = x - y$. When computing $z_i \in \{0, 1\}$, one adds $x_i - y_i \in \{-1, 0, 1\}$ and the carry $c_i \in \{-1, 0, 1\}$ and from the resulting number $z_i' \in \{-2, -1, 0, 1, 2\}$ writes down the digit z_i and gets the next carry c_{i+1}.

Now for (a), let $m = \min\{i : x_i \neq y_i\}$. Then for each $i < m$, $c_i = 0$ and $z_i' = 0$, so $z_i = 0$.

For (b), let $n = \max\{i : x_i \neq y_i\}$. If $z_n' \in \{0, 1\}$, $\phi(z) \leq n$. If $z_n' \in \{-2, -1\}$, then $c_{n+1} = 1$, so $\phi(z) = n + 1$. If $z_n' = 2$, then $c_{n+1} = -1$, consequently $c_{n+2} = 1$, and so $\phi(z) = n + 2$.

For (c), let $i \in \sigma(x-y)$. By (a), $i \geq \min(\sigma(x) \cup \sigma(y))$. Assume on the contrary that $i \notin \bigcup_{j=0}^{2}((\sigma(x) \cup \sigma(y)) + j)$. Then $i \geq 2$ and $x_{i-j} = y_{i-j} = 0$ for each $j \in \{0, 1, 2\}$. But then the same argument as in (b) shows that $z_i = 0$, a contradiction. □

Let $\mathcal{T}_0$ denote the topology on $\mathbb{Z}$ induced by the direct sum topology on $\bigoplus_\omega\{0, 1\}$ with respect to the mapping $x \mapsto (x_i)_{i<\omega}$. It follows from Lemma 3.1 (a) that $\mathcal{T}_0$ is a group topology. It is easy to see that $\mathcal{T}_0$ is the 2-adic topology, that is, one with a neighborhood base at 0 consisting of subgroups $2^n\mathbb{Z}$, where $n < \omega$.

Proposition 3.2. *Let $\mathcal{T}$ be a nondiscrete group topology on $\mathbb{Z}$ finer than $\mathcal{T}_0$. Suppose $(\mathbb{Z}, \mathcal{T})$ contains no discrete subset with a unique accumulation point which is a convergent sequence in $\mathcal{T}_0$. Then for every nonprincipal ultrafilter p on G converging to 0 in $\mathcal{T}$, the ultrafilter $\phi(p)$ on ω is rapid.*

Proof. It is similar to that of Proposition 2.1.

Given an increasing $f : \omega \to \omega$, define $U \subseteq \mathbb{Z}$ by

$$U = \{x : f(\theta(x)) \le \phi(x) - 2\} \cup \{0\}.$$

It then follows that U is a neighborhood of 0 in $\mathcal{T}$. Pick a neighborhood V of 0 in $\mathcal{T}$ such that $V - V \subseteq U$. We claim that for every $n < \omega$,

$$|\phi(V) \cap f(n)| \le 2^n.$$

To see this, assume the contrary. Then there exist distinct nonzero $x, y \in V$ such that $\phi(x), \phi(y) < f(n)$ and $h(x)(i) = h(y)(i)$ for each $i < n$. Let $z = x - y$. Then $0 \neq z \in U$, and by Lemma 3.1 ((a) and (b)), $\theta(z) \ge n$ and $\phi(z) \le f(n) + 1$. Hence, $f(n) \le f(\theta(z)) \le \phi(z) - 2 < f(n)$, a contradiction. □

As in section 2 from Proposition 3.2 and [3] we obtain that

Corollary 3.3. *It is consistent with ZFC that for every nondiscrete group topology $\mathcal{T}$ on $\mathbb{Z}$ finer than $\mathcal{T}_0$, $(\mathbb{Z}, \mathcal{T})$ contains a discrete nonclosed subset which is a convergent sequence in $\mathcal{T}_0$.*

However, in order to prove Theorem 1.1 for $\mathbb{Z}$ using Proposition 3.2, we need to be more elaborate than in section 2.

For every filter $\mathcal{F}$ on ω, let $\mathcal{N}(\mathcal{F})$ denote the filter on $\mathbb{Z}$ with a base consisting of subsets $U_A = \{x \in \mathbb{Z} : \sigma(x) \subseteq A\}$, where $A \in \mathcal{F}$.

Proposition 3.4. *Let $\mathcal{F}$ be a filter on ω with $\bigcap \mathcal{F} = \emptyset$ and let $I = \bigcap_{A \in \mathcal{F}} \overline{A}$. Then the following statements are equivalent:*
(i) there is a group topology $\mathcal{T}(\mathcal{F})$ on $\mathbb{Z}$ for which $\mathcal{N}(\mathcal{F})$ is the neighborhood filter of 0,
(ii) $\mathcal{F} \subseteq 1 + \mathcal{F}$,
(iii) I is a left ideal of $\beta\omega$.

Here, ω is considered as a discrete semigroup and $\beta\omega$ is the Stone-Čech compactification of ω. The points of $\beta\omega$ are the ultrafilters on ω and the topology is generated by taking as a base the subsets $\overline{A} = \{p \in \beta\omega : A \in p\}$, where $A \subseteq G$. For any $p, q \in \beta\omega$, the ultrafilter $p + q$ has a base consisting of subsets $\bigcup_{x \in A}(x + B_x)$, where $A \in p$ and $B_x \in q$ for every $x \in A$. See [1] and [8] for more information about $\beta\omega$ and, more generally, βS.

Proof. (i) $\Rightarrow$ (ii) It suffices to show that for every $A \in \mathcal{F}$, there is $B \in \mathcal{F}$ such that $B + 1 \subseteq A$. Since $\mathcal{T}(\mathcal{F})$ is a group topology, there is $B \in \mathcal{F}$ such that $-U_B - U_B \subseteq U_A$. To see that this B is as required, let $n \in B$. Pick $x \in \mathbb{Z}$ with $\sigma(x) = \{n\}$. Then $x \in U_B$, and so $(-2)x \in U_A$. But $\sigma((-2)x) = \sigma(x) + 1 = \{n + 1\}$. Hence, $n + 1 \in A$.

(ii) $\Rightarrow$ (i) We have to show that for every $A \in \mathcal{F}$, there is $B \in \mathcal{F}$ such that $U_B - U_B \subseteq U_A$. Since $\overline{\mathcal{F}}$ is a left ideal, there is $B \in \mathcal{F}$ such that $B + j \subseteq A$ for each $j \in \{0, 1, 2\}$.

To see that this B is as required, let $x, y \in U_B$. Then $\sigma(x), \sigma(y) \subseteq B$. By Lemma 3.1 (c), $\sigma(x-y) \subseteq \bigcup_{j=0}^{2}((\sigma(x) \cup \sigma(y)) + j)$. Hence, $\sigma(x-y) \subseteq \bigcup_{j=0}^{2}(B+j) \subseteq A$, and so $x - y \in U_A$.

(*ii*) $\Leftrightarrow$ (*iii*) is the corollary of the next lemma. □

Lemma 3.5. *Let S be a semigroup, let $\mathcal{F}$ be a filter on S, and let $I = \bigcap_{U \in \mathcal{F}} \overline{U}$. Then I is a left ideal of βS if and only if for every $x \in S$, $\mathcal{F} \subseteq x\mathcal{F}$.*

Proof. Necessity. Let $U \in \mathcal{F}$. For every $q \in I$, pick $B_q \in q$ such that $xB_q \subseteq U$. The subsets $\overline{B_q}$, where $q \in I$, form an open cover of I, so there is a finite subcover $\{\overline{B_{q_i}} : i < n\}$. Let $V = \bigcup_{i<n} B_{q_i}$. Then $V \in \mathcal{F}$ and $xV \subseteq U$. Hence, $U \in x\mathcal{F}$.

Sufficiency. Let $q \in I$. For every $x \in S$, $xq \in I$, because $\mathcal{F} \subseteq x\mathcal{F}$. It then follows for every $p \in \beta S$, $pq \in I$. □

Corollary 3.6. *Let $\mathcal{F}$ be a filter on ω and let $I = \bigcap_{U \in \mathcal{F}} \overline{U}$. Then I is a left ideal of $\beta\omega$ if and only if $\mathcal{F} \subseteq 1 + \mathcal{F}$.*

Proof. From $\mathcal{F} \subseteq 1 + \mathcal{F}$ it follows, by induction, that for every $n < \omega$, $\mathcal{F} \subseteq n + \mathcal{F}$. □

Now, in order to prove Theorem 1.1 for $\mathbb{Z}$, we need a minimal left ideal of $\beta\omega$ which contains a nonrapid ultrafilter.

Lemma 3.7. *Let p be an accumulation point of a subset $\{p_n : n < \omega\} \subseteq \omega^*$.*

(a) If all p_n are rapid, then p is rapid.

(b) If all p_n are nonrapid, then p is nonrapid.

Proof. (a) Let $f : \omega \to \omega$ be given. Without loss of generality one may suppose that f is increasing. For every $n < \omega$, pick $A_n \in p_n$ such that $|A_n \cap f(m)| \leq m$ for each $m < \omega$ and $A_n \cap f(n) = \emptyset$. Define $A \in p$ by $A = \bigcup_{n<\omega} A_n$. Then for each $m < \omega$, $A \cap f(m) = \bigcup_{n<m}(A_n \cap f(m))$, and consequently, $|A \cap f(m)| \leq m^2$.

(b) For every $n < \omega$, pick $f_n : \omega \to \omega$ such that whenever $A_n \in p_n$, there are infinitely many $m < \omega$ such that $|A_n \cap f_n(m)| > m$. Define $f : \omega \to \omega$ by $f(m) = \max\{f_n(m) : n < m\}$, so $f(m) \geq f_n(m)$ for all $n < m < \omega$. Now let $A \in p$. Pick $n < \omega$ and $A_n \in p_n$ such that $A_n \subseteq A$, and pick $m > n$ such that $|A_n \cap f_n(m)| > m$. Then $|A \cap f(m)| > m$. □

Recall that a semigroup is *left cancellative* if left translations are injective.

Corollary 3.8. *Let S be a countable left cancellative semigroup and let $p, q \in S^*$. Then pq is rapid if and only if q is rapid.*

Proof. Indeed, pq is an accumulation point of $\{sq : s \in S\}$, and each sq is rapid if and only if q is rapid. □

Sufficiency of Corollary 3.8 for $\mathbb{N}$ was established in [2].

Now to prove Theorem 1.1 for $\mathbb{Z}$, pick a nonprincipal ultrafilter u on ω which is not rapid and let $I = \beta\omega + u$. By Corollary 3.8, every ultrafilter from I is not rapid. Let $\mathcal{F}$ be the filter on ω such that $I = \bigcap_{A\in\mathcal{F}} \overline{A}$. By Proposition 3.4, there is a group topology $\mathcal{T}(\mathcal{F})$ on $\mathbb{Z}$ for which $\mathcal{N}(\mathcal{F})$ is the neighborhood filter of 0. Let $\mathcal{T}$ be any nondiscrete group topology on $\mathbb{Z}$ finer that $\mathcal{T}(\mathcal{F})$. For every ultrafilter p on $\mathbb{Z}$ converging to 0 in $\mathcal{T}$, $\phi(p) \in I$. Hence, by Proposition 3.2, $(\mathbb{Z}, \mathcal{T})$ contains a discrete nonclosed subset which is a convergent sequence in $\mathcal{T}_0$.

4 Quasicyclic groups

For every prime $p \neq 2$, we consider $\mathbb{Z}_{p^\infty}$ as the group generated by elements a_n, where $1 \leq n < \omega$, such that $pa_1 = 0$ and $pa_{n+1} = a_n$, so that every $x \in \mathbb{Z}_{p^\infty}$ can be uniquely written as $x = \sum_{i=0}^{n} x_i a_{i+1}$, where $n < \omega$, $x_i \in \{-1, 0, 1, \dots, p-2\}$ for each $i \leq n$, and $x_n \neq 0$ unless $n = 0$. We consider $\mathbb{Z}_{2^\infty}$ as the group generated by elements a_n, where $1 \leq n < \omega$, such that $4a_1 = 0$ and $4a_{n+1} = a_n$, so that every $x \in \mathbb{Z}_{2^\infty}$ can be uniquely written as $x = \sum_{i=0}^{n} x_i a_{i+1}$, where $n < \omega$, $x_i \in \{-1, 0, 1, 2\}$ for each $i \leq n$, and $x_n \neq 0$ unless $n = 0$.

For every prime p and for every $x \in \mathbb{Z}_{p^\infty}$, extend the expansion $x = \sum_{i=0}^{n} x_i a_{i+1}$ to $x = \sum_{i<\omega} x_i a_{i+1}$ by putting $x_i = 0$ for all $i > n$. Let

$$\sigma(x) = \{i < \omega : x_i \neq 0\},\ \phi(x) = \max \sigma(x),\ \text{ and } \theta(x) = \min \sigma(x).$$

Lemma 4.1. *For any distinct nonzero $x, y \in \mathbb{Z}_{p^\infty}$,*
(a) $\theta(x-y) \geq \min\{i : x_i \neq y_i\} - 1$, *in particular,*
$\theta(x-y) \geq \min\{\theta(x), \theta(y)\} - 1$,
(b) $\phi(x-y) = \max\{i : x_i \neq y_i\}$, *in particular,*
$\phi(x-y) \leq \max\{\phi(x), \phi(y)\}$, *and*
(c) $\sigma(x-y) \subseteq \bigcup_{j=0}^{1}((\sigma(x) \cup \sigma(y)) - j)$.

Proof. Let $z = x - y$. When computing $z_i \in \{-1, 0, 1, \dots, p-2\}$ ($z_i \in \{-1, 0, 1, 2\}$ if $p = 2$), one adds $x_i - y_i \in \{1-p, \dots, p-1\}$ ($x_i - y_i \in \{-3, \dots, 3\}$ if $p = 2$) and the carry $c_i \in \{-1, 0, 1\}$ and from the resulting number $z_i' \in \{-p, \dots, p\}$ ($z_i' \in \{-4, \dots, 4\}$ if $p = 2$) writes down the digit z_i and gets the next carry c_{i-1}. □

Let $\mathcal{T}_0$ denote the topology on $\mathbb{Z}_{p^\infty}$ induced by the direct sum topology on $\bigoplus_\omega\{-1, 0, 1, \dots, p-2\}$ ($\bigoplus_\omega\{-1, 0, 1, 2\}$ if $p = 2$) with respect to the mapping $x \mapsto (x_i)_{i<\omega}$. It follows from Lemma 4.1 (a) that $\mathcal{T}_0$ is a group topology.

Proposition 4.2. *Let $\mathcal{T}$ be a nondiscrete group topology on $\mathbb{Z}_{p^\infty}$ finer than $\mathcal{T}_0$. Suppose $(\mathbb{Z}_{p^\infty}, \mathcal{T})$ contains no discrete subset with a unique accumulation point which is a convergent sequence in $\mathcal{T}_0$. Then for every nonprincipal ultrafilter q on $\mathbb{Z}_{p^\infty}$ converging to 0 in $\mathcal{T}$, the ultrafilter $\phi(q)$ on ω is rapid.*

Proof. Given an increasing $f : \omega \to \omega$, define $U \subseteq \mathbb{Z}_{p^\infty}$ by

$$U = \{x : f(\theta(x)) \leq \phi(x)\} \cup \{0\}.$$

It then follows that U is a neighborhood of 0 in $\mathcal{T}$. Pick a neighborhood V of 0 in $\mathcal{T}$ such that $V - V \subseteq U$. We claim that for every $n < \omega$,

$$|\phi(V) \cap f(n)| \leq p^{n+1} \quad (|\phi(V) \cap f(n)| \leq 4^{n+1} \text{ if } p = 2).$$

To see this, assume the contrary. Then there exist distinct nonzero $x, y \in V$ such that $\phi(x), \phi(y) < f(n)$ and $h(x)(i) = h(y)(i)$ for each $i \leq n$. Let $z = x - y$. Then $0 \neq z \in U$, and by Lemma 4.1 ((a) and (b)), $\theta(z) \geq n$ and $\phi(z) < f(n)$. Hence, $f(n) \leq f(\theta(z)) \leq \phi(z) < f(n)$, a contradiction. □

For every filter $\mathcal{F}$ on ω, let $\mathcal{N}(\mathcal{F})$ denote the filter on $\mathbb{Z}_{p^\infty}$ with a base consisting of subsets $U_A = \{x \in \mathbb{Z} : \sigma(x) \subseteq A\}$, where $A \in \mathcal{F}$.

Proposition 4.3. *Let $\mathcal{F}$ be a filter on ω with $\bigcap \mathcal{F} = \emptyset$ and let $I = \bigcap_{A \in \mathcal{F}} \overline{A}$. Then the following statements are equivalent:*
(i) there is a group topology $\mathcal{T}(\mathcal{F})$ on $\mathbb{Z}_{p^\infty}$ for which $\mathcal{N}(\mathcal{F})$ is the neighborhood filter of 0,
(ii) $\mathcal{F} \subseteq -1 + \mathcal{F}$,
(iii) $(-\beta\omega) + I \subseteq I$.

Proof. It is similar to that of Proposition 3.4. We restrict ourselves to proving (*i*) ⇒ (*ii*) for $p \neq 2$. We have to show that for every $A \in \mathcal{F}$, there is $B \in \mathcal{F}$ such that $B - 1 \subseteq A$. Since $\mathcal{T}(\mathcal{F})$ is a group topology, there is $B \in \mathcal{F}$ such that $pU_B \subseteq U_A$. To see that this B is as required, let $n \in B$. Since $\sigma(a_{n+1}) = \{n\}$, $a_{n+1} \in U_B$, so $pa_{n+1} \in U_A$. But $\sigma(pa_{n+1}) = \sigma(a_n) = \{n - 1\}$. Hence, $n - 1 \in A$. □

Now to prove Theorem 1.1 for $\mathbb{Z}_{p^\infty}$, pick a nonprincipal ultrafilter u on ω which is not rapid and let $I = \beta\mathbb{Z} + u$. By Corollary 3.8, every ultrafilter from I is not rapid. Let $\mathcal{F}$ be the filter on ω such that $I = \bigcap_{A \in \mathcal{F}} \overline{A}$. By Proposition 4.3, there is a group topology $\mathcal{T}(\mathcal{F})$ on $\mathbb{Z}_{p^\infty}$ for which $\mathcal{N}(\mathcal{F})$ is the neighborhood filter of 0. Let $\mathcal{T}$ be any nondiscrete group topology on $\mathbb{Z}_{p^\infty}$ finer that $\mathcal{T}(\mathcal{F})$. For every ultrafilter q on $\mathbb{Z}_{p^\infty}$ converging to 0 in $\mathcal{T}$, $\phi(q) \in I$. Hence, by Proposition 4.2, $(\mathbb{Z}_{p^\infty}, \mathcal{T})$ contains a discrete nonclosed subset which is a convergent sequence in $\mathcal{T}_0$.

5 Proof of Theorem 1.2

Given an Abelian group G and a sequence $(a_n)_{n=1}^\infty$ in G,

(1) let $\mathcal{T}((a_n)_{n=1}^\infty)$ denote the largest group topology on G (not necessarily Hausdorff) in which $(a_n)_{n=1}^\infty$ converges to 0, and

(2) for every sequence $(m_n)_{n=1}^{\infty}$ of positive integers, let

$$U((m_n)_{n=1}^{\infty}) = \bigcup_{n=1}^{\infty} \sum_{i=1}^{n} B_{m_i}^n,$$

where $B_m^n = \{0, \pm a_m, \pm a_{m+1}, \dots, \pm a_n\}$.

Proposition 5.1. *The subsets $U((m_n)_{n=1}^{\infty})$, where $(m_n)_{n=1}^{\infty}$ runs over the sequences of positive integers, form a neighborhood base at 0 in $\mathcal{T}((a_n)_{n=1}^{\infty})$.*

Proof. See [8, Theorem 1.17 and Definition 1.18]. □

We shall write $U(m_1, m_2, \dots)$ instead of $U((m_n)_{n=1}^{\infty})$. Notice that $U(1, 2, \dots)$ consists of all $x \in G$ which can be written as $x = 0$ or

$$x = c_1 a_{i_1} + \dots + c_k a_{i_k},$$

where $k \geq 1$, $1 \leq i_1 < \dots < i_k$, and for each $j \in \{1, \dots, k\}$, $c_j \in \mathbb{Z} \setminus \{0\}$ and $|c_1| + \dots + |c_j| \leq i_j$. If for every $x \in U(1, 2, \dots)$, such a decomposition is unique, we say that $U(1, 2, \dots)$ has the *unique decomposition property.*

Suppose $U = U(1, 2, \dots)$ has the unique decomposition property. The *direct sum topology* on U is one induced by the direct sum topology on $\bigoplus_{n<\omega}\{0, \pm 1, \dots, \pm n\}$. For every $x \in U$, write $x = c_1 a_{i_1} + \dots + c_k a_{i_k}$ and, for every $n < \omega$, define $x_n \in \{0, \pm 1, \dots, \pm n\}$ by putting $x_n = c_j$ if $n = i_j$ for some $j \in \{1, \dots, k\}$ and $x_n = 0$ otherwise. Let

$$\sigma(x) = \{n < \omega : x_n \neq 0\}, \ \phi(x) = \max \sigma(x), \text{ and } \theta(x) = \min \sigma(x)$$

($\phi(0) = -1$ and $\theta(0) = \omega$). Notice that for every $x \in U$ and $n < \omega$, x can be uniquely written as $x = y + z$, where $y, z \in U$, $\phi(y) < n$, and $\theta(z) \geq n$.

Recall that a topological group is metrizable if and only if it is first countable and Hausdorff.

Proposition 5.2. *For every nondiscrete metrizable Abelian topological group $(G, \mathcal{T}_0)$, either $(G, \mathcal{T}_0)$ contains a copy of $\bigoplus_{\omega} \mathbb{Z}_p$ with the direct sum topology for some prime p or there is a sequence $(a_n)_{n=1}^{\infty}$ in G such that $U = U(1, 2, \dots)$ has the unique decomposition property and $\mathcal{T}_0$ induces the direct sum topology on U.*

Proof. Suppose that every neighborhood of 0 in $\mathcal{T}_0$ contains elements of arbitrarily big (possibly infinite) order. We show that the second possibility holds.

Construct inductively a sequence $(a_n)_{n=1}^{\infty}$ in G and a decreasing sequence $(V_n)_{n=1}^{\infty}$ of open neighborhoods of 0 forming a neighborhood base at 0 such that

(i) a_n has order $> n$ and $ca_n \in V_n$ for each $c \in \{\pm 1, \pm 2, \dots, \pm n\}$, and

(ii) $ca_n + V_{n+1} \subseteq V_n$ and $(ca_n + V_{n+1}) \cap (c' a_n + V_{n+1}) = \emptyset$ for all distinct $c, c' \in \{0, \pm 1, \dots, \pm n\}$.

Let $U = U(1, 2, \ldots)$. Then

(1) whenever $x \in U$ and $x = c_1 a_{i_1} + \cdots + c_k a_{i_k}$, one has $x \in V_{i_1}$, and

(2) whenever $x, y \in U$, $x \neq y$, $x = c_1 a_{i_1} + \cdots + c_k a_{i_k}$, $y = d_1 a_{j_1} + \cdots + d_l a_{j_l}$, and $n = \max\{i_k, j_l\} + 1$, one has $(x + V_n) \cap (y + V_n) = \emptyset$.

It suffices to check (2). Without loss of generality one may suppose that $i_1 \leq j_1$, and if $i_1 = j_1$, then $c_1 \neq d_1$. But then $x + V_n \subseteq c_1 a_{i_1} + V_{i_1+1}$ and $y + V_n \subseteq d_1 a_{i_1} + V_{i_1+1}$ if $i_1 = j_1$ and $y + V_n \in V_{i_1+1}$ otherwise.

It is immediate from (2) that $U = U(1, 2, \ldots)$ has the unique decomposition property. To see that $\mathcal{T}_0$ induces the direct sum topology on U, let $u \in U$, $n > \phi(u)$, and $W_n = \{x \in U : x_i = u_i \text{ for each } i < n\}$. We show that $W_n = U \cap (u + V_n)$ (put $V_0 = G$).

To see that $W_n \subseteq U \cap (u + V_n)$, let $x \in W_n$. Write $x = u + y$, where $y \in U$ and $\theta(y) \geq n$. Then by (1), $y \in V_n$. Conversely, let $y \in V_n$ and $x = u + y \in U$. Write $x = v + z$, where $v, z \in U$, $\phi(v) < n$, and $\theta(z) \geq n$, so $z \in V_n$. Then by (2), from $u + y = v + z$ it follows that $u = v$, and so $y = z \in U$. Hence, $x = u + y \in W_n$.

If not every neighborhood of 0 in $\mathcal{T}_0$ contains elements of arbitrarily big order, then the first possibility holds. Indeed, there is a prime p such that every neighborhood of 0 contains an element of order p. Construct inductively a sequence $(a_n)_{n<\omega}$ of elements of order p and a decreasing sequence $(V_n)_{n<\omega}$ of open neighborhoods of 0 forming a neighborhood base at 0 such that $\langle a_n \rangle + V_{n+1} \subseteq V_n$ and $(x + V_{n+1}) \cap (y + V_{n+1}) = \emptyset$ for all distinct $x, y \in \langle a_n \rangle$. Then $\bigoplus_{n<\omega} \langle a_n \rangle$ is a copy of $\bigoplus_\omega \mathbb{Z}_p$. □

Now to prove Theorem 1.2, let $(G, \mathcal{T}_0)$ be a nondiscrete metrizable Abelian topological group. By Proposition 5.2, we have two possibilities. The first one falls under section 2. Therefore, we may assume that the second one holds. Let $(a_n)_{n=1}^\infty$ be the corresponding sequence in G and $U = U(1, 2, \ldots)$.

Proposition 5.3. *Let $\mathcal{T}$ be a group topology on G finer than $\mathcal{T}_0$ and endow U with the topology induced by $\mathcal{T}$. Suppose U contains no discrete nonclosed subset which is a convergent sequence in $\mathcal{T}_0$. Then for every nonprincipal ultrafilter p on G containing U and converging to 0 in $\mathcal{T}$, the ultrafilter $\phi(p)$ on ω is rapid.*

Proof. Given an increasing $f : \omega \to \omega$, define $V \subseteq U$ by

$$V = \{x : f(\theta(x)) \leq \phi(x)\} \cup \{0\}.$$

It then follows that V is a neighborhood of 0 in $\mathcal{T}$.

Pick a neighborhood W of 0 in $\mathcal{T}$ such that $W - W \subseteq V$. Then for every $n < \omega$,

$$|\phi(W) \cap f(n)| \leq \prod_{i<n} (2i + 1).$$

□

Now to finish the proof of Theorem 1.2, pick a nonprincipal ultrafilter u on ω which is not rapid. For every $A \subseteq \omega$, let $U_A = \{x \in U : \sigma(x) \subseteq A\}$. The subsets $V \cap U_A$, where V

runs over the neighborhoods of 0 in $\mathcal{T}_0$ and $A \in u$, form a neighborhood base at 0 in some group topology $\mathcal{T}^u$ on G. Let $\mathcal{T}$ be any nondiscrete group topology on G finer than $\mathcal{T}^u$. For every nonprincipal ultrafilter p on G converging to 0 in $\mathcal{T}$, one has $\phi(p) = u$. Hence by Proposition 5.3, the subset U of $(G, \mathcal{T})$ contains a discrete nonclosed subset which is a convergent sequence in $\mathcal{T}_0$.

Remark 5.4. *Suppose that topology $\mathcal{T}$ in Proposition 5.3 contains no discrete subset with a unique accumulation point and satisfies the following condition: the filter on ω with a base consisting of subsets $\phi(W)$, where W runs over the neighborhoods of 0 in $\mathcal{T}$, is an ultrafilter. Then this ultrafilter is a P-point. The proof is similar to that of [8, Theorem 5.19]. Of course, the same holds for direct sums as well. It follows that using P-points one can prove the following a little bit weaker version of Theorem 1.2: every nondiscrete metrizable group topology on an Abelian group G can be refined to a maximally nondiscrete group topology $\mathcal{T}$ such that $(G, \mathcal{T})$ contains a countable discrete subset with a unique accumulation point.*

Acknowledgment: The second and the third authors were supported by NRF grants IFR2011033100072 and IFR1202220164, respectively.

Bibliography

[1] Hindman, N. and Strauss, D. *Algebra in the Stone-Čech compactification*, De Gruyter, Berlin, 1998.
[2] Krautzberger, P. *On rapid idempotent ultrafilters*, Semigroup Forum (2014), doi 10.1007/s00233-014-9598-8.
[3] Miller, A. *There are no Q-points in Laver's model for the Borel conjecture*, Proc. Amer. Math. Soc. 78 (1980), 103-106.
[4] Protasov, I. and Zelenyuk, Y. *Topologies on groups determined by sequences*, Lviv: VNTL Publishers, 1998.
[5] Ross, D. and Zelenyuk, Y. *Topological groups and $P^-_{\kappa^+}$-points*, Topology Appl. 157 (2010), 2467-2475.
[6] Shelah, S. *Proper and improper forcing*, Springer, Berlin, 1998.
[7] Sipacheva, O. *Nonexistence of countable extremally disconnected groups with many open subgroups*, Topology Appl., accepted, arXiv:1405.6356.
[8] Zelenyuk, Y. *Ultrafilters and topologies on groups*, De Gruyter, Berlin, 2011.

Camillo Trapani and Francesco Tschinke

Faithfully representable topological $\star$-algebras: some spectral properties

Abstract: A faithfully representable topological *-algebra (FR*-algebra) $\mathfrak{A}_0$ is characterized by the fact that it possesses sufficiently many *-representations. Some spectral properties are examined, by constructing a convenient quasi *-algebra $\mathfrak{A}$ over $\mathfrak{A}_0$, starting from the order bounded elements of $\mathfrak{A}_0$.

Keywords: topological *-algebras; bounded elements

MSC: 46K05; 46K10

1 Introduction

This note is devoted to a particular family of topological *-algebras that share with C*-algebras a crucial feature: they possess *sufficiently many* *-representations. Throughout this paper, we term *topological *-algebra* a *-algebra $\mathfrak{A}_0$ equipped with a locally convex topology $\mathfrak{t}$ such that, for each $a \in \mathfrak{A}_0$, the mappings $x \mapsto ax$, $a \mapsto xa$ and the involution * are continuous in $\mathfrak{A}_0[\mathfrak{t}]$. A **-representation* of $\mathfrak{A}_0$ is a *-homomorphism of $\mathfrak{A}_0$ into the *-algebra $\mathcal{L}^\dagger(\mathcal{D})$ of all weakly continuous endomorphisms of a pre-Hilbert space $\mathcal{D}$. In [16] one of us considered the case where $\mathfrak{A}_0$ is *faithfully representable*, for short an FR*-algebra, in the sense that, for every nonzero element $a \in \mathfrak{A}_0$, there exists a continuous (in a sense that will be specified later) *-representation π which does not vanish at a. The main scope of that paper was to identify elements of $\mathfrak{A}_0$ that are mapped into bounded operators by any *-representation of $\mathfrak{A}_0$ and for this reason they can deserve the name of *bounded elements* of $\mathfrak{A}_0$.

Bounded elements of topological *-algebras have a long story. This notion was first introduced by Allan [1] whose goal was to develop a spectral theory for topological *-algebras. The need of introducing bounded elements was suggested by the successful spectral analysis for closed operators in Hilbert space $\mathcal{H}$: a complex number λ is in the spectrum $\sigma(T)$ of a closed (possibly, unbounded) operator T if $T - \lambda I$ has no inverse in the *-algebra $\mathcal{B}(\mathcal{H})$ of bounded operators. Allan's definition sounds as follows: an element x of the topological *-algebra $\mathfrak{A}_0[\mathfrak{t}]$ is *Allan bounded* if there exists $\lambda \neq 0$ such that the set $\{(\lambda^{-1}x)^n;\ n = 1, 2, \dots\}$ is a bounded subset of $\mathfrak{A}_0[\mathfrak{t}]$. Of course one would expect that Allan bounded elements of some topological *-algebra of (possibly) unbounded operators (e.g., an O*-algebra) are its bounded operators; but this is not the

Camillo Trapani, Dipartimento di Matematica e Informatica, Università di Palermo, I-90123 Palermo, Italy, e-mail: camillo.trapani@unipa.it
Francesco Tschinke, Dipartimento di Matematica e Informatica, Università di Palermo, I-90123 Palermo, Italy, e-mail: francesco.tschinke@unipa.it
https://doi.org/10.1515/9783110413557-016

case as we shall see. Thus Allan bounded elements need not be realized by bounded operators in some $\star$-representation of $\mathfrak{A}_0$ in Hilbert space, as it would be natural to expect.

Bounded elements in purely algebraic terms have been considered by Vidav [18] and Schmüdgen [11] with respect to some (positive) wedge. This purely algebraic definition can be extended by considering as strongly positive elements those belonging to the $\mathfrak{t}$-closure in $\mathfrak{A}_0$ of the, say, algebraic cone of positive elements of a $\star$-algebra. The main result is that *order bounded* elements, as we call them, allow equivalent characterizations in terms of continuous positive linear functionals and also in terms of $\star$-representations, that, if the positive wedge is a cone, are sufficiently many to separate points of $\mathfrak{A}_0$.

Of course, several other possibilities for defining bounded elements can be considered. For instance, one may say that x is *left* $\mathfrak{t}$*-bounded*, if there exists $\gamma_x > 0$ such that

$$p_\alpha(xy) \le \gamma_x p_\alpha(y), \quad \forall \alpha \in \Delta; \forall y \in \mathfrak{A}_0,$$

where $\{p_\alpha; \alpha \in \Delta\}$ is a directed family of seminorms defining the topology $\mathfrak{t}$ of $\mathfrak{A}_0$ [6]; or *spectrally bounded* if its spectrum is a bounded subset of the complex plane. Moreover some attempts to extend this notion to the larger set-up of locally convex quasi $\star$-algebras [12, 14, 15] or locally convex partial $\star$-algebras [4, 5] has been done. But in these cases, Allan's notion cannot be adopted, since powers of a given element x need not be defined.

The class of FR*-algebras considered here enjoys a series of nice properties. In particular, order bounded elements are exactly those realized by bounded operators by *any* continuous $\star$-representation of $\mathfrak{A}_0$ in Hilbert space. In [16], a study of some spectral properties of an FR*-algebra was undertaken, under the additional assumption that the set of bounded elements, which is often a C*-normed algebra, is in fact a C*-algebra. This assumption is quite strong and not fulfilled in general by topological unbounded operator algebras, which are in a sense the model to which the whole study is aimed to. Removing this assumption requires to invoke the theory of partial $\star$-algebras or quasi $\star$-algebras [3] on which some similar studies have been performed in a recent past [2, 5, 8].

The paper is organized as follows. After some preliminaries (Section 2), we review in Section 3 the main properties of FR-algebras we will need in the sequel. In Section 4 we revisit the spectral analysis begun in [16] by considering the topological $\star$-algebra $\mathfrak{A}_0$ where we started from as a subspace of a conveniently constructed topological quasi $\star$-algebra $\mathfrak{A}$. With this approach, we can avoid the assumption that bounded elements of $\mathfrak{A}_0$ constitute a C*-algebra contained in $\mathfrak{A}_0$, but, as a drawback, we are forced to go beyond the original framework of topological $\star$-algebras. This is not so surprising if we consider that a spectral theory for unbounded operator algebras cannot be successfully formulated within the same algebra, without dismissing the traditional spectral theory of closed (or closable) operators, considered as single objects.

Topological algebras have been, as well-known, the main research subject of the late Prof. Anastasios Mallios. His monography [9] remains one of the cornerstones of the literature on this topic. One of the authors of this note (C.T.) had many occasions of discussing Mathematics with him in Athens and serves both a human and scientific indelible memory of him.

2 Preliminaries and basic results

The following preliminary definitions will be needed in the sequel. For more details we refer to [3, 10].

2.1 Basics

Let $\mathcal{H}$ be a complex Hilbert space and $\mathcal{D}$ a dense subspace of $\mathcal{H}$. We denote by $\mathcal{L}^\dagger(\mathcal{D},\mathcal{H})$ the set of all (closable) linear operators X such that $D(X) = \mathcal{D}$, $D(X^\star) \supseteq \mathcal{D}$ and by $\mathcal{L}^\dagger(\mathcal{D})$ the subspace consisting of all operators $X \in \mathcal{L}^\dagger(\mathcal{D},\mathcal{H})$ such that $X\mathcal{D} \subset \mathcal{D}$, $X^\star\mathcal{D} \subset \mathcal{D}$. The set $\mathcal{L}^\dagger(\mathcal{D})$ is a $\star$-algebra with respect to the ordinary operations of addition, multiplication by scalars, multiplication (defined as composition of maps) and involution $X \mapsto X^\dagger := X^\star \upharpoonright_{\mathcal{D}}$. We denote by $I_{\mathcal{D}}$ the identity operator in $\mathcal{D}$. Clearly, $I_{\mathcal{D}}$ is the unit of $\mathcal{L}^\dagger(\mathcal{D})$. A $\star$-subalgebra $\mathfrak{M}$ of $\mathcal{L}^\dagger(\mathcal{D})$, containing the identity $I_{\mathcal{D}}$, is called an *O*-algebra* [10].

Let $\mathfrak{M}$ be a †-invariant subset of $\mathcal{L}^\dagger(\mathcal{D},\mathcal{H})$ (i.e., an O*-family on $\mathcal{D}$), containing the identity $I_{\mathcal{D}}$. The space $\mathcal{D}$ endowed with the graph topology $t_{\mathfrak{M}}$, defined by the seminorms

$$\xi \in \mathcal{D} \to \|X\xi\|, \quad X \in \mathfrak{M},$$

will be denoted by $\mathcal{D}_{\mathfrak{M}}$. If $\mathfrak{M}$ is a closed O*-family, the collection of bounded subsets of $\mathcal{D}[t_{\mathfrak{M}}]$ is the same of that corresponding to $t_{\mathcal{L}^\dagger(\mathcal{D},\mathcal{H})}$ and it is denoted by $\mathcal{B}$.

A (closed) O*-family $\mathfrak{M}$ on $\mathcal{D}$ can be endowed with some *traditional* operator topologies:

- the *weak operator topology* τ_w defined by the seminorms

$$X \in \mathfrak{M} \to p_{\xi,\eta}(X) := |\langle X\xi|\eta\rangle|, \quad \xi, \eta \in \mathcal{D};$$

- the *strong operator topology* τ_s defined by the seminorms

$$X \in \mathfrak{M} \to p_\xi(X) := \|X\xi\|, \quad \xi \in \mathcal{D};$$

- the *strong* operator topology* τ_{s^*} defined by the seminorms

$$X \in \mathfrak{M} \to p^*_\xi(X) := \max\{\|X\xi\|, \|X^\dagger\xi\|\}, \quad \xi \in \mathcal{D},$$

that mimick the corresponding topologies of bounded operator algebras.

The following *uniform* topologies τ_u, τ^u and τ^u_* generalize the norm of $\mathcal{B}(\mathcal{H})$.

- τ_u defined by the seminorms

$$p_{\mathcal{M}}(X) = \sup_{\xi,\eta\in\mathcal{M}} |\langle X\xi|\eta\rangle|; \quad \mathcal{M} \in \mathcal{B};$$

- τ^u defined by the seminorms

$$p^{\mathcal{M}}(X) = \sup_{\xi\in\mathcal{M}} \|X\xi\|; \quad \mathcal{M} \in \mathcal{B}$$

- τ^u_* defined by the seminorms

$$p^{\mathcal{M}}_*(X) = \max\{p^{\mathcal{M}}(X), p^{\mathcal{M}}(X^\dagger)\}; \quad \mathcal{M} \in \mathcal{B}.$$

It is easily seen that $\mathfrak{M}[\tau_w]$ and $\mathfrak{M}[\tau_u]$ are topological $\star$-algebras.

We list here some easy properties of the topologies defined above. For every bounded subset $\mathcal{M}$ of $\mathcal{D}[t]$, one has

- there exists $\gamma_{\mathcal{M}} > 0$ such that $p_{\mathcal{M}}(X) \le \gamma_{\mathcal{M}}\, p^{\mathcal{M}}(X), \quad \forall X \in \mathfrak{M}$.
- $p_{\mathcal{M}}(X^\dagger X) = p^{\mathcal{M}}(X)^2, \quad \forall X \in \mathfrak{M}$.

As shown in [3, Prop. 4.2.3] $\mathcal{L}^\dagger(\mathcal{D},\mathcal{H})[\tau^u_*]$ is complete. It is also complete when endowed with the topology τ_{s^*}.

Let $\mathfrak{A}_0$ be a $\star$-algebra and $\mathcal{D}_\pi$ a dense domain in a certain Hilbert space $\mathcal{H}_\pi$. A linear map π from $\mathfrak{A}_0$ into $\mathcal{L}^\dagger(\mathcal{D}_\pi)$ such that:

(i) $\pi(a^*) = \pi(a)^\dagger, \quad \forall a \in \mathfrak{A}_0$,

(ii) if $a, b \in \mathfrak{A}_0$, then $\pi(ab) = \pi(a)\pi(b)$,

is called a *$\star$-representation* of $\mathfrak{A}_0$. Moreover, if $\mathfrak{A}_0$ has a unit $e \in \mathfrak{A}_0$, we assume $\pi(e) = I_{\mathcal{D}_\pi}$, the identity of $\mathcal{D}_\pi$.

A $\star$-representation π of a topological $\star$-algebra $\mathfrak{A}_0[\mathrm{t}]$ is said to be a $(\mathrm{t}, \tau_{\mathrm{w}})$-continuous if, for every $\xi, \eta \in \mathcal{D}_\pi$, there exists a t-continuous seminorm p on $\mathfrak{A}_0$ such that

$$|\langle \pi(a)\xi|\eta\rangle| \le p(a), \quad \forall a \in \mathfrak{A}_0.$$

A linear functional ω on $\mathfrak{A}_0$ is called positive if $\omega(a^*a) \ge 0$, for every $a \in \mathfrak{A}_0$. To every positive linear functional ω on $\mathfrak{A}_0$ there corresponds a Hilbert space $\mathcal{H}_\omega$ and a linear map λ_ω from $\mathfrak{A}_0$ into a dense subspace $\lambda_\omega(\mathfrak{A}_0) \subset \mathcal{H}_\omega$ and a $\star$-representation π_ω acting on a dense domain $\mathcal{D}_{\pi_\omega}$ such that $\lambda_\omega(\mathfrak{A}_0) \subset \mathcal{D}_{\pi_\omega} \subset \mathcal{H}_\pi$ and

$$\omega(b^*xa) = \langle \pi_\omega(x)\lambda_\omega(a)|\lambda_\omega(b)\rangle, \quad \forall a, b, x \in \mathfrak{A}_0.$$

The representation π_ω can be taken to be closed [10]. If $\mathfrak{A}_0$ has a unit e, then there exists a vector ξ_ω such that $\lambda_\omega(\mathfrak{A}_0) = \{\pi_\omega(a)\xi_\omega,\ a \in \mathfrak{A}_0\}$ and

$$\omega(x) = \langle \pi_\omega(x)\xi_\omega|\xi_\omega\rangle, \quad \forall x \in \mathfrak{A}_0.$$

We will refer to π_ω as to the GNS $\star$-representation of $\mathfrak{A}_0$ defined by ω.

2.2 Left $\mathfrak{t}$-boundness vs. Allan boundness

It is worth comparing the two notions of boundness already considered. We have the following

Proposition 2.1. *Let $\mathfrak{A}_0[\mathfrak{t}]$ be a topological *-algebra. If $x \in \mathfrak{A}_0[\mathfrak{t}]$ is left $\mathfrak{t}$-bounded, then x is Allan bounded.*

Proof. Let $\{p_\alpha; \alpha \in \Delta\}$ be a directed family of seminorms defining the topology $\mathfrak{t}$, and let us consider a *left $\mathfrak{t}$-bounded* element $x \in \mathfrak{A}_0[\mathfrak{t}]$, that is:

$$\exists \gamma_x > 0 : p_\alpha(xy) \leq \gamma_x p_\alpha(y), \forall \alpha \in \Delta, \forall y \in \mathfrak{A}_0.$$

In particular, we can put $y = x^{n-1}$: in this way, we have: $p_\alpha(x^n) \leq \gamma_x p_\alpha(x^{n-1})$, then, iterating, $p_\alpha(\frac{x^n}{\gamma_x^n}) \leq \frac{1}{\gamma_x} p_\alpha(x)$. Hence, the set $\{(\lambda^{-1}x)^n;\ n = 1, 2, \dots\}$ is a bounded subset of $\mathfrak{A}_0[\mathfrak{t}]$ with $\lambda = \gamma_x$. □

The converse is not necessarily true, as shown in the following example.

Example 2.2. *Let us consider the topological *-algebra $\mathcal{L}^\dagger(\mathcal{D})[\tau_w]$, where τ_w is the weak operator topology. It is simple to verify that if $X \in \mathcal{L}^\dagger(\mathcal{D})[\tau_w]$ is left τ_w-bounded then X is a bounded operator. In fact, if X is left τ_w-bounded, then there exists $\gamma_x > 0$: $|\langle XY\xi|\eta\rangle| \leq \gamma_x |\langle Y\xi|\eta\rangle|$, for all $\xi, \eta \in \mathcal{D}$ and for all $Y \in \mathcal{L}^\dagger(\mathcal{D})$. If we choose $Y = 1$, we have*

$$|\langle X\xi|\eta\rangle| \leq \gamma_x |\langle \xi|\eta\rangle| \leq \gamma_x \|\xi\| \|\eta\|, \quad \xi, \eta \in \mathcal{D}.$$

This implies that X is a bounded operator.

It is clear that if X is an unbounded operator of $\mathcal{L}^\dagger(\mathcal{D})$ such that $X^2 = 0$, then it is Allan bounded (as any nilpotent operator), but, X being unbounded, cannot be left τ_w-bounded. In order to construct an operator $X \in \mathcal{L}^\dagger(\mathcal{D})$ such that $X^2 = 0$ we proceed as follows. Let $\{e_k\}_{k \in \mathbb{N}^+}$ be an ortonormal basis of the separable Hilbert space $\mathcal{H}$. Let $\mathcal{D}$ be the following subspace of $\mathcal{H}$:

$$\mathcal{D} := \left\{ f = \sum_{k=1}^{\infty} f_k e_k \in \mathcal{H} : \quad \sum_{k=1}^{\infty} |f_k|^2 k^{2p} < \infty, \quad \forall p \in \mathbb{N} \right\}$$

and define

$$Xf := \sum_k k f_{2k} e_{2k+1}, \quad f \in \mathcal{D}.$$

Then X is an unbounded operator of $\mathcal{L}^\dagger(\mathcal{D})$ and $X^2 f = 0$.

3 FR*-algebras: an overlook

In this section we give a short overview to FR*-algebras, already studied in [16].

Definition 3.1. *A topological* **-algebra* $\mathfrak{A}_0[\mathfrak{t}]$ *is called* faithfully representable, *shortly an* FR*-algebra, *if for every* $x \in \mathfrak{A}_0 \setminus \{0\}$ *there exists a* $(\mathfrak{t}, \tau_w)$*-continuous* **-representation* π *of* $\mathfrak{A}_0$ *such that* $\pi(x) \neq 0$.

Let $\mathrm{Rep}_c(\mathfrak{A}_0)$ denote the family of all $(\mathfrak{t}, \tau_w)$-continuous *-representation of $\mathfrak{A}_0$. Then, as shown in [16], the following statements are equivalent:
(i) $\mathfrak{A}_0$ is an FR*-algebra;
(ii) For every $x \in \mathfrak{A}_0 \setminus \{0\}$, there exists a $\mathfrak{t}$-continuous positive linear functional ω such that $\omega(x^* x) > 0$.

The set of all $\mathfrak{t}$-continuous positive linear functionals on $\mathfrak{A}_0$ will be denoted by $\mathcal{P}_c(\mathfrak{A}_0)$.

Remark 3.2. *It is well known that in a C*-algebra every positive linear functional is continuous. But, apart from C*-algebras, there exist nonnormed topological algebras* $\mathfrak{A}_0[\mathfrak{t}]$ *that share the same property. For instance, every Fréchet Q-algebra with unit has this property [7, Corollary 17.9]. It is easily seen that there is equivalence between the following two statements:*
a) Every **-representation is* $(\mathfrak{t}, \tau_w)$*-continuous.*
b) Every positive linear functional is $\mathfrak{t}$*-continuous.*

3.1 Order structure

Let $\mathfrak{A}_0$ be a *-algebra with unit e. We denote by

$$(\mathfrak{A}_0)^+_{\mathrm{alg}} = \left\{ \sum_{k=1}^{n} x_k^* x_k,\ x_k \in \mathfrak{A}_0,\ n \in \mathbb{N} \right\}$$

the set (wedge) of positive elements of $\mathfrak{A}_0$.

If $\mathfrak{A}_0[\mathfrak{t}]$ is a topological *-algebra, *strongly positive* elements of $\mathfrak{A}_0$ are then defined as members of $\overline{(\mathfrak{A}_0)^+_{\mathrm{alg}}}^{\mathfrak{t}}$. We put $(\mathfrak{A}_0)^+ := \overline{(\mathfrak{A}_0)^+_{\mathrm{alg}}}^{\mathfrak{t}}$.

The set $(\mathfrak{A}_0)^+$ is an *m-admissible wedge* in the sense of Schmüdgen [10, Sect. 1.4]; i.e.,
(1) $e \in (\mathfrak{A}_0)^+$;
(2) $x + y \in (\mathfrak{A}_0)^+, \quad \forall x, y \in (\mathfrak{A}_0)^+$;
(3) $\lambda x \in (\mathfrak{A}_0)^+, \quad \forall x \in (\mathfrak{A}_0)^+,\ \lambda \geq 0$;
(4) $a^* x a \in (\mathfrak{A}_0)^+, \quad \forall x \in (\mathfrak{A}_0)^+,\ a \in \mathfrak{A}_0$.

The wedge $(\mathfrak{A}_0)^+$ defines an order on the real vector space $(\mathfrak{A}_0)_h = \{x \in \mathfrak{A}_0 : x = x^*\}$ by $x \leq y \Leftrightarrow y - x \in (\mathfrak{A}_0)^+$.

This order can be used to introduce a further notion of *bounded* element.

Let $x \in \mathfrak{A}_0$; put $\mathfrak{R}(x) = \frac{1}{2}(x + x^*)$, $\mathfrak{I}(x) = \frac{1}{2i}(x - x^*)$. Then $\mathfrak{R}(x), \mathfrak{I}(x) \in (\mathfrak{A}_0)_h$ (the set of selfadjoint elements of $\mathfrak{A}_0$) and $x = \mathfrak{R}(x) + i\mathfrak{I}(x)$.

An element $x \in \mathfrak{A}_0$ is called *order bounded* if there exists $\gamma \geq 0$ such that

$$\pm\Re(x) \leq \gamma e; \qquad \pm\Im(x) \leq \gamma e.$$

We denote by $(\mathfrak{A}_0)_b$ the family of order bounded elements. It can be proved that $(\mathfrak{A}_0)_b$ is a *-algebra.

For $x \in (\mathfrak{A}_0)_h$, put

$$\|x\|_b := \inf\{\gamma > 0 : -\gamma e \leq x \leq \gamma e\}.$$

Then, $\|\cdot\|_b$ is a seminorm on the real vector space $((\mathfrak{A}_0)_b)_h$ and, if $\mathfrak{A}_0 \cap (-(\mathfrak{A}_0)^+) = \{0\}$, $\|\cdot\|_b$ is a norm on $((\mathfrak{A}_0)_b)_h$.

FR*-algebras can be characterized as follows.

Proposition 3.1. *Let $\mathfrak{A}_0[t]$ be a topological *-algebra with unit e. The following statements are equivalent.*
(i) $(\mathfrak{A}_0)^+ \cap (-(\mathfrak{A}_0)^+) = \{0\}$, i.e. $(\mathfrak{A}_0)^+$ is a cone.
(ii) $\mathcal{P}_c(\mathfrak{A}_0)$ is sufficient; i.e., for every $a \in \mathfrak{A}_0 \setminus \{0\}$, there exists $\omega \in \mathcal{P}_c(\mathfrak{A}_0)$ such that $\omega(a^*a) > 0$.
(iii) $\mathfrak{A}_0[t]$ is an FR-algebra.*

Remark 3.3. *The notion of FR*-algebra looks very much close to that of *-semisimple topological *-algebra. It is convenient to compare the two definitions. We remind that, if $\mathfrak{A}_0[t]$ is a topological *-algebra, one can consider the set $R'(\mathfrak{A}_0)$ of all* bounded *topologically irreducible *-representations π on a Hilbert space $\mathcal{H}$ that are continuous from $\mathfrak{A}_0[t]$ into $\mathcal{B}(\mathcal{H})[\|\cdot\|]$. The *-radical of $\mathfrak{A}_0$ is then defined by*

$$R^*_{\mathfrak{A}_0} := \bigcap\{Ker(\pi);\ \pi \in R'(\mathfrak{A}_0)\}.$$

*A topological *-algebra $\mathfrak{A}_0[t]$ is called* *-semisimple *if $R^*_{\mathfrak{A}_0} = \{0\}$.*

Following a similar path we can consider the set

$$\mathfrak{R}^*_{\mathfrak{A}_0} := \bigcap\{Ker(\pi);\ \pi \in Rep_c(\mathfrak{A}_0)\}.$$

*Clearly a topological *-algebra is an FR*-algebra if, and only if, $\mathfrak{R}^*_{\mathfrak{A}_0} = \{0\}$.*

*Since $R'(\mathfrak{A}_0) \subseteq Rep_c(\mathfrak{A}_0)$, then $\mathfrak{R}^*_{\mathfrak{A}_0} \subseteq R^*_{\mathfrak{A}_0}$. Thus every *-semisimple topological *-algebra is an FR*-algebra. The converse is false in general. Indeed, let $\mathfrak{M}$ be an O*-algebra on a domain $\mathcal{D}$ of Hilbert space $\mathcal{H}$, endowed with the weak operator topology. The identical *-representation $\iota_{\mathfrak{M}}$, $\iota_{\mathfrak{M}} : X \in \mathfrak{M} \to X \in \mathfrak{M}$, is clearly faithful, so that if $\mathfrak{M}$ has no nontrivial invariant subspaces, then $\mathfrak{R}^*_{\mathfrak{M}}$ reduces to $\{0\}$. Thus $\mathfrak{M}$ is an FR*-algebra. This applies in particular to the O*-algebra $\mathfrak{M}(q,p)$ on the Schwartz space $\mathcal{S}(\mathbb{R})$, generated by the operator q of multiplication by the real variable x and the operator $p = -i\frac{d}{dx}$. This algebra does not admit any* bounded *representation; thus the notion of *-semisimplicity becomes meaningless in this case.*

Proposition 3.2. *Let $\mathfrak{A}_0[\mathsf{t}]$ be an* FR*-*algebra with unit e.*

Assume that the following condition (P) holds

$$\text{(P)}\quad \textit{If } y \in \mathfrak{A}_0 \textit{ and } \omega(y) \geq 0, \textit{for every } \omega \in \mathcal{P}_c(\mathfrak{A}_0), \textit{then } y \in (\mathfrak{A}_0)^+.$$

Then, for an element $x \in \mathfrak{A}_0$, the following statements are equivalent.

(i) $x \in (\mathfrak{A}_0)^+$;

(ii) $\omega(x) \geq 0$, *for every* $\omega \in \mathcal{P}_c(\mathfrak{A}_0)$

(iii) $\pi(x) \geq 0$, *for every* $\pi \in \mathrm{Rep}_c(\mathfrak{A}_0)$.

Theorem 2. *Let $\mathfrak{A}_0[\mathsf{t}]$ be a topological *-algebra with unit e and assume that condition (P) holds. For $x \in \mathfrak{A}_0$, the following statements are equivalent.*

(i) *x is order bounded.*

(ii) *There exists $\gamma_x > 0$ such that*

$$|\omega(a^*xa)| \leq \gamma_x \omega(a^*a), \quad \forall \omega \in \mathcal{P}_c(\mathfrak{A}_0),\ a \in \mathfrak{A}_0.$$

(iii) *There exists $\gamma_x > 0$ such that*

$$|\omega(b^*xa)| \leq \gamma_x \omega(a^*a)^{1/2}\omega(b^*b)^{1/2}, \quad \forall \omega \in \mathcal{P}_c(\mathfrak{A}_0),\ a, b \in \mathfrak{A}_0.$$

(iv) *$\pi(x)$ is a bounded operator, for every $\pi \in \mathrm{Rep}_c(\mathfrak{A}_0)$, and*

$$\sup\{\|\overline{\pi(x)}\|,\ \pi \in \mathrm{Rep}_c(\mathfrak{A}_0)\} < \infty.$$

Let x be order bounded and define

$$\mathsf{q}(x) = \sup\{|\omega(b^*xa)|;\ \omega \in \mathcal{P}_c(\mathfrak{A}_0),\ a, b \in \mathfrak{A}_0;\ \omega(a^*a) = \omega(b^*b) = 1\}.$$

Then, $\mathsf{q}(x) = \|x\|_b$, for every $x = x^* \in (\mathfrak{A}_0)_b$. Since q extends $\|\cdot\|_b$, we adopt the notation $\|\cdot\|_b$ for both. By (iv) it follows easily that, for every $x \in (\mathfrak{A}_0)_b$,

$$\|x\|_b = \sup\{\|\overline{\pi(x)}\|,\ \pi \in \mathrm{Rep}_c(\mathfrak{A}_0)\}. \tag{1}$$

It is easy to see that $\|\cdot\|_b$ is a norm on $(\mathfrak{A}_0)_b$ such that, for every $x, y \in (\mathfrak{A}_0)_b$,

(i) $\|x^*\|_b = \|x\|_b$;

(ii) $\|xy\|_b \leq \|x\|_b\,\|y\|_b$;

(iii) $\|x^*x\|_b = \|x\|_b^2$.

4 Spectral properties revisited

In what follows a crucial role will be played by an *auxiliary* topology which stems from the family $\mathcal{P}_c(\mathfrak{A}_0)$ of t-continuous positive linear functionals. We define in fact the *strong** topology t_{s^*}, on $\mathfrak{A}_0$ by the family of seminorms

$$x \in \mathfrak{A}_0 \to \omega(x^*x)^{1/2} + \omega(xx^*)^{1/2}, \quad \omega \in \mathcal{P}_c(\mathfrak{A}_0).$$

From the definition itself it follows that $\mathfrak{t}_{s^*}$ is coarser than $\mathfrak{t}$ and that every $\omega \in \mathcal{P}_c(\mathfrak{A}_0)$ is also $\mathfrak{t}_{s^*}$-continuous. Moreover, by an easy application of the Cauchy-Schwarz inequality and by the definition itself of the topology $\mathfrak{t}_{s^*}$, the sesquilinear form φ_ω defined by

$$\varphi_\omega(x, y) = \omega(y^* x), \quad x, y \in \mathfrak{A}_0. \tag{1}$$

is *jointly* continuous for $\mathfrak{t}_{s^*}$.

Proposition 4.1. *Let $\mathfrak{A}_0$ be a* FR**-algebra with unit e. Then $\|\cdot\|_b$ is a C^*-norm on $(\mathfrak{A}_0)_b$. Moreover if $\mathfrak{A}_0$ is $\mathfrak{t}_{s^*}$-complete, then $(\mathfrak{A}_0)_b$ is a C*-algebra with norm $\|\cdot\|_b$.*

4.1 Constructing a locally convex quasi *-algebra over $(\mathfrak{A}_0)_b$

In [16] a notion of spectrum was proposed, under the assumption that $(\mathfrak{A}_0)_b$ is a C*-algebra contained in $\mathfrak{A}_0$. We remove this condition, but we suppose that $(\mathfrak{A}_0)_b$ is $\mathfrak{t}_{s^*}$-dense in $\mathfrak{A}_0$. For doing this, as announced in the Introduction, we need to build up a larger structure (actually, a quasi*-algebra [3]), having $\mathfrak{A}_0$ as a subspace. The construction runs as follows.

Lemma 4.2. *For every $x \in (\mathfrak{A}_0)_b$, the multiplications $a \mapsto ax$, $a \mapsto xa$, $a \in \mathfrak{A}_0$, are $\mathfrak{t}_{s^*}$-continuous.*

Proof. We first observe that, if $\omega \in \mathcal{P}_c(\mathfrak{A}_0)$ and $y \in \mathfrak{A}_0$, the linear functional ω_y on $\mathfrak{A}_0$ defined by $\omega_y(b) = \omega(y^* by)$, $b \in \mathfrak{A}_0$, is $\mathfrak{t}$-continuous, since the multiplication in $\mathfrak{A}_0$ is separately continuous; i.e. $\omega_y \in \mathcal{P}_c(\mathfrak{A}_0)$. If $x \in (\mathfrak{A}_0)_b$ we have, using (ii) of Theorem 2,

$$\begin{aligned}\omega(x^* a^* ax)^{1/2} + \omega(axx^* a)^{1/2} &= \omega_x(a^* a)^{1/2} + \omega(axx^* a^*)^{1/2}\\ &\le (\omega_x(a^* a)^{1/2} + \omega_x(aa^*)^{1/2}) + \gamma_{x^*x}(\omega(a^* a)^{1/2} + \omega(aa^*)^{1/2})\end{aligned}$$

which proves that the map $a \mapsto ax$ is $\mathfrak{t}_{s^*}$-continuous in $\mathfrak{A}_0$. The continuity of the map $a \mapsto xa$ follows easily by taking adjoints. □

Let us denote by $\mathfrak{A} := \widetilde{\mathfrak{A}_0}[\mathfrak{t}_{s^*}]$, the completion of $\mathfrak{A}_0[\mathfrak{t}_{s^*}]$. The assumption that $(\mathfrak{A}_0)_b$ is $\mathfrak{t}_{s^*}$-dense in $\mathfrak{A}_0$, implies, obviously, that $\mathfrak{A} := \widetilde{(\mathfrak{A}_0)_b}[\mathfrak{t}_{s^*}]$. Then, if $a \in \mathfrak{A}$ there exists a net (a_α), of elements of $(\mathfrak{A}_0)_b$, $\mathfrak{t}_{s^*}$-converging to a. Hence, taking into account Lemma 4.2, we can define multiplications as follows

$$a \cdot x := \lim_\alpha a_\alpha x, \qquad x \cdot a := \lim_\alpha x a_\alpha, \quad x \in (\mathfrak{A}_0)_b, a \in \mathfrak{A}$$

and the involution by

$$a^* := \lim_\alpha a_\alpha^*.$$

It is easily seen that these maps are well-defined (i.e., they do not depend on the particular net chosen for approximating a).

Proposition 4.3. *Let $\mathfrak{A}_0$ be an* FR*-algebra, such that $(\mathfrak{A}_0)_b$ is t_{s^*}-dense in $\mathfrak{A}_0$. Then, $\mathfrak{A}[\mathrm{t}_{s^*}]$ is a locally convex quasi *-algebra over $(\mathfrak{A}_0)_b$.*

Proof. By the definition itself of the topology of the completion $\mathfrak{A}$, it follows that both the continuity properties stated in Lemma 4.2 are preserved when a runs over $\mathfrak{A}$. Similarly, the involution $a \mapsto a^*$ is continuous. □

Remark 4.1. *As a quasi *-algebra $\mathfrak{A}$, over $(\mathfrak{A}_0)_b$, $\mathfrak{A}$ is a particular* partial *-algebra *[3]. The latter is characterized by the two lattices $\mathcal{F}^L$, $\mathcal{F}^R$ of, respectively, left- and right-multipliers, respectively $L(S)$, $R(S)$, of a subset $S \in \mathfrak{A}$. For the quasi *-algebra $\mathfrak{A}$ over $(\mathfrak{A}_0)_b$ constructed here, we simply have, for $a \in \mathfrak{A}$*

$$L(a) = R(a) = \begin{cases} \mathfrak{A} & \text{if } a \in (\mathfrak{A}_0)_b \\ (\mathfrak{A}_0)_b & \text{if } a \in \mathfrak{A} \setminus (\mathfrak{A}_0)_b. \end{cases}$$

4.2 Ips-forms

In order to explore spectral properties of $\mathfrak{A}_0[\mathrm{t}]$ we need to introduce a particular class of positive sesquilinear forms on $\mathfrak{A} = \widetilde{\mathfrak{A}_0}[\mathrm{t}_{s^*}]$.

Let φ be a positive sesquilinear form on $\mathfrak{A} \times \mathfrak{A}$. Then we have

$$\varphi(x, y) = \overline{\varphi(y, x)}, \quad \forall\, x, y \in \mathfrak{A}, \tag{2}$$

$$|\varphi(x, y)|^2 \leqslant \varphi(x, x)\varphi(y, y), \quad \forall\, x, y \in \mathfrak{A}. \tag{3}$$

We put

$$N_\varphi = \{x \in \mathfrak{A} : \varphi(x, x) = 0\}.$$

By (3), we have

$$N_\varphi = \{x \in \mathfrak{A} : \varphi(x, y) = 0, \quad \forall\, y \in \mathfrak{A}\},$$

and so N_φ is a subspace of $\mathfrak{A}$ and the quotient space $\mathfrak{A}/N_\varphi := \{\lambda_\varphi(x) \equiv x + N_\varphi; x \in \mathfrak{A}\}$ is a pre-Hilbert space with respect to the inner product $\langle \lambda_\varphi(x) | \lambda_\varphi(y) \rangle = \varphi(x, y)$, $x, y \in \mathfrak{A}$. We denote by $\mathcal{H}_\varphi$ the Hilbert space obtained by completion of $\mathfrak{A}/N_\varphi$.

Let us consider an *ips-form* φ with core $(\mathfrak{A}_0)_b$, that is φ is a positive sesquilinear form on $\mathfrak{A} \times \mathfrak{A}$ satisfying
(ips$_1$)$\lambda_\varphi((\mathfrak{A}_0)_b)$ is dense in $\mathcal{H}_\varphi$;
(ips$_2$)$\varphi(ax, y) = \varphi(x, a^\star y)$, $\forall\, a \in \mathfrak{A}, \forall\, x, y \in (\mathfrak{A}_0)_b$;
(ips$_3$)$\varphi(a^\star x, by) = \varphi(x, (ab)y)$, $\forall\, a \in L(b), \forall\, x, y \in (\mathfrak{A}_0)_b$.

In other words, an ips-form is an *everywhere defined* biweight with core $(\mathfrak{A}_0)_b$, in the sense of [3].

To every ips-form φ on $\mathfrak{A}$, with core $(\mathfrak{A}_0)_b$, there corresponds a triple $(\pi_\varphi, \lambda_\varphi, \mathcal{H}_\varphi)$, where $\mathcal{H}_\varphi$ is a Hilbert space, λ_φ is a linear map from $(\mathfrak{A}_0)_b$ into $\mathcal{H}_\varphi$ and π_φ is a *-representation on $\mathfrak{A}$ in the Hilbert space $\mathcal{H}_\varphi$. We refer to [3] for more details on this celebrated GNS construction.

If $\omega \in \mathcal{P}_c(\mathfrak{A}_0)$, the positive sesquilinear form φ_ω on $\mathfrak{A}_0 \times \mathfrak{A}_0$, defined as in (1), is, as we already remarked, jointly continuous with respect to t_{s^*}. Then φ_ω extends by continuity to $\mathfrak{A} \times \mathfrak{A}$. Let us call $\widetilde{\varphi}_\omega$ this extension.

Lemma 4.4. *If $\omega \in \mathcal{P}_c(\mathfrak{A}_0)$, then $\widetilde{\varphi}_\omega$ is an ips-form on $\mathfrak{A} \times \mathfrak{A}$ with core $\mathfrak{A}_0$.*

Proof. It is easy to check that the conditions (ips$_2$) and (ips$_3$) are satisfied and since each φ_ω, $\omega \in \mathcal{P}_c(\mathfrak{A}_0)$ is jointly t_{s^*}-continuous, then (ips$_1$) is also satisfied, so that every $\widetilde{\varphi}_\omega$, $\omega \in \mathcal{P}_c(\mathfrak{A}_0)$ is an ips-form (see [4, Sect.4]). □

Let us denote by $\mathcal{M}(\mathfrak{A})$ the family of ips-forms on $\mathfrak{A} \times \mathfrak{A}$ defined as above starting from $\mathcal{P}_c(\mathfrak{A}_0)$; i.e.,

$$\mathcal{M}(\mathfrak{A}) = \{\varphi = \widetilde{\varphi}_\omega;\ \omega \in \mathcal{P}_c(\mathfrak{A}_0)\}.$$

As in [4], we extend to families of ips-forms, the notion of *sufficiency* in the following way.

Definition 4.2. *A family $\mathcal{M}$ of ips-forms on $\mathfrak{A} \times \mathfrak{A}$ is sufficient if $x \in \mathfrak{A}$ and $\varphi(x, x) = 0$, for every $\varphi \in \mathcal{M}$, imply $x = 0$.*

By Proposition 3.1, we know that, for a FR-*-algebra $\mathfrak{A}_0$, $\mathcal{P}_c(\mathfrak{A}_0)$ is sufficient. However the corresponding family of ips-forms $\mathcal{M}(\mathfrak{A})$ need not be sufficient in the sense of Definition 4.2.

Definition 4.3. *We say that the FR-*-algebra $\mathfrak{A}_0$ is* full *if the family $\mathcal{M}(\mathfrak{A})$ is sufficient.*

It is not always simple to determine the whole set $\mathcal{M}(\mathfrak{A})$. But it is clear that if some subset $\mathcal{N}$ of $\mathcal{M}(\mathfrak{A})$ is sufficient so is $\mathcal{M}(\mathfrak{A})$. So that in what follows we suppose that $\mathcal{M}(\mathfrak{A})$ has a sufficient subset $\mathcal{N}$ which is *balanced* in the following sense [13, 17]: if $\varphi \in \mathcal{N}$ then $\varphi_a \in \mathcal{N}$, for every $a \in (\mathfrak{A}_0)_b$ where $\varphi_a(x, y) = \varphi(xa, ya)$, $x, y \in \mathfrak{A}$. This choice reveals to be more flexible for examining examples. The term *balanced* is borrowed by Yood, who gave a similar definition for positive functionals [19].

Remark 4.4. *Let $\mathcal{N}$ be a balanced subset of $\mathcal{M}(\mathfrak{A})$. The sesquilinear forms of $\mathcal{N}$ can be used to define on $\mathfrak{A}$ several topologies. We will only consider the topology $\mathsf{t}_{s^*}^{\mathcal{N}}$ defined by the following family of seminorms:*

$\mathsf{t}^{\mathcal{N}}_{s^*}$: $x \mapsto \varphi(x,x)^{1/2} + \varphi(x^*,x^*)^{1/2}, \quad \varphi \in \mathcal{N}.$
It is easily seen that $\mathsf{t}^{\mathcal{N}}_{s^*}$ *induces on* $\mathfrak{A}_0$ *the topology* t_{s^*}.

For full FR*-algebras we can define *extensions* of the multiplication in the following way. We prevent the raeder that what follows is an application of results obtained in [5] to the present situation. Proofs which are simple adaptations of those given in the quoted paper are therefore omitted.

Definition 4.5. *Let* $\mathfrak{A}_0$ *be a full* FR**-algebra with unit* e *and* $x, y \in \mathfrak{A}$ *and* $\mathcal{N}$ *a sufficient balanced subset of* $\mathcal{M}(\mathfrak{A})$.

- *We say that the weak multiplication* $x \square y$ *is well-defined if there exists* $z \in \mathfrak{A}$ *such that:*

$$\varphi(ya, x^* b) = \varphi(za, b), \ \forall a, b \in (\mathfrak{A}_0)_b, \forall \varphi \in \mathcal{N}.$$

  In this case, we put $x \square y := z$.
- *We say that the strong multiplication* $x \bullet y$ *is well-defined (and* $x \in L^s(y)$ *or* $y \in R^s(x)$*) if there exists* $w \in \mathfrak{A}$ *such that:*

$$\varphi(wa, z^* b) = \varphi(ya, (x^* z^*)b) \ \text{ whenever } \ z \in L(x), \ \forall \varphi \in \mathcal{N}, \ \forall a, b \in (\mathfrak{A}_0)_b,$$

  and

$$\varphi(w^* a, vb) = \varphi(x^* a, (yv)b) \ \text{ whenever } \ v \in R(y), \ \forall \varphi \in \mathcal{N}, \ \forall a, b \in \mathfrak{A}_0.$$

  In this case, we put $x \bullet y := w$.

It is obvious that the these defifinitions depend on the choice of $\mathcal{N}$. The following result is immediate.

Proposition 4.5. *If the* FR*-*-algebra* $\mathfrak{A}_0$ *is full, then* $\mathfrak{A}$ *is a partial *-algebra with respect to the weak multiplication defined by any sufficient balanced subset* $\mathcal{N}$ *of* $\mathcal{M}(\mathfrak{A})$.

Remark 4.6. *The uniqueness of each element* z, w *in the previous proposition results from the sufficiency of the family* $\mathcal{N}$. *Moreover, it is clear that if* $x \bullet y$ *is well-defined, then* $x \square y$ *is well-defined too. In particular, if* $x, y \in (\mathfrak{A}_0)_b$ *then,* $x \bullet y$ *and* $x \square y$ *are both well defined and* $x \bullet y = x \square y = xy$.

An easy consequence of the previous definitions is the following form of the *associative law.*

Proposition 4.6. *Let* $x, y, z \in \mathfrak{A}$. *Assume that* $x \square y$, $(x \square y) \square z$ *and* $y \bullet z$ *are all well-defined. Then* $x \in L(y \bullet z)$ *and*

$$x \square (y \bullet z) = (x \square y) \square z.$$

An element x has a *strong inverse* if there exists $x^{-1} \in \mathfrak{A}$ such that $x \bullet x^{-1} = x^{-1} \bullet x = e$. The mixed associativity of Proposition 4.6 implies that, if a strong inverse of x exists, then it is unique.

Definition 4.7.
An element $x \in \mathfrak{A}$ is called $\mathcal{N}$-bounded *if there exists $\gamma > 0$ such that:*

$$|\varphi(xa, b)| \le \gamma\, \varphi(a, a)^{1/2} \varphi(b, b)^{1/2}, \ \forall \varphi \in \mathcal{N},\ a, b \in \mathfrak{A}_0\,.$$

Remark 4.8. *It is clear that if, $x \in \mathfrak{A}_0$, then x is $\mathcal{M}(\mathfrak{A})$-bounded, if and only if it is order bounded, by Theorem 2.*

Let us now define:

$$\begin{aligned} q_{\mathcal{N}}(x) &:= \inf\{\gamma > 0 : \varphi(xa, xa) \le \gamma^2 \varphi(a, a),\ \forall \varphi \in \mathcal{N},\ \forall a \in (\mathfrak{A}_0)_b\} \\ &= \sup\{\varphi(xa, xa)^{1/2} : \varphi \in \mathcal{N},\ a \in (\mathfrak{A}_0)_b,\ \varphi(a, a)^{1/2} = 1\} \end{aligned}$$

and

$$\mathcal{D}(q_{\mathcal{N}}) := \{x \in \mathfrak{A} : x \text{ is } \mathcal{N}\text{-bounded}\}.$$

It is clear that $(\mathfrak{A}_0)_b \subseteq \mathcal{D}(q_{\mathcal{N}})$.

Then the following holds:

Proposition 4.7. *Let x, y be $\mathcal{N}$-bounded elements of $\mathfrak{A}$. The following statements hold:*
(i) x^* is $\mathcal{N}$-bounded also, and $q_{\mathcal{N}}(x) = q_{\mathcal{N}}(x^*)$;
(ii) If $x \square y$ is well-defined, then $x \square y$ is $\mathcal{N}$-bounded and

$$q_{\mathcal{N}}(x \square y) \le q_{\mathcal{N}}(x)\, q_{\mathcal{N}}(y).$$

The proof of (ii) is a consequence of the following inequality (4).

Let $x, y \in \mathcal{D}(q_{\mathcal{N}})$ then, for every $\varphi \in \mathcal{N}$ and for every $a, b \in (\mathfrak{A}_0)_b$ we get

$$\begin{aligned} |\varphi(ya, x^* b)| &\le \varphi(ya, ya)^{1/2} \varphi(x^* b, x^* b)^{1/2} \\ &\le q_{\mathcal{N}}(x)^{1/2} q_{\mathcal{N}}(y)^{1/2} \varphi(a, a)^{1/2} \varphi(b, b)^{1/2}. \end{aligned} \tag{4}$$

As we have seen at the beginning of this section, to every $\varphi \in \mathcal{N}$ there corresponds a Hilbert space $\mathcal{H}_\varphi$ constructed from cosets $\lambda_\varphi(a)$, $a \in \mathfrak{A}$. Then (4) can be read as follows. Let us define, for $x, y \in \mathfrak{A}$ fixed, the sesquilinear form on $\lambda_\varphi((\mathfrak{A}_0)_b)$

$$\Theta_{x,y}(\lambda_\varphi(a), \lambda_\varphi(b)) := \varphi(ya, x^* b), \quad a, b \in (\mathfrak{A}_0)_b.$$

Then, by (4), $\Theta_{x,y}$ is a bounded sesquilinear form on $\lambda_\varphi((\mathfrak{A}_0)_b) \times \lambda_\varphi((\mathfrak{A}_0)_b)$ and it extends to $\mathcal{H}_\varphi \times \mathcal{H}_\varphi$. Hence, there exist a bounded operator $B_{\varphi,x,y}$ such that

$$\Theta_{x,y}(\lambda_\varphi(a), \lambda_\varphi(b)) = \varphi(ya, x^* b) = \langle B_{\varphi,x,y} \lambda_\varphi(a) | \lambda_\varphi(b) \rangle, \quad \forall a, b \in (\mathfrak{A}_0)_b).$$

Assume now that the following condition hold

(wb)there exists $z \in \mathfrak{A}$ such that, for every $\varphi \in \mathcal{N}$, $B_{\varphi,x,y}\lambda_\varphi(a) = \lambda_\varphi(za)$ for every $a \in (\mathfrak{A}_0)_b$.

In this case, $x \Box y$ is well defined and equals z.

If the condition (wb) is satisfied for every $x, y \in \mathcal{D}(q_\mathcal{N})$, then we say that $\mathcal{N}$ is *well-behaved*. In this case, $\mathcal{D}(q_\mathcal{N})$ is a C*-algebra with the weak or strong multiplication and the norm $q_\mathcal{N}$. This has been proved in [5], under a stronger condition.

The above discussion makes clear that requiring that $\mathcal{D}(q_\mathcal{N})$ is a C*-algebra, as we did in [5], is a quite strong condition, rarely fulfilled in concrete examples. But in any case we have:

Proposition 4.8. *$\mathcal{D}(q_\mathcal{N})$ is complete under the norm $q_\mathcal{N}$.*

Proof. Let $\{x_n\}$ be a Cauchy sequence with respect to the norm $q_\mathcal{N}$. Then $\{x_n^*\}$ is Cauchy too. Since, for every $\varphi \in \mathcal{N}$ and $a \in (\mathfrak{A}_0)_b$ the sesquilinear form φ_a, with $\varphi_a(x, y) = \varphi(xa, ya)$, belongs to $\mathcal{N}$, we have

$$\varphi((x_n - x_m)a, (x_n - x_m)a) \to 0, \text{ as } n, m \to \infty$$

and

$$\varphi((x_n^* - x_m^*)a, (x_n^* - x_m^*)a) \to 0, \text{ as } n, m \to \infty.$$

Therefore, $\{x_n\}$ is Cauchy also with respect to $\mathfrak{t}_{s^*}^\mathcal{N}$. Then, there exists $x \in \mathfrak{A}$ such that $x_n \xrightarrow{\mathfrak{t}_{s^*}^\mathcal{N}} x$. Since

$$\varphi(xa, xa) = \lim_{n\to\infty} \varphi(x_n a, x_n a) \le \limsup_{n\to\infty} q_\mathcal{M}(x_n)^2 \varphi(a, a)$$

and $\limsup_{n\to\infty} q_\mathcal{M}(x_n)^2 < \infty$ (by the boundedness of the sequence $\{q_\mathcal{M}(x_n)\}$), we conclude that x is $\mathcal{N}$-bounded. Finally, by the Cauchy condition, for every $\epsilon > 0$, there exists $n_\epsilon \in \mathbb{N}$ such that, for every $n, m > n_\epsilon$, $q_\mathcal{M}(x_n - x_m) < \epsilon$. This implies that

$$\varphi((x_n - x_m)a, (x_n - x_m)a) < \epsilon\varphi(a, a), \quad \forall \varphi \in \mathcal{N},\ a \in \mathfrak{A}_0.$$

Then, for fixed $n > n_\epsilon$ and $m \to \infty$, we obtain

$$\varphi((x_n - x)a, (x_n - x)a) \le \epsilon\varphi(a, a), \quad \forall \varphi \in \mathcal{N},\ a \in \mathfrak{A}_0.$$

This implies that $q_\mathcal{N}(x_n - x) \le \epsilon$. □

Since $q_\mathcal{N}(a) = \|a\|_b$, for every $a \in (\mathfrak{A}_0)_b$, for making notations lighter, we adopt the notation $\|\cdot\|_b$ for both of them.

Corollary 1. *If $(\mathfrak{A}_0)_b$ is dense in $\mathcal{D}(q_\mathcal{N})[\|\cdot\|_b]$, then $\mathcal{D}(q_\mathcal{N})[\|\cdot\|_b]$ is a C*-algebra, possibly not contained in $\mathfrak{A}_0$.*

4.3 The spectrum

Let now $\mathfrak{A}_0$ be an FR*-algebra with unit e and such that $(\mathfrak{A}_0)_b$ is t_{s^*}-dense in $\mathfrak{A}_0$. We assume, in addition that $\mathcal{D}(q_{\mathcal{N}})[\|\cdot\|_b]$ is a C*-algebra.

Then we define the *resolvent* $\rho_\circ(x)$ of $x \in \mathfrak{A}_0$ is defined by

$$\rho_\circ(x) = \left\{\lambda \in \mathbb{C} : \text{ the strong inverse } (x - \lambda e)^{-1} \text{exists in } \mathcal{D}(q_{\mathcal{N}})\right\}.$$

The *spectrum* of x is defined as $\sigma_\circ(x) := \mathbb{C} \setminus \rho_\circ(x)$.

In similar way as in [12] it can be proved that: (a) $\rho_\circ(x)$ is an open subset of the complex plane; (b) the map $\lambda \in \rho_\circ(x) \mapsto (x - \lambda e)^{-1} \in (\mathfrak{A}_0)_b$ is analytic in each connected component of $\rho_\circ(x)$.

As usual, we define the *spectral radius* of $x \in \mathfrak{A}_0$ by

$$r_\circ(x) := \sup\{|\lambda| : \lambda \in \sigma_\circ(x)\}.$$

Now we conclude by proving that bounded elements of $\mathfrak{A}_0$ can be characterized in terms of the spectrum. We have, in fact

Theorem 3. *Let $\mathfrak{A}_0$ be an FR*-algebra with unit e and such that $(\mathfrak{A}_0)_b$ is t_{s^*}-dense in $\mathfrak{A}_0$. Assume, in addition, that $\mathcal{D}(q_{\mathcal{N}})[\|\cdot\|_b]$ is a C*-algebra. Let $x \in \mathfrak{A}_0$. Then, $r_\circ(x) < \infty$ if and only if $x \in (\mathfrak{A}_0)_b$.*

Proof. If $x \in (\mathfrak{A}_0)_b$, then $\sigma_\circ(x)$ coincides with the spectrum of x as an element of the C*-algebra $\mathcal{D}(q_{\mathcal{N}})[\|\cdot\|_b]$ and so $\sigma_\circ(x)$ is compact. Conversely, assume that $r_\circ(x) < \infty$. Then the function $\lambda \mapsto (x - \lambda e)^{-1}$ is $\|\cdot\|_b$-analytic in the region $|\lambda| > r_\circ(x)$. Therefore it can be expanded in a $\|\cdot\|_b$-convergent Laurent series

$$(x - \lambda e)^{-1} = \sum_{k=1}^{\infty} \frac{a_k}{\lambda^k}, \qquad |\lambda| > r^{\mathcal{M}}(x),$$

with $a_k \in \mathcal{D}(q_{\mathcal{N}})$ for each $k \in \mathbb{N}$. As usual

$$a_k = \frac{1}{2\pi i} \int_\gamma \frac{(x - \lambda e)^{-1}}{\lambda^{-k+1}} d\lambda, \qquad k \in \mathbb{N},$$

where $\gamma := \{\lambda \in \mathbb{C} : |\lambda| = R : R > r_\circ(x)\}$ and the integral on the r.h.s. is meant to converge with respect to $\|\cdot\|_b$.

Using the previous integral representation and the continuity, for every $\varphi \in \mathcal{N}$ and $b, b' \in (\mathfrak{A}_0)_b$, we have

$$\varphi(xa_k b, b') = \varphi(a_{k+1} b, b').$$

This implies that $xa_k = a_{k+1}$.

Similarly, one shows that $xa_1 = -x$. Thus, in conclusion, $x = -a_2 \in \mathcal{D}(q_{\mathcal{N}}) \cap \mathfrak{A}_0 = (\mathfrak{A}_0)_b$. □

Corollary 2. *Let $\mathfrak{A}_0$ be an* FR*-*algebra* $\mathfrak{A}_0$, *with unit* e. *Then*

$$\begin{cases} r_\circ(x) \leq \|x\|_b & \text{if } x \in (\mathfrak{A}_0)_b \\ r_\circ(x) = +\infty & \text{if } x \notin (\mathfrak{A}_0)_b. \end{cases}$$

Example 4.9. *The maximal O^*-algebra $\mathcal{L}^\dagger(\mathcal{D})$ on a domain $\mathcal{D}$, endowed the topology $\mathfrak{t} = \tau_u$ is an* FR*-algebra. For every $\xi \in \mathcal{D}$ the positive linear functional ω_ξ defined by $\omega_\xi(X) = \langle X\xi|\xi\rangle$ is τ_u-continuous. Let $\mathcal{N}$ be the family of extensions of the set of forms $\{\varphi_{\omega_\xi}, \xi \in \mathcal{D}\}$, defined as in (1). The topology $\mathfrak{t}_{s^*}^{\mathcal{N}}$ coincides with τ_{s^*}. The *-algebra $(\mathcal{L}^\dagger(\mathcal{D}))_b$ is exactly the *-algebra $\mathcal{L}_b^\dagger(\mathcal{D}) = \{X \in \mathcal{L}^\dagger(\mathcal{D}) : \overline{X} \in \mathcal{B}(\mathcal{H})\}$ of all elements of $\mathcal{L}^\dagger(\mathcal{D})$ that are bounded operators. This *-algebra is τ_{s^*} – dense in $\mathcal{L}^\dagger(\mathcal{D})$. The completion of $\mathcal{L}_b^\dagger(\mathcal{D})$ is the partial *-algebra $\mathcal{L}^\dagger(\mathcal{D}, \mathcal{H})$ whose bounded part consists of the restrictions to $\mathcal{D}$ of the elements of $\mathcal{B}(\mathcal{H})$. The weak- and strong multiplications defined here are nothing but the weak- and strong multiplications of $\mathcal{L}^\dagger(\mathcal{D}, \mathcal{H})$ (see, e.g. [3, 5]). Finally, as shown in [4], the spectrum of an element $Y \in \mathcal{L}^\dagger(\mathcal{D})$ as defined in this section coincides with the ordinary spectrum of the closed operator $\overline{Y}$.*

Bibliography

[1] Allan, G.R. *A spectral theory for locally convex algebras*, Proc. London. Math. Soc. 15 (1965) 399–421.

[2] Antoine, J-P., Bellomonte, G., and Trapani, C. *Fully representable and *-semisimple topological partial *-algebras*, Studia Mathematica, 208 (2012), 167–194.

[3] J.-P. Antoine, A. Inoue, C. Trapani, *Partial *-algebras and their operator realizations*, Kluwer, Dordrecht, 2002.

[4] Antoine, J-P., Trapani, C., and Tschinke, F. *Spectral properties of partial *-algebras*, Mediterr. j. math. 7 (2010) 123–142.

[5] J.-P. Antoine, C.Trapani and F. Tschinke, *Bounded elements in certain topological partial *-algebras*, Studia Math. 203 (2011) 223–251.

[6] Bhatt, S.J. *Quotient bounded elements in locally convex algebras*, J. Austral. Math. Soc. (Series A) 33 (1982), 102–113.

[7] Fragoulopoulou, M. *Topological algebras with involution*, NorthHolland, 2005.

[8] Fragoulopoulou, M., Trapani, C., and Triolo, S. *Locally convex quasi *-algebras with sufficiently many *-representations*, J. Math. Anal. Appl. 388 (2012), 1180–1193.

[9] Mallios, A. *Topological algebras. Selected topics*, NorthHolland, 1986.

[10] Schmüdgen, K. *Unbounded operator algebras and representation theory*, Birkhäuser Verlag, Basel, 1990.

[11] Schmüdgen, K. *A strict Positivstellensatz for theWeyl algebra*, Math. Ann. 331 (2005) 779–794.

[12] Trapani, C. *Bounded elements and spectrum in Banach *-algebras*, Studia Mathematica 172 (2006) 249–273.

[13] Trapani, C. Unbounded C*-seminorms, biweights and *-representations of partial *-algebras: a review, International J. Math. Math.Sci., Volume 2006 (2006), Article ID 79268, 34 pages.

[14] Trapani, C. *Bounded and strongly bounded elements of Banach *-algebras*, Contemporary Math. 427 (2007) 417–424.

[15] Trapani, C. **-Representations, seminorms and structure properties of normed *-algebras*, Studia Mathematica, Vol. 186, 47–75 (2008).
[16] Trapani, C. *Order bounded elements of topological *-algebras*, Proceedings of the Conference ICATTA 2011 (Vera Cruz, Mexico), Commun. Math. Appl., 3 (2012) , 99–108.
[17] Trapani, C. and Tschinke, F. *Unbounded C*-seminorms and biweights on partial *-algebras* Mediterr. j. math. 2 (2005) 301–313.
[18] Vidav, I. *On some *regular rings*, Acad. Serbe Sci. Publ. Inst. Math. 13 (1959) 73–80.
[19] Yood, B. *C*-seminorms*, Studia Math., 118 (1996), 19–26.

Konstantinos Tzironis

On continuity of complementors in topological algebras

Abstract: In this paper, we introduce *fundamental* (*left, right*) *complemented algebras* with a complementor, say $\perp$. This notion is hereditary to the closed two-sided ideals for preannihilator algebras. Using the notion of the *axially closed elements* in a topological algebra, we characterize the continuity of the complementor $\perp$ via the respective ones on the minimal closed two-sided ideals of a certain fundamental left complemented algebra.

Keywords: (Left) preannihilator algebra, fundamental (left) complemented algebra, complementor, axially closed element

MSC: 46H05, 46H10

1 Introduction and Preliminaries

Complementarity of non-normed topological algebras was studied in [4]. This led to precomplemented (in particular complemented) topological algebras. Here, we deal with the continuity of complementors, in the context of fundamental complemented algebras. The respective problem has been faced, as well, in the classical case of Banach algebras (see [2]). All algebras, employed below, are taken over the field $\mathbb{C}$ of complexes. A *topological algebra* E is an algebra which is a topological vector space and the ring multiplication is separately continuous (see e.g., [7]). $\overline{S}$ stands for the topological closure of a subset S in a topological algebra.

An element x of an algebra E, is called *idempotent* if $x = x^2$. Throughout we employ non-zero idempotents. A non-zero element x of E is called *primitive*, if it can not be expressed as the sum of two orthogonal idempotents; viz. there are no non-zero idempotents $y, z \in E$ with $x = y + z$ and $yz = zy = 0$.

We denote by $\mathcal{A}_l(S)$ (resp. $\mathcal{A}_r(S)$) the left (right) annihilator of a (non empty) subset S of an algebra E, being a left (resp. right) ideal of E. If S is a left (resp. right) ideal of E, then the ideals $\mathcal{A}_l(S)$ and $\mathcal{A}_r(S)$ are two-sided. If $\mathcal{A}_l(E) = (0)$ (resp. $\mathcal{A}_r(E) = (0)$) we say that E is a *left* (resp. *right*) *preannihilator algebra*. For a right preannihilator algebra it is also used the term *proper algebra*. E is named *preannihilator* if it is both left and right preannihilator. Besides, a topological algebra E is called *annihilator*, if it is preannihilator and satisfies the conditions

$$\text{If } \mathcal{A}_l(I) = (0) \text{ with } I \in \mathcal{L}_r, \text{ then } I = E$$

Konstantinos Tzironis, Department of Mathematics, University of Athens, Panepistimioupolis, Athens 15784, Greece. e-mail: tzirk@math.uoa.gr

https://doi.org/10.1515/9783110413557-017

and

$$\text{if } \mathcal{A}_r(I) = (0) \text{ with } I \in \mathcal{L}_l, \text{ then } I = E.$$

$m_l(E)$ (resp. $m_r(E)$, $m(E)$) denotes the set of all minimal left (right, two-sided) ideals of an algebra E. Taking a topological algebra E, $\mathcal{L}_l(E) \equiv \mathcal{L}_l$ (resp. $\mathcal{L}_r(E) \equiv \mathcal{L}_r$, $\mathcal{L}(E) \equiv \mathcal{L}$) stands for the set of all closed left (right, two-sided) ideals of E, while $m_l(E)$ (resp. $m_r(E)$, $m(E)$) denotes the set of all minimal closed left (resp. right, two-sided) ideals of E. If for $I \in \mathcal{L}$ the relation $I^2 = (0)$ implies $I = (0)$, then E is called *topologically semiprime*.

The next notion was introduced in [4, p. 3723, Definition 2.1]. See also [3].

A topological algebra is called *left complemented* if there exists a mapping $\perp : \mathcal{L}_l \longrightarrow \mathcal{L}_l : I \mapsto I^\perp$, such that

$$\text{if } I \in \mathcal{L}_l, \text{ then } E = I \oplus I^\perp. \tag{1}$$

$I^\perp$ is called a *complement* of I.

$$\text{If } I, J \in \mathcal{L}_l, I \subseteq J, \text{ then } J^\perp \subseteq I^\perp. \tag{2}$$

$$\text{If } I \in \mathcal{L}_l, \text{ then } (I^\perp)^\perp = I. \tag{3}$$

$\perp$ as before is called a *left complementor* on E. In what follows we denote by $(E, \perp)$ a left complemented algebra with a left complementor $\perp$.

A *right complemented algebra* is defined analogously and we talk about a *right complementor*. A left and right complemented algebra is simply called a complemented algebra.

A topological algebra E is named *left precomplemented*, if for every $I \in \mathcal{L}_l$, there exists $I' \in \mathcal{L}_l$ such that $E = I \oplus I'$. Similarly a *right precomplemented algebra* is defined. A left and right precomplemented algebra is called a *precomplemented algebra* (see [4, p. 3725, Definition 2.7]).

All the results appeared in this paper hold true by interchanging "left" by "right".

For the next basic result, which a realization of precomplementors is given, we refer to [4, p. 3725, Theorem 2.9].

Theorem 1.1. *Let E be a left preannihilator, left precomplemented algebra. Then, for each $I \in \mathcal{L}$, there exists a unique $I' \in \mathcal{L}$, such that $E = I \oplus I'$ and $I' = \mathcal{A}_l(I) = \mathcal{A}_r(I)$. Thus $\mathcal{L}_l(I) \subseteq \mathcal{L}_l$ and $\mathcal{L}_r(I) \subseteq \mathcal{L}_r$ for all $I \in \mathcal{L}$. Furthermore, if $J, K \in \mathcal{L}$ and $J \cap K = (0)$, then $K \subseteq J'$ and $J \subseteq K'$ where J', K' are complements of J, K respectively.*

As a direct consequence of the previous result, we get that *if a topological algebra is left preannihilator and left precomplemented, then it is preannihilator.*

For our presentation, we shall also need the next results.

Lemma 1.2. *Let E be a vector space and M, N_1, N_2 subspaces of E, such that $E = M \oplus N_1, M \cap N_2, = (0)$ and $N_1 \subseteq N_2$. Then $N_1 = N_2$.*

Proof. See [6, p. 154, Lemma 3.17]. □

For the proof of the next two results we refer to [4, p. 3727, Theorem 3.1].

Theorem 1.3. *Let $(E, \perp)$ be a left preannihilator, left complemented algebra. If I_λ is a closed two-sided ideal in E, then the mapping*

$$\perp_\lambda : \mathcal{L}_l(I_\lambda) \longrightarrow \mathcal{L}_l(I_\lambda) \text{ with } I \mapsto I^{\perp_\lambda} = I^\perp \cap I_\lambda$$

defines a complementor on the topological algebra I_λ.

The next result concerns properties inherited on closed two-sided ideals (see also Theorem 1.1).

Corollary 1.4. *Any closed two-sided ideal of a left preannihilator, left complemented algebra is a preannihilator, left complemented algebra.*

2 Fundamental left complemented algebras. The main theorem

This section concerns fundamental (left) complemented algebras (Definition 2.5). The defining properties of a topological algebra, as before, are inherited to the closed two-sided ideals (Proposition 2.11). For a certain fundamental left complemented algebra and using axially closed elements (Definition 2.9) we characterize the continuity of the complementor of the algebra concerned via the respective ones on its minimal closed two-sided ideals (Theorem 2.19).

The next result generalizes Lemma 2.8.8 in [8, p. 98].

Lemma 2.1. *Let E be a topologically semiprime algebra. Then, for any minimal left (resp. right) ideal I of E, there exists a minimal closed two-sided ideal containing I.*

Proof. Consider the set $G = \{J_\lambda : \lambda \in \Lambda,\ J_\lambda \in \mathcal{L} \text{ with } I \subseteq J_\lambda\}$, which obviously is nonempty. If $J = \bigcap_{\lambda\in\Lambda} J_\lambda$, then J is a closed two-sided ideal of E containing J and actually, it is the least closed two-sided ideal with this property. Claim that J is minimal, as well. Let J' be a closed two-sided ideal with $J' \subseteq J$. Then $J' \cap I$ is a left ideal with $J' \cap I \subseteq I$. By the minimality of the left ideal I, we get either $J' \cap I = (0)$ or $J' \cap I = I$. In the latter case, $I \subseteq J'$ and hence $J' \in G$, whence $J \subseteq J'$. Therefore $J' = J$. If $J' \cap I = (0)$, then $J'I \subseteq J'$ and $J'I \subseteq I$. Namely, $J'I \subseteq J' \cap I$, that yields $J'I = (0)$. Thus, $I \subseteq \mathcal{A}_r(J')$, and as $\mathcal{A}_r(J')$ is a closed two-sided ideal, we get $\mathcal{A}_r(J') \in G$ and hence $J \subseteq \mathcal{A}_r(J')$. Therefore

$J'^2 \subseteq J'J \subseteq J'\mathcal{A}_r(J') = (0)$. So, since E is topologically semiprime, $J' = (0)$, and this completes the assertion. □

The next result generalizes Lemma 3 in [1, p. 39], stated for Banach algebras. See also the comment after Theorem 2.4, in [5].

Lemma 2.2. *Let E be a right preannihilator, left precomplemented algebra. Then, for any $x \in E$, $x \in \overline{Ex}$.*

Proof. From the left precomplementation of E, for the (closed) left ideal $\overline{Ex}$ there is $I \in \mathcal{L}_l$ such that $E = \overline{Ex} \oplus I$. Thus, for $x \in E$, $x = y + z$ with $y \in \overline{Ex}$ and $z \in I$. Hence, for any $t \in E$, we get $tz = tx - ty$, that yields $tz \in \overline{Ex}$. So, since $tz \in I$, we also have $tz \in \overline{Ex} \cap I = (0)$ for all $t \in E$. But, E is right preannihilator, thus $z = 0$ and hence $x = y \in \overline{Ex}$. □

Lemma 2.3. *Let E be a left preannihilator, left precomplemented algebra. If $I \in \mathcal{L}_l$ with $I \subseteq K_\alpha$ for some $\alpha \in A$ (where $\{K_\alpha\}_{\alpha \in A}$ is the family of all minimal closed two-sided ideals of E), and I' is a complement of I, then $K_\beta \subseteq I'$ with $\alpha \neq \beta$.*

Proof. By Theorem 1.1, we get

$$E = K_\alpha \oplus K'_\alpha \text{ where } K'_\alpha = \mathcal{A}_l(K_\alpha) = \mathcal{A}_r(K_\alpha) \text{ and } K_\beta \subseteq K'_\alpha,\ \alpha \neq \beta. \tag{1}$$

Claim that $K'_\alpha \subseteq I'$ where I' is a (pre)complement of I (in E). So, take $x \in K'_\alpha$. Then $x = y + z$ with $y \in I$, $z \in I'$. For any $t \in K_\alpha$, we get

$$ty = tx - tz. \tag{2}$$

Since $x \in K'_\alpha$, the form of K'_α implies $tx = 0$, and by (2), $ty = -tz$. But, obviously $ty = -tz \in I \cap I' = (0)$ for all $t \in K_\alpha$, and thus $y \in \mathcal{A}_r(K_\alpha) = K'_\alpha$. But, $y \in I \subseteq K_\alpha$, as well. Thus $y = 0$ and hence $x = z \in I'$. Therefore $K'_\alpha \subseteq I'$ that, in view of (1), yields the assertion. □

Remark 2.4. *Let E be a left precomplemented algebra and $I, I' \in \mathcal{L}_l$ with $E = I \oplus I'$, then there is a linear mapping $T : E \to E$ with $T^2 = T$ (projection) such that $ImT = I$ and $\ker T = I'$. Indeed, take $x \in E$, then there are unique y, z with $y \in I$ and $z \in I'$ such that $x = y + z$. We define $T(x) = y$. Then T is a well defined linear mapping with $T^2 = T$, $ImT = I$ and $\ker T = I'$. Moreover, for $y \in ImT$ there is $x \in E$ with $T(x) = y$, and thus*

$$T(y) = T(T(x)) = T^2(x) = y \text{ namely } T(y) = y \text{ for all } y \in ImT.$$

If there was another linear map $T' : E \to E$ with the respective properties, as T has, then for an element $x \in E$, there are unique w, t with $w \in I$ and $t \in I'$ such that $x = w + t$, thus

$$T(x) = T(w) + T(t) = w = T'(w + t) = T'(x).$$

The last relation assures the uniqueness of T*. So, we can use the symbol* $T = T(I, I')$*. If in particular, the map in question is continuous, we set the next.*

Definition 2.5. *(M. Haralampidou) A left precomplemented algebra* E *is called* fundamental *if, for any* $I \in \mathcal{L}_l$*, and a (pre)complement of* I*, say* $I' \in \mathcal{L}_l$ *(viz.* $E = I \oplus I'$*) there is a continuous linear mapping* $T : E \to E$ *such that* $T^2 = T$*,* $\operatorname{Im} T = I$ *and* $\ker T = I'$*.*

A fundamental right precomplemented algebra *is defined analogously. A fundamental left and right precomplemented algebra is simply called a* fundamental precomplemented algebra.

Since a left complemented algebra is left precomplemenetd, the analogous notion for a *fundamental (left) complemented algebra* can be defined.

We now give two realizations of Definition 2.5.

Proposition 2.6. Any precomplemented Banach algebra is fundamental.

Proof. See [2, p. 386]. □

Proposition 2.7. Any left precomplemented algebra E in which all $I \in \mathcal{L}_l$ are locally convex spaces and the sum $E = I \oplus I'$, $I \in \mathcal{L}_l$ is the topological direct sum of the I, I', in the sense of [9, p. 90], is a fundamental algebra.

Proof. See [9, p. 95, Proposition 29]. □

Definition 2.8. *Let* $(E, \perp)$ *be a fundamental left complemented algebra. A net* $(I_\delta)_{\delta \in \Delta}$ *of minimal closed left ideal (of* E*) is* $\perp$*-convergent to* $I_0 \in \mathcal{L}_l$ *if*

$$T_\delta \equiv T_\delta(I_\delta, I_\delta^\perp) \underset{\text{uniformly}}{\longrightarrow} T_0(I_0, I_0^\perp)$$

on any minimal right ideal of E*.*

For certain left complemented algebras the closedness of minimal left ideals in the previous definition is redundant (see [3, p. 969, Theorem 3.2]).

Definition 2.9. *(M. Haralampidou) An element* x *in a topological algebra* E *is said to be* axially closed *if the left ideal* Ex *is minimal closed.*

In particular, a subset of E *is named* axially closed *if each of its elements is axially closed.*

Concerning the previous notion, we note that if $x \in E$ is a primitive idempotent and the (closed) left ideal Ex is a left precomplemented algebra, then, in view of [3, p. 964, Theorem 2.1], the ideal concerned is minimal closed (namely, the element x is axially closed).

Definition 2.10. *Let $(E, \perp)$ be a fundamental left complemented algebra. The mapping $\perp$ is named* continuous *whenever for each convergent, axially closed net $(a_\delta)_{\delta\in\Delta}$ with $a_\delta \underset{\delta}{\longrightarrow} a_0 \in E$, $a_0 \neq 0$, and such that $Ea_0 \in \mathcal{L}_l$, the net $(Ea_\delta)_{\delta\in\Delta}$ is $\perp$-convergent in Ea_0. Namely,*

$$T_\delta \equiv T_\delta(Ea_\delta, Ea_\delta^\perp) \underset{uniformly}{\longrightarrow} T_0(Ea_0, Ea_0^\perp)$$

on any minimal right ideal of E.

In view of the next result, the notion "fundamental left complemented algebra" is inherited to the closed two-sided ideals.

Proposition 2.11. *Let $(E, \perp)$ be a left preannihilator, fundamental left complemented algebra. Then any closed two-sided ideal I_λ in E is a fundamental left complemented algebra with a complementor $\perp_\lambda$, as in Theorem 1.3.*

Proof. Take $J \in \mathcal{L}_l(I_\lambda)$, then by Theorem 1.1, $J \in \mathcal{L}_l(E)$. By hypothesis, there is a continuous linear mapping $T : E \to E$ with $T^2 = T$, $ImT = J$ and $\ker T = J^\perp$. Denote by $T_\lambda = T|_{I_\lambda}$ the restriction of T on I_λ. Actually, $T_\lambda : I_\lambda \to I_\lambda$ is well defined, linear and continuous. Indeed, for $x \in I_\lambda$, $x = y + z$ with $y \in J$ and $z \in J^\perp$. Therefore, taking in mind that the map T is the identity one on $J = ImT$, we get

$$T_\lambda(x) = T(x) = T(y) + T(z) = T(y) = y \in J.$$

Linearity and continuity of T_λ is obvious. Moreover $T_\lambda^2 = T_\lambda$. Furthermore,

$$ImT_\lambda = T_\lambda(I_\lambda) = T(I_\lambda) \subseteq ImT = J.$$

Besides, for any $y \in J$, $T(y) = y$. From here, $T_\lambda(y) = y$. Thus, $J \subseteq ImT_\lambda$ and hence $ImT_\lambda = J$. Finally,

$$\ker T_\lambda = \{x \in I_\lambda, T_\lambda(x) = 0\} = I_\lambda \cap \ker T = I_\lambda \cap J^\perp = J^{\perp_\lambda}$$

and this completes the proof. □

Lemma 2.12. *Let E be a topological algebra and $I \in \mathcal{L}(E)$ with $\mathcal{L}_l(I) \subseteq \mathcal{L}_l(E)$. Then the following hold.*

(i) If J is a minimal closed left ideal of the topological algebra I, then J is a minimal closed left ideal of E.

(ii) If x is an axially closed element in I, then Ix is a minimal closed left ideal in E.

Proof. (i) Since $\mathcal{L}_l(I) \subseteq \mathcal{L}_l(E)$, $J \in \mathcal{L}_l(E)$. If J as closed left ideal is not minimal in E, then there exists $J' \in \mathcal{L}_l(E)$ with $(0) \neq J' \subseteq J$ and $J' \neq J$. Since $IJ' \subseteq EJ' \subseteq J'$, we get $J' \in \mathcal{L}_l(I)$. Thus, J is not a minimal closed left ideal in I, that is a contradiction.

(ii) By hypothesis, Ix is a minimal closed left ideal in I, thus (i) yields the assertion. □

In the sequel, we refer to conditions under which axially elements of a topological algebra E remain axially ones in closed two-sided ideals of E.

Lemma 2.13. *Let E be a left preannihilator left precomplemented algebra. If $I \in \mathcal{L}(E)$, $a \in E$ such that $Ea \subseteq I$ and the left ideal Ea is minimal closed in E, then it is a minimal closed left ideal in I.*

Proof. It is trivial to show that $Ea \in \mathcal{L}_l(I)$. Let now $J \in \mathcal{L}_l(I)$ with $(0) \neq J \subseteq Ea$. Then $EJ = (I \oplus I')J \subseteq IJ + I'J = IJ$ for the latter equality, we also use the fact that $I' = \mathcal{A}_l(I)$ (see Theorem 1.1). Thus, $EJ \subseteq IJ \subseteq J$ and finally, $J \in \mathcal{L}_l(E)$. The minimality of Ea in E implies $J = Ea$, and this completes the assertion. □

Lemma 2.14. *Let E be a left preannihilator left precomplemented algebra and $I \in \mathcal{L}(E)$. If $(a_\delta)_{\delta\in\Delta}$ is an axially closed net in E with $Ea_\delta \subseteq I$ for every $\delta \in \Delta$, then the following hold.*

(i) $a_\delta \in I$ for every $\delta \in \Delta$.

(ii) The net $(a_\delta)_{\delta\in\Delta}$ is axially closed in I.

Proof. By lemma 2.2 and the comment after Theorem 1.1, we have $a_\delta \in \overline{Ea_\delta}$. But, Ea_δ is closed in E, and hence (i) is fulfilled.

(ii) We first note that the left ideal Ea_δ is minimal (cf. Lemma 2.13). Obviously, Ia_δ is a left ideal of I and $Ia_\delta \subseteq Ea_\delta$. Thus either $Ia_\delta = Ea_\delta$ or $Ia_\delta = (0)$. In the latter case, and since I is (right) preannihilator (see Theorem 1.4), we get $a_\delta = 0$ (see also (i)) and hence $Ea_\delta = (0)$, a contradiction. The previous leads to $Ia_\delta = Ea_\delta$ being a minimal closed left ideal of I. This assures that the net $(a_\delta)_{\delta\in\Delta}$ is axially closed in I. □

Lemma 2.15. *Let E be a left preannihilator left precomplemented algebra and $I \in \mathcal{L}(E)$. If $a \in I$, and Ia is a minimal closed left ideal of I, containing a, then Ea is a minimal closed left ideal of E.*

Proof. In view of Theorem 1.1 and Lemma 2.12, Ia is a minimal closed left ideal of E. Since $Ia \subseteq Ea \subseteq EIa \subseteq Ia$, we get $Ea = Ia$, thus the assertion. □

Lemma 2.16. *Let $(E, \perp)$ be a left preannihilator left complemented algebra and $I_0 \in \mathcal{L}(E)$.*

(i) If $a \in I_0$ and I_0a is a minimal closed left ideal in I_0, then $a \in I_0a$ and the left ideal Ea is minimal closed.

(ii) If $(a_\delta)_{\delta\in\Delta}$ is an axially closed net in I_0, then it is axially closed in E, as well.

Proof. (i) By Theorem 1.3 and the comment after Theorem 1.1, I_0 is a preannihilator left complemented algebra with a left complementor $\perp_0 : \mathcal{L}_l(I_0) \longrightarrow \mathcal{L}_l(I_0)$ with $I \mapsto$

$I^{\perp_0} = I^{\perp} \cap I_0$. Thus, in view of Lemma 2.2, $a \in \overline{I_0 a} = I_0 a$ and hence (Lemma 2.15) Ea is a minimal closed left ideal in E.

(ii) It follows from (i).

□

Lemmas 2.14, 2.15 and 2.16 lead to

Corollary 2.17. Let E be a left preannihilator left complemented algebra and $I \in \mathcal{L}(E)$. Consider the assertions:

(1) The net $(a_\delta)_{\delta \in \Delta}$ is axially closed in I.

(2) The net $(a_\delta)_{\delta \in \Delta}$ is axially closed in E.

Then (1) $\Rightarrow$ (2). If in particular, $Ea_\delta \subseteq I$ for every $\delta \in \Delta$, then (2) $\Rightarrow$ (1).

Corollary 2.18. Let E be a left preannihilator left complemented algebra. If $I \in \mathcal{L}(E)$ and $(a_\delta)_{\delta \in \Delta}$ is a net in I, then the assertions (1) and (2) of Corollary 2.17 are equivalent.

Theorem 2.19. *Let $(E, \perp)$ be a left preannihilator, topologically semiprime fundamental left complemented algebra. Suppose that the set of all minimal closed two-sided ideals $m(E) = \{I_\lambda\}_{\lambda \in \Lambda}$ of E satisfies the relation $\bigcap_{\lambda \in \Lambda} I_\lambda^{\perp} = (0)$, then the following assertions are equivalent:*

(1) The mapping $\perp$ is continuous.

(2) For every $\lambda \in \Lambda$ the mapping $\perp_\lambda$ is continuous. Here, $\perp_\lambda$ denotes the complementor on I_λ, defined in Theorem 1.3.

Proof. We first note that $I_\lambda, \lambda \in \Lambda$ is a fundamental left complemented algebra (see Proposition 2.11).

(1) $\Rightarrow$ (2) : Consider $I_\lambda \in m(E)$ and $(a_\delta)_{\delta \in \Delta}$ a convergent axially closed net in I_λ with $a_\delta \underset{\delta}{\longrightarrow} a_0 \in I_\lambda,\ a_0 \neq 0$, then (by definition) $I_\lambda a_\delta \in m_l(I_\lambda)$, and $I_\lambda a_0 \in \mathcal{L}_l(I_\lambda)$. We show that $I_\lambda a_\delta \underset{\perp_\lambda, \delta}{\longrightarrow} I_\lambda a_0$ over any minimal right ideal of I_λ. Consider a minimal right ideal R of I_λ. By the assumption for $(a_\delta)_{\delta \in \Delta}$, we get $I_\lambda a_\delta = Ea_\delta \underset{\perp, \delta}{\longrightarrow} Ea_0 = I_\lambda a_0$ (see Lemma 2.15 its proof, and Lemma 2.16). Since $I_\lambda^{\perp} = \mathcal{A}_r(I_\lambda)$, we get $RI_\lambda^{\perp} \subseteq I_\lambda I_\lambda^{\perp} = I_\lambda \mathcal{A}_r(I_\lambda) = (0)$, and hence $RE = R(I_\lambda \oplus I_\lambda^{\perp}) \subseteq RI_\lambda + RI_\lambda^{\perp} = RI_\lambda \subseteq R$. Thus R is a right ideal of E, as well. In particular, R is minimal. Indeed, if J is a non zero right ideal of E, contained in R, then $JI_\lambda \subseteq JE \subseteq J$. Namely, J is a right ideal in I_λ. Thus, $J = R$.

Now, by the continuity of $\perp$, we have $T_\delta(Ea_\delta, Ea_\delta^{\perp}) \underset{\text{uniformly}}{\longrightarrow} T_0(Ea_0, Ea_0^{\perp})$ in R or $T_\delta(I_\lambda a_\delta, I_\lambda a_\delta^{\perp}) \underset{\text{uniformly}}{\longrightarrow} T_0(I_\lambda a_0, I_\lambda a_0^{\perp})$ in R. Since $R \subseteq I_\lambda$, we finally get

$$T_\delta(I_\lambda a_\delta, I_\lambda a_\delta^{\perp})|_R = T_\delta(I_\lambda a_\delta, I_\lambda a_\delta^{\perp_\lambda})|_R,\ \delta \in \Delta$$

and

$$T_0(I_\lambda a_0, I_\lambda a_0^{\perp})|_R = T_0(I_\lambda a_\delta, I_\lambda a_0^{\perp_\lambda})|_R.$$

Hence $\perp_\lambda$ is continuous.

(2) $\Rightarrow$ (1) : Consider an axially closed net $(a_\delta)_{\delta\in\Delta}$ in E such that $a_\delta \underset{\delta}{\longrightarrow} a_0 \in E$, $Ea_0 \in \mathcal{L}_l(E)$ and $a_0 \neq 0$. We are going to prove that $Ea_\delta \underset{\perp}{\longrightarrow} Ea_0$. For each $\delta \in \Delta$, we denote by I_δ the minimal closed two-sided ideal of E that contains Ea_δ (see Lemma 2.1). Claim that there is $I_0 \in m(E)$ with $Ea_\delta \subseteq I_0$ for every $\delta > \delta_0$, where δ_0 is an appropriate index in Δ. Namely, $(Ea_\delta)_{\delta\in\Delta}$ is finally contained in some $I_0 \in m(E)$. Otherwise, for any $I_0 \in m(E)$, we consider a subnet $(Ea_{\delta'})_{\delta'\in\Delta'}$ such that $Ea_{\delta'}$ is not contained in I_0 for every $\delta' \in \Delta'$. By the previous argumentation, $Ea_{\delta'} \subseteq I_{\delta'}, I_{\delta'} \in m(E)$ and $I_0 \cap I_{\delta'} = (0)$ provided that both $I_0, I_{\delta'} \in m(E)$. Therefore $Ea_{\delta'} \cap I_0 = (0)$. Furthermore, $EI_0a_{\delta'} \subseteq I_0a_{\delta'}$. Thus, $I_0a_{\delta'}$ is a left ideal of E with $I_0a_{\delta'} \subseteq Ea_{\delta'}$ and $Ea_{\delta'}$ is a minimal closed left ideal of E. Therefore either $I_0a_{\delta'} = (0)$ or $I_0a_{\delta'} = Ea_{\delta'}$. In the latter case, and since I_0 is a right ideal of E, we get $I_0a_{\delta'} \subseteq I_0$ and hence $Ea_{\delta'} \subseteq I_0$, a contradiction. Thus, $I_0a_{\delta'} = (0)$ that yields $a_{\delta'} \in \mathcal{A}_r(I_0) = I_0^\perp$ (see also Theorem 1.1). But, $a_{\delta'} \underset{\delta'}{\longrightarrow} a_0$ and $I_0^\perp \in \mathcal{L}(E)$, thus $a_0 \in I_0^\perp$ and hence $a_0 \in \bigcap_{\lambda\in\Lambda} I_\lambda^\perp = (0)$ that still is a contradiction. The previous argumentation assures the existence of an ideal $I_0 \in m(E)$ such that $Ea_\delta \subseteq I_0$ for every $\delta > \delta'$ with $\delta' \in \Delta$, appropriately chosen. Without any loss of generality, we may assume that there is $I_0 \in m(E)$ such that $Ea_\delta \subseteq I_0$ for every $\delta \in \Delta$. Since Ea_δ, $\delta \in \Delta$ is a minimal closed left ideal in E, we trivially have $a_\delta \neq 0$, for every $\delta \in \Delta$. In view of Lemma 2.13, Ea_δ is a minimal closed left ideal of I_0. Besides, by Lemma 2.14, $a_\delta \in I_0$ and $I_0a_\delta = Ea_\delta$ for every $\delta \in \Delta$ and the net $(a_\delta)_{\delta\in\Delta}$ is actually closed in I_0. So, since $a_\delta \underset{\delta}{\longrightarrow} a_0$ and I_0 is closed, we get that $a_0 \in I_0$.

Since,

$$I_0a_0 = I_0(\lim a_\delta) = \lim(Ea_\delta) = Ea_0.$$

I_0a_0 is closed. Therefore $Ea_\delta = I_0a_\delta \underset{\perp_0,\delta}{\longrightarrow} I_0a_0 \subseteq I_0$. Namely,

$$T_\delta(I_0a_\delta, I_0a_\delta^{\perp_0}) \underset{\text{uniformly}}{\longrightarrow} T_0(I_0a_0, I_0a_0^{\perp_0})$$

over any minimal right ideal of I_0. Now, take any $J \in m_l(E)$ with $J \subseteq I_0$. Then for any minimal right ideal R of E, we denote by I_λ the minimal closed two-sided ideal that contains R. We have two cases.

If $I_0 \neq I_\lambda$, then $T(J, J^\perp) = 0$ in R provided that $R \subseteq I_\lambda \subseteq J^\perp$ (see Lemma 2.3). Moreover, by hypothesis, $\ker T = J^\perp$. Thus,

$$T_\delta(Ea_\delta, Ea_\delta^\perp) = 0 = T_0(Ea_\delta, Ea_\delta^\perp) \ \text{(in } R). \ \text{Hence}$$

$$T_\delta(Ea_\delta, Ea_\delta^{\perp_0}) \underset{\text{uniformly}}{\longrightarrow} T_0(Ea_0, Ea_0^{\perp_0}).$$

In the case $I_0 = I_\lambda$, one gets $T(J, J^\perp) = T|_{I_0}(J, J^{\perp_0}) = 0$ in R. Thus, the claim that "the net $(Ea_\delta)_{\delta\in\Delta}$ is $\perp$-convergent to Ea_0" is equivalent to the claim "the net $(Ea_\delta)_{\delta\in\Delta}$ is $\perp_0$-convergent to Ea_0". Therefore $\perp$ is continuous. □

Acknowledgment: This work is based on the author's Doctoral Thesis (in process), University of Athens, written under the supervision of Professor Marina Haralampidou

Bibliography

[1] Alexander, F.E. *On complemented and annihilator algebras*, Glasgow Math. J., 10(1969), 38–45.
[2] Alexander, F.E. *Representation theorems for complemented algebras*, Trans. Amer. Math. Soc., 148(1970), 385–398.
[3] Haralampidou, M. *Structure theorems for complemented topological algebras*, Boll. U.M.I., 7, (1993), 961–971.
[4] Haralampidou, M. *On complementing topological algebras*, Journal of Mathematical Sciences, Vol. 96, No 6 (1999), 3722–3734. Contemp. Math. 341 (2004), 63–71.
[5] Haralampidou, M. *Dual complementors in topological algebras*, Banach Center Publications, Institute of Math. Polish Academy of Sci. 67(2005), 219–233.
[6] Haralampidou, M. *On the Krull property in topological algebras,* Comment. Math. Prace Mat. 46(2006), no. 2, 141–162.
[7] Mallios, A. *Topological Algebras. Selected Topics.* North-Holland, Amsterdam, 1986.
[8] Rickart, C.E. *General theory of Banach algebras*, R.E. Krieger, Huntington, N.Y., 1974.
[9] Robertson, A.P. and Robertson, W. *Topological Vector Spaces*, Cambridge Univ. press, 1973.

Shukhrat M. Usmanov

Dominated ergodic theorem for isometries of non-commutative L_p-spaces, $1 < p < \infty$, $p \neq 2$

Abstract: We prove the dominated ergodic theorem for positive invertible, or surjective, isometries of the non-commutative L_p-spaces associated with finite von Neumann algebras. As a corollary, we obtain the individual ergodic theorem for such isometries.

Keywords: Non-commutative L_p-spaces; Isometries; Von Neumann algebras; Dominated ergodic theorem; Non-commutative ergodic theory

MSC: Primary 46L55, Secondary 37A55

1 Introduction

In this work we study behavior of ergodic averages for positive invertible (or surjective) isometries of the non-commutative L_p-spaces associated with finite von Neumann algebras.

In classical ergodic theory, the dominated ergodic theorem for isometries of L_p-spaces, $1 < p < \infty$, $p \neq 2$, is of great importance for development of the theory. This theorem was obtained by A. Bellow in 1964 [1]. The main idea of her proof was to reduce it to well-known case of the measure preserving transformations (see [9], Theorem 6.3, Chapter 1.). She approximated a positive invertible isometry by non-singular isometries induced by periodic transformations, and then transferred a dominated estimate for ergodic averages from measure preserving case to non-singular case, from which she obtained the dominated ergodic theorem (see also [5], Theorem 2.1).

In non-commutative ergodic theory, one usually considers a particular class of operators, the weakly closed algebras of operators in Hilbert spaces (von Neumann algebras) and seeks for conditions of convergence of ergodic averages for transformations of those algebras. Those studies are important for construction of models in quantum dynamics as well as in non-commutative probability theory. The first important result in this area was obtained by C. Lance in [10]. He proved a non-commutative version of individual ergodic theorem for *-automorphisms of von Neumann algebras. After that, a large number of results in this direction have been obtained. In particular, F. J. Yeadon developed a technique to obtain non-commutative maximal inequalities for contractions of von Neumann algebras [22], [23]. Another contribution was made by S. Litvinov who generalized the classical Banach principle into the von Neumann

Shukhrat M. Usmanov, Waldorf University, Department of Mathematics, 106 South 6th Street, Forest City, Iowa 50436, U.S.A.; E-mail: Shukhrat.Usmanov@waldorf.edu

https://doi.org/10.1515/9783110413557-018

algebra setting in his Ph.D. dissertation [11]; this result was further developed by M. Goldstein and S. Litvinov [3] and V. Chilin, S. Litvinov and A. Skalski [2].

In this work we study ergodic averages for positive surjective isometries of non-commutative Lp-spaces ($1 < p < \infty$), $p \neq 2$. For such isometries we prove the dominated ergodic theorem. This result is a non-commutative analog of the Bellow's dominated ergodic theorem for isometries of classical L_p-spaces. Using this result and a non-commutative Banach principle, recently obtained in [2], we prove the individual ergodic theorem for positive surjective isometries of non-commutative L_p-spaces ($1 < p < \infty$), $p \neq 2$, associated with finite von Neumann algebras.
This work is devoted to study of the case of the finite von Neumann algebras. According to the opinion of the author, it is possible to generalize the main result of the work (Theorem 5.3) on semi-finite case (for technical approach of that, see, for example, an interesting work [18]).

2 Preliminaries

Let H be a (separable) Hilbert space, $B(H)$ be a *-algebra of all bounded linear operators on H. A *-subalgebra M of $B(H)$ is called a *von Neumann algebra*, if M is closed in weak operator topology and the identity operator $\mathbf{1}$ belongs to M. The commutative von Neumann subalgebra $Z(M) = \{x \in M : xy = yx, \forall y \in M\}$ of M is called a *center* of M.

A continuous linear functional τ on M is called *normal*, if $\tau(lub_\alpha x_\alpha) = lub_\alpha \tau(x_\alpha)$ for all increasing nets $\{x_\alpha\}$ in M_+ with an upper bound; *finite*, if $\tau(\mathbf{1}) < \infty$; *faithful*, if $\tau(x) = 0$ implies $x = 0$ for any positive $x \in M$. A continuous linear functional τ on M_+ is called a *trace* if $\tau(xx^*) = \tau(x^*x)$ for any $x \in M_+$.

A von Neumann algebra M is a *finite* if it possesses at least one faithful normal finite trace τ. We assume that $\tau(\mathbf{1}) = 1$ (a *normalized* trace).

A linear map $\theta : M \to M$ is called a **-automorphism* if $\theta(xy) = \theta(x)\theta(y)$ and $\theta(x^*) = \theta(x)^*$ for all $x, y \in M$. An automorphism θ of M is called *properly outer* if for every $e \in M, e^2 = e, e \neq 0, \theta(e) = e$, the reduced automorphism on eMe is not inner (is not generated by some unitary from M). An automorphism of M is called *aperiodic* if each θ^n, $n \in N$ is properly outer.

A linear map $\theta : M \to M$ is called a **-anti-automorphism* if $\theta(xy) = \theta(y)\theta(x)$ and $\theta(x^*) = \theta(x)^*$ for all $x, y \in M$. A *-anti-automorphism θ of M is called *properly outer* if each θ^2 is a properly outer *-automorphism. A *-anti-automorphism θ of M is called *aperiodic* if each θ^n, $n \in N$ is properly outer.

The Jordan *-automorphism of the von Neumann algebra, M, is a complex-linear map $\alpha : M \to M$ satisfying $\alpha(x^2) = \alpha(x)^2, x \in M$. R. Kadison [8] shows that any Jordan *-automorphism is isometric and that there exists a projection f (i.e. $f^2 = f$)

in the center of M such that $x \to \alpha(x)f$ is a $\star$-automorphism and $x \to \alpha(x)(\mathbf{1} - f)$ is a $\star$-anti-automorphism.

Let M be a finite von Neumann algebra acting on H. Assume that M is equipped with a finite normal faithful trace τ. For given $1 < p < \infty$ define on M a norm

$$\|x\|_p = [\tau(|x|^p)]^{1/p}, x \in M,$$

where $|x| = (x^*x)^{1/2}$ is a modulus of x. Then $(M, \|.\|_p)$ is a normed space, whose completion is the non-commutative L_p-space associated with $(M, \tau$. Denote it as $L_p(M, \tau)$. Elements of $L_p(M, \tau)$ can be considered as closed densely defined operators on H which are affiliated to M.

We also set $L_\infty(M, \tau) = M$ and $L_0(M, \tau)$ - the space of all closed densely defined operators which are affiliated to M and are measurable with respect to (M, τ).

All the results stated in this section can be found in [19], [20], [15] and [16].

3 Positive isometries generated by the trace-preserving automorphisms

Let M be a von Neumann algebra with a finite, normal, faithful trace τ, and let T be an invertible isometry T on non-commutative L_p-space $L_p(M, \tau)$, associated with M and τ. In [21], F. J. Yeadon provided a description of the action of an arbitrary isometry T on the elements of M via $T(x) = WB\theta(x)$ for all $x \in M$, where W is an unitary operator from M, B is a positive operator affiliated to center $Z(M)$ of M and θ is a Jordan $\star$-automorphism of M. If T is a positive isometry, then $T(x) = B\theta(x)$. We will say in this case, that T is induced by B and θ.

Lemma 3.1 *Assume that T is a positive isometry of L. If $\tau \circ \theta = \tau$, then $B = \mathbf{1}$.*

Proof. The proof easy follows from the Radon-Nikodym Theorem for traces on finite von Neumann algebras (see [P], Sec. 5.3) and from faithfullness of the trace τ. □

Let T be a positive isometry of $L_p(M, \tau)$, induced by B and τ-preserving Jordan $\star$-automorphism θ. By Lemma 3.1, it follows that $B = \mathbf{1}$ and, hence, $T(x) = \theta(x)$, where $x \in M$. Now, we state and a non-commutative version of the dominated ergodic theorem for such an isometry T.

By $A_n^T(x)$, we will denote the *Cesaro n-sums* for T:

$$A_n^T(x) = \frac{1}{n} \sum_{k=0}^{n-1} T^k(x), \quad x \in M.$$

Definition 3.2. We say that *the dominated ergodic theorem holds* for an isometry, T, on $L_p(M, \tau)$ if, for every positive $x \in L_p(M, \tau)$, there exists a positive element, $a \in$

$L_p(M, \tau)$, and a constant, $C_p > 0$ (which does not depend on x), such that

$$0 \leq A_n^T(x) \leq a, \text{ and } \|a\|_p \leq C_p \|x\|_p.$$

In fact, the element $a \in L_p(M, \tau)$ from Definition 1 is a non-commutative analogy of the maximal function of Hardy and Littlewood for Cesaro averages in classical ergodic theory. Importance of maximal functions and maximal inequalities in commutative ergodic theory is well-known. Unfortunately, generally speaking, there does not exist the maximal function for non-commuting averages, and it is the main impediment to the development of the non-commutative ergodic theory. In recent work [7] M. Junge and Q. Xu applied the technique of non-commutative interpolation to construct the maximal element for Cesaro averages with non-commuting $\{T^n, n \in \mathbf{N}\}$. For more details, they have shown that if M is a semi-finite von Neumann algebra with a normal semi-finite faithful trace τ, $L_p(M, \tau)$ is the associated non-commutative L_p-space $(1 < p < \infty)$, and T is a positive L_p-contraction on $L_p(M, \tau)$, such that a restriction T on M is a L_∞-contraction on M with $\tau(T(x)) \leq \tau(x)$ for all $x \in L_1(M, \tau) \cap M_+$, then for any $x \in L_p(M, \tau)$ with $x \geq 0$ there is a $a \in L_p(M, \tau$ such that $A_n^T(x) \leq a, \forall n \in \mathbf{N}$, and $||a||_p \leq C_p||x||_p$, where $C_p > 0$ is a constant depending only on p (see [7], Theorem 0.1).

The following Proposition is a non-commutative version of the classical Dominated Ergodic Theorem for measure preserving transformations.

Proposition 3.3. *Let T be a positive, surjective isometry of $L_p(M, \tau)$, induced by τ-preserving Jordan *-automorphism θ. Then the dominated ergodic theorem holds for T on $L_p(M, \tau)$.*

Proof. The statement of the Proposition follows immediately from Theorem 0.1 of [7]. □

4 Cesaro sums for positive isometries

Consider two finite, normal, faithful traces τ and τ_1 on M. By the Radon-Nikodym Theorem for finite von Neumann algebras (see [P], Sec. 5.3), there exists a positive operator, h, affiliated with $Z(M)$, such that $\tau(x) = \tau_1(hx)$ for all $x \in M$.

Let T and S be isometries on $L_p(M, \tau)$ and $L_p(M, \tau_1)$, respectively, induced by the same Jordan *-automorphism, θ, of M, $T(x) = B\theta(x)$, $S(x) = B_1\theta(x)$, $x \in M$, where B and B_1 are the positive operators affiliated to $Z(M)$. Consider also the traces $\tau \circ \theta$ and $\tau_1 \circ \theta$ and the corresponding Radon-Nikodym derivatives ρ and δ, also affiliated with $Z(M)$, such that $\tau \circ \theta = \tau(\rho.)$, $\tau_1 \circ \theta = \tau_1(\delta.)$. Let us note, that if $x \in L_p(M, \tau)$, then $\tau(x^p) < \infty$ implies $\tau_1(hx^p) < \infty$. So, $\tau_1((h^{1/p}x)^p) < \infty$, and hence $h^{1/p}x \in L_p(M, \tau_1)$.

Consider now the actions of the Jordan *-automorphism θ on the operators ρ and δ, affiliated to $Z(M)$. Consider, for example, the spectral decomposition of $\delta = \int_0^\infty \lambda de_\lambda$, where all spectral projectors e_λ are contained in $Z((M)$, and consider $\theta(\delta)$ as a weak limit of

$$\lim_{M\to\infty} \int_0^M \theta(\lambda) de_\lambda.$$

Since θ is acting as an *-automorphism on $Z(M)$, then $\theta(\delta)$ (as well as $\theta(\rho)$ or $\theta(B)$ for any Radon-Nikodym derivative operator B, affiliated to $Z(M)$, is well-defined.

Next lemma describes the operators B and B_1 in terms of operators ρ and δ.

Lemma 4.1. For B and B_1, defined as above, we have

(i) $B = (\theta(\rho^{-1}))^{1/p}$;

(ii) $B_1 = (\theta(\delta^{-1}))^{1/p}$.

Proof. Since, for $x \in M$,

$$\|T(x)\|_{p,\tau}^p = \tau(|T(x)|^p) = \tau(B^p\theta(x^p)) = \tau(\theta(\theta^{-1}(B^p)x^p)) = \tau(\rho\theta^{-1}(B^p)x^p),$$

and since $\|T(x)\|_{p,\tau} = \|x\|_{p,\tau}$, we have

$$\tau(\rho\theta^{-1}(B^p)x^p) = \tau(x^p).$$

Therefore, it follows from the fact that τ is faithful that $\rho\theta^{-1}(B^p) = \mathbf{1}$. Hence, $B = (\theta(\rho^{-1}))^{1/p}$. Similarly one can prove that $B_1 = (\theta(\delta^{-1}))^{1/p}$. □

Lemma 4.2. With the same hypothesis as in Lemma 4.1, $B_1 = h^{1/p}\theta(h^{-1/p})B$.

Proof. Let $x \in M$ be arbitrary. As in the proof of Lemma 3.1, $\tau(x^p) = \tau(B^p\theta(x^p))$. By the Radon-Nikodym Theorem, $\tau(B^p\theta(x^p)) = \tau_1(hB^p\theta(x^p))$ and $\tau(x^p) = \tau_1(hx^p)$. Since S is an isometry, we have

$$\tau_1((h^{1/p}x)^p) = \|h^{1/p}x\|_{p,\tau_1}^p = \|S(h^{1/p}x)\|_{p,\tau_1}^p = \tau_1(B_1^p\theta(h)\theta(x^p)).$$

So, $\tau_1(hB^p\theta(x^p)) = \tau_1(B_1\theta(h)\theta(x^p))$. Since τ_1 is faithful, we have:

$$B^p h = B_1^p\theta(h).$$

Therefore, it follows that

$$B_1 = h^{1/p}\theta(h^{-1/p})B. \quad \square$$

Proposition 4.3. *Let T and S be positive isometries on $L_p(M, \tau)$ and $L_p(M, \tau_1)$, respectively, such that T is induced by B and θ and S is induced by B_1 and θ. Then the* *dominated ergodic theorem holds for T, if and only if the dominated ergodic theorem holds for S.*

Proof. The action of S^n on $L_p(M, \tau_1)$ can be expressed as

$$S^n(y) = B_1\theta(B_1)\theta^2(B_1)...\theta^{n-1}(B_1)\theta^n(y)$$

for $y \in L_p(M, \tau_1)$. Since $B_1 = h^{1/p}\theta(h^{-1/p})B$ by Lemma 4.2, we can replace B_1 with $h^{1/p}\theta(h^{-1/p}B$ and simplify to obtain

$$S^n(y) = B\theta(B)...\theta^{n-1}(B)h^{1/p}\theta(h^{1/p})...\theta^{n-1}(h^{1/p})\theta(h^{-1/p})...\theta^n(h^{-1/p})\theta^n(y) =$$

$$= h^{1/p}B\theta(B)...\theta^{n-1}(B)\theta^n(h^{-1/p}y) = h^{1/p}T^n(h^{-1/p}y).$$

Therefore, we can express T in the form

$$T^n(x) = h^{-1/p}S^n(xh^{1/p}), \text{ for } x \in L_p(M, \tau).$$

Thus,

$$A_n^T(x) = \frac{1}{n}\sum_{k=0}^{n-1} T^k(x) = \frac{1}{n}\sum_{k=0}^{n-1} h^{-1/p}S^k(xh^{1/p}) = h^{-1/p}A_n^S(xh^{1/p}).$$

Assume, now, that the Dominated Ergodic Theorem holds for T with a and C_p, i.e.

$$0 \le A_n^T(x) \le a, \text{ and } \|a\|_p \le C_p\|x\|_p.$$

Therefore, $0 \le A_n^T(x) = h^{-1/p}A_n^S(xh^{1/p}) \le a$, implies that

$$0 \le A_n^S(xh^{1/p}) \le h^{1/p}a, \text{ and } \|a\|_{p,\tau} \le C_p\|x\|_{p,\tau}.$$

Since for $x \in L_p(M, \tau)$, we have $h^{1/p}x \in L_p(M, \tau_1)$, it follows that

$$\|h^{1/p}a\|_{p,\tau_1} \le C_p\|h^{1/p}x\|_{p,\tau_1}.$$

Thus, the dominated ergodic theorem holds for S with $h^{1/p}a$ and C_p, if it holds for T with a and C_p. By symmetry, the converse implication follows. □

5 Dominated ergodic theorem for positive isometries

The main result of this section is a Dominated ergodic theorem for positive surjective isometries of non-commutative L_p-spaces (Theorem 5.3). In the beginning we state two technical lemmas.

Lemma 5.1. *Let M be a finite vNA, τ be a normal, finite, faithful trace on M with $\tau(\mathbf{1}) = 1$, and θ be a $*$-automorphism of M. Assume that $\theta = id$ on $Z(M)$. Then $\tau \circ \theta = \tau$.*

Proof. Assume that M is a factor. If τ is a normal, faithful, finite trace on M, then $\tau \circ \theta$ is also the normal, faithful, finite trace on M. By the Radon-Nikodym Theorem for finite von Neumann algebras (see [P], Sec. 5.3), there exists a positive operator h affiliated with $Z(M)$, such that $\tau\theta(x) = \tau(hx)$ for all $x \in M$. Since M is the factor, the center $Z(M)$ is trivial ($Z(M) = \mathbf{C1}$). So, $\tau \circ \theta = \lambda\tau$ for some scalar $\lambda \in \mathbf{C}$. Thus, it follows that $\tau(\theta(\mathbf{1})) = \tau(\mathbf{1})$; hence, $\lambda = 1$.

Assume, now, that $Z(M)$ is a commutative von Neumann algebra of type I, i.e. $Z(M) = \sum_{i \in I} \mathbf{C} e_i$, where $\{e_i\}_{i \in I}$ is a family of orthogonal projections such that $e_i M$ is a factor for each $i \in I$. Then, $\theta = id$ on $Z(M)$ implies that $\theta(e_i) = e_i$ for each i. Consequently, $\theta(e_i M) = e_i M$ for each factor $e_i M$, which implies that $\tau \circ \theta = \tau$.

Consider, now, the case when $Z(M)$ is a continuous commutative von Neumann algebra. Let

$$M = \int_{\{\Gamma, m\}}^{\oplus} M(\lambda) dm(\lambda)$$

be a decomposition of M in a direct integral of factors on a standard σ-finite measure space (Γ, m) (see [19], Chapter 4, Theorem 8.21). Since $\theta = id$ on $Z(M)$, for almost all λ (with respect to measure m) the restricted trace $\theta_\lambda = \theta|_{M(\lambda)}$ is well-defined. Consider the representation of the trace τ in terms of a direct integral of traces

$$\tau = \int_{\Gamma}^{\oplus} \tau_\lambda dm(\lambda)$$

(see [19], Chapter 4). Then,

$$\tau_\lambda \circ \theta_\lambda = \tau_\lambda$$

for almost all λ (with respect to m). Hence, $\tau \circ \theta = \tau$. Since the center of M is a direct sum of the discrete and continuous parts, the proof of lemma is completed. □

Let M be a finite von Neumann algebra M with a finite, normal, faithful trace τ and let $L_p(M, \tau)$ be a non-commutative L_p-space associated with M and τ, $1 < p < \infty$.

Lemma 5.2. Let M be a finite von Neumann algebra. Then for any Jordan $*$-automorphism θ of M and any $\varepsilon > 0$ there exist an integer number $n > 0$ and a projection $c \in Z(M)$ such that $\tau(\theta^n(x)) = \tau(x)$ for all $x \in cM$ and $\tau(\mathbf{1} - c) < \varepsilon$.

Proof. Let us consider a contraction $\theta' = \theta|_{Z(M)}$ of θ on the center $Z(M)$ of M. Since $Z(M)$ is a commutative subalgebra of M, θ' is acting on $Z(M)$ as a $*$-automorphism. Then there exists a resolution $\{c_i\}_{i=0}^{\infty}$ of identity $\mathbf{1}$ in $Z(M)$, $\sum_{i=0}^{\infty} c_i = \mathbf{1}$, such that the contraction of θ' on $c_0 Z(M)$ is an aperiodic automorphism, the contraction of θ' on

$c_1Z(M)$ is an automorphism of period 1, the contraction of θ' on $c_2Z(M)$ is the periodic automorphism of period 2,..., the contraction of θ' on $c_iZ(M)$ is the periodic automorphism of period i,..., etc.

Consider a commutative von Neumann algebra $c_0Z(M)$. Since the contraction $\theta'|_{c_0Z(M)}$ of the automorphism θ' on $c_0Z(M)$ is the aperiodic automorphism, we can apply to $\theta'|_{c_0Z(M)}$ the classical Linderholm approximation theorem for aperiodic transformations (see [12], Sec. 2, Chapter 1). Then there exist a central projection $c' \in c_0Z(M)$ and a natural number n_0 such that $\theta'|_{c'Z(M)}$ is periodic with period n_0 and $\tau(c_0 - c') < \frac{\varepsilon}{2}$.

Consider now the commutative von Neumann algebra $(\mathbf{1} - c_o)Z(M)$. Since M ia finite, then for given $\varepsilon > 0$ there exists an integer $N > 0$ such that

$$\tau(\sum_{i=N+1}^{\infty} c_i) < \frac{\varepsilon}{2}.$$

Consider the central projection $c'' = \sum_{i=1}^{N} c_i$ and a contraction of θ' on $c''Z(M)$ is the periodic automorphism with period n_1 which is less then $N!$. Set $c = c' + c''$, $n = max\{n_0, n_1\}$. Then a contraction θ'' of θ' on $cZ(M)$ is a periodic automorphism with period n, and

$$\tau(\mathbf{1} - c) < \frac{\varepsilon}{2} + \frac{\varepsilon}{2} = \varepsilon.$$

Thus completes the proof of Lemma 5.2. □

Theorem 5.3. *(An non-commutative analogy of commutative dominated ergodic theorem for positive surjective isometries). Let T be a positive surjective isometry on non-commutative L_p-space $L_p(M, \tau)$. Then given $x \in L_p(M, \tau)$ and $\varepsilon > 0$ there exist a projection $e \in M$, a positive element $a \in L_p(M, \tau)$, and a positive constant C_p, such that*

(i) $\tau(\mathbf{1} - e) < \varepsilon$;

(ii) $0 \leq A_n^T(exe) \leq a$ *for all positive integers n;*

(iii) $\|a\|_p \leq C_p\|exe\|_p$.,

where the constant C_p does not depend on x and ε.

Proof. Assume that T is induced by B and Jordan *-automorphism θ as $T(x) = B\theta(x)$ for all $x \in M$. By Lemma 5.2, for given $\varepsilon > 0$, there exists a projection, e, and a positive integer, $n > 0$, such that $\theta(e) = e$, $\tau(\mathbf{1} - e) < \varepsilon$ and $\theta|_{eMe}$ is periodic with period n. Consider a new trace, τ_1, on eMe given by

$$\tau_1(x) = \tau(\frac{1}{n}\sum_{k=0}^{n-1} \theta^k(x)), \text{ where } x \in eMe.$$

Then it follows that $\tau_1 \circ \theta(x) = \tau_1(x)$ for all $x \in eMe$. Consider, now, the L_p-space $L_p(eMe, \tau_1)$. We denote the norm of $L_p(eMe, \tau_1)$, induced by the trace τ_1, by $\|.\|_{p,\tau_1}$. By the Radon-Nikodym theorem for finite von Neumann algebra, the traces $\tau|_{eMe}$ and

τ_1 are connected via an element $h \in L_p(eMe, \tau_1)$ by

$$\tau(x) = \tau_1(hx) \text{ for all } x \in eMe$$

(see [P], Sec. 5.3). Consider a positive operator $B_1 = h^{1/p}\theta(h^{-1/p})B$, affiliated with $Z(eMe)$. Using B_1, we construct a positive linear transformation $S(.) = B_1\theta(.)$ on eMe, induced by B_1 and θ. Then, S is a positive invertible transformation on cM. We need to show that S is an isometry on eMe in $\|.\|_{p,\tau_1}$-norm. Using the techniques, developed in Lemma 3.1 and Lemma 4.2, we obtain:

$$\|S(x)\|_{p,\tau_1} = \tau_1((S(x))^p) = \tau_1(B_1^p(\theta(x))^p) = \tau_1(h\theta(h^{-1})B^p(\theta(x))^p) =$$

$$= \tau(B^p\theta(h^{-1})\theta(x^p)) = \tau((B\theta(h^{-1/p}x))^p) = \|T(h^{-1/p}x)\|_p = \|h^{-1/p}x\|_p =$$

$$= \tau(h^{-1}x^p) = \tau_1(x^p) = \|x\|_{p,\tau_1}.$$

Therefore, we obtain that, for all $x \in eMe$,

$$\|S(x)\|_{p,\tau_1} = \|x\|_{p,\tau_1}.$$

Since S is a $\|.\|_\infty$ and $\|.\|_{p,\tau_1}$-contraction, we can extend it from eMe to $L_p(eMe, \tau_1)$. Therefore, we have a positive surjective isometry S on $L_p(eMe, \tau_1)$, induced by B_1 and τ_1-preserving Jordan *-automorphism θ. It was shown in Lemma 3.1, that for τ_1-preserving θ, we have $B_1 = \mathbf{1}$. So, it follows that $S(y) = \theta(y)$ for all $y \in eMe$. By Proposition 3.3, the Dominated Ergodic Theorem holds for the positive surjective isometry S on $L_p(eMe, \tau_1)$ with a and C_p. Then, by Proposition 4.3, the Dominated Ergodic Theorem holds for T on $L_p(eMe, \tau)$ with $h^{-1/p}a$ and C_p, and $\tau(\mathbf{1} - e) < \varepsilon$. This completes the proof of the Theorem 5.3. □

6 Individual ergodic theorem for positive isometries

In this section we prove the individual ergodic theorem for positive surjective isometries of non-commutative L_p-space $L = L_p(M, \tau)$, $1 < p < \infty$, associated with a finite von Neumann algebra, M, and a finite, normal, faithful trace, τ.

Definition 6.1. A sequence $\{x_n\} \subset L$ is said to converge *bilaterally almost uniformly* (in short, b.a.u.) to x, if for every $\varepsilon > 0$, there is a projection $e \in M$, such that $\tau(\mathbf{1} - e) < \varepsilon$, and $\lim_{n\to\infty} \|e(x_n - x)e\| = 0$.

In the proof of the main result of this section (Theorem 6.3) we will utilize the following fundamental result due by V. Chilin, S. Litvinov and A. Skalski in [2]. A subset $X_0 \subset X_+$, where X is an ordered real Banach space with the closed convex cone X_+ and $X = X_+ - X_+$, is called *minorantly dense* in X_+, if, for every $x \in X_+$, there is a sequence

$\{x_n\} \subset X_0$, such that $x_n \le x$ for each n, and $\|x - x_n\| \to 0$ as $n \to \infty$.

Theorem *(non-commutative Banach principle for b.a.u. convergence)([2], Theorem 2.7). Let X be an ordered real Banach space with the closed convex cone X_+, $X = X_+ - X_+$. Let $a_n : X \to S(M)$ be a sequence of positive continuous linear maps satisfying the condition*

(i) for every $x \in X_+$ and $\varepsilon > 0$ there is $b \in M_+$, such that $b \ne 0$, $b \le \mathbf{1}$, $\tau(\mathbf{1} - b) < \varepsilon$ and

$$\sup_n \|ba_n(x)b\| < \infty.$$

If, for every x from a minorantly dense subset $X_0 \subset X_+$ we have

(ii) $a_m(x) - a_n(x) \to$ b.a.u., $m, n \to \infty$,

then (ii) holds for all of X.

Let S be a positive, surjective isometry on $L = L_p(M, \tau)$, $1 < p < \infty$, induced by trace preserving $\star$-automorphism θ. Let $\overline{N_S}$ be a closure of the linear subspace $N_S = \{x - Sx : x \in L_p(M, \tau)\}$ of L in $\|.\|_p$-norm.

Lemma 6.2. *For every $x_0 \in \overline{N_S}$, the sequence $\frac{1}{n}S^n(x_0)$ converges to 0 bilaterally almost uniformly.*

Proof. The isometry S is a contraction in the $\|.\|_\infty$-norm on M (because S is a $\star$-automorphism of M), and in the $\|.\|_1$-norm (because $\tau \circ \theta = \tau$ on M). Then, using the Mean Ergodic Theorem (see [9],Chapter 2, Theorem 1.1), we can see that $A_n^S(x_0)$ converges to 0 b.a.u., if and only if $\frac{1}{n}S^n(x_0)$ converges to 0 b.a.u. Hence, we need to prove that $A_n^S(x_0)$ converges to 0 b.a.u.

For given $\varepsilon > 0$, let us consider the sequences $\{\varepsilon_k\}_{k=1}^\infty$ and $\{\delta_k\}_{k=1}^\infty$ of positive numbers, such that $\varepsilon_k \to 0$, $\delta_k \to 0$ and

$$\sum_{k=1}^\infty \frac{\delta_k^p}{\varepsilon_k^p} < \varepsilon.$$

For every $x \in M$, there exist the sequences $\{a_k\} \subset M$, $\{b_k\} \subset M$ and $\{c_k\} \subset L_p(M, \tau)$, such that $x_0 = b_k + c_k$, where $b_k = a_k - S(a_k) \in M$ (because $S = \theta$ on M) and $\|c_k\|_p < \delta_k$. If x is self-adjoint, then we can choose $\{b_k\}$, $\{b_k\}$ and $\{c_k\}$ also self-adjoint ($a_k^* = a_k$, $b_k^* = b_k$, $c_k^* = c_k$ for all $k \in \mathbf{Z}$). We can write:

$$A_n^S(x_0) = A_n^S(b_k) + A_n^S(c_k).$$

We can see then that

$$\|A_n^S(b_k)\|_\infty \le \frac{2}{n+1}\|a_k\|_\infty.$$

Consider, now, the sequence $\{c_k\}$. By Yeadon's inequality (see [22]), there exists a sequence of self-adjoint projections $\{e_k\} \subset M$ such that $\sup_k \|e_k A_n^S(c_k^p)e_k\|_\infty \le \varepsilon_k^p$, and $\tau(e_k^\perp) \le \frac{\tau(c_k^p)}{\varepsilon_k^p} \le \frac{\delta_k^p}{\varepsilon_k^p}$.

Let $e = \wedge_{k=1}^{\infty} e_k$, then $\tau(e^{\perp}) \le \sum_{k=1}^{\infty} \frac{\delta_k^p}{\varepsilon_k^p} < \varepsilon$.

Therefore, we see that

$$\|e(A_n^S(x_))e\|_{\infty} \le \frac{2}{n+1}\|a_k\|_{\infty} + \|eA_n^S(c_k)e\|_{\infty} \le \frac{2}{n+1}\|a_k\|_{\infty} + \varepsilon_k \to 0$$

as $k \to \infty$.

Hence, $A_n^S(x_0) \to 0$ b.a.u., which implies that $\frac{1}{n}S^n(x_0) \to 0$ b.a.u. □

Theorem 6.3. *Let T be a positive surjective isometry on $L_p(M,\tau)$, $1 < p < \infty$. For every $x \in L_p(M,\tau)$, the limit of $\frac{1}{n}\sum_{k=0}^{n-1} T^k(x)$ exists bilateral almost uniformly.*

Proof. Consider the subspaces $F_T = \{x \in L_p(M,\tau) : T(x) = x\}$ and $N_T = \{x - T(x) : x \in L_p(M,\tau)\}$. Denote as $\overline{N_T}$ the closure of N_T in $\|.\|_p$-norm. Then, there exists a canonical splitting

$$L_p(M,\tau) = F_T \oplus \overline{N_T}$$

(see [9], Chapter 2, Theorem 2.1.3). Denote by P the projection from $L_p(M,\tau)$ onto F_T. Let x an arbitrary element of M with a unique decomposition $x = x_1 \oplus x_0$, where $x_1 \in F_T$, $x_0 \in \overline{N_T}$. Then $A_n^T(x) = A_n^T(x_1) + A_n^T(x_0) = x_1 + A_n^T(x_0)$. Hence, we need to prove that $A_n^T(x_0) \to 0$ b.a.u.

Assume that T is induced by B and θ. In the proof of Theorem 5.3 we constructed an isometry, S, on $L_p(eMe,\tau_1)$, which was induced by B_1 and θ and preserves $\tau_1(.) = \tau(h^{-1}(.))$. Since $B_1 = \mathbf{1}$ (see Lemma 3.1), then the isometry $S = \theta$ on eMe. By Lemma 6.2, $\frac{1}{n}S^n(x_0) \to 0$ b.a.u.

We know that $T^n(x) = h^{-1/p}S^n(xh^{-1/p})$ for all $x \in L_p(M,\tau)$ (see the proof of Proposition 4.3). If $\frac{1}{n}S^n(x_0) \to 0$ b.a.u., then $\frac{1}{n}T^n(x_0) \to 0$ b.a.u. By the Mean Ergodic Theorem ([9], Chapter 2, Theorem 2.1.1), $A_n^S(x_0)$ converges to 0 b.a.u. if and only if $\frac{1}{n}S^n(x_0)$ converges to 0 b.a.u. Hence, $A_n^T(x_0)$ converges to 0 b.a.u. for $x_0 \in \overline{N_T}$.

We assumed that $x \in M$, but the element $x_1 = P(x)$ need not to be in M. Assume that $x_1 \in M$. Then, $\frac{1}{n}T^n(x_1) \to 0$ b.a.u. Hence, $\frac{1}{n}T^n(x) = \frac{1}{n}T^n(x_1) + \frac{1}{n}T^n(x_0) \to 0$ b.a.u. Then, the limit of $A_n^T(x)$ exists b.a.u. for such $x \in L_p(M,\tau)$ that $x_1 = P(x) \in M$.

To complete the proof of the theorem, we need to prove that the limit of $A_n^T(x)$ exists b.a.u. for $x \in L_p(M,\tau)$. To show this, we apply the Non-commutative Banach Principle to $M_+ \subset L_p(M,\tau)_+$. We consider $L_p(M,\tau)$ as a real Banach space with the norm $\|.\|_p$ and $L_p(M,\tau)_+$ as the closed convex cone generating $L_p(M,\tau)$, that is

$$L_p(M,\tau) = L_p(M,\tau)_+ - L_p(M,\tau)_+.$$

Then, M_+ is minorantly dense in $L_p(M,\tau)_+$. Now, $A_n^T(x)$ is a sequence of positive, continuous, linear maps on $L_p(M,\tau)$. By Theorem 5.3(ii), $0 \le A_n^T(x) \le a$, and by Theorem 5.3(iii), $\|a\|_p \le C_p\|exe\|_p$. Then $\sup_n\{\{A_n^T(x)\}\} \le C_p < \infty$, and then the sequence $\{A_n^T(x)\}$ (where $x \in L_p(M,\tau)_+$) satisfies the condition (i) of the Non-commutative Banach Principle. Since $\{A_n^T(x)\}$ converges b.a.u. for all $x \in M_+$, then condition (ii) of the

Non-commutative Banach Principle holds for $A_n^T(x)$. Hence, the limit of $A_n^T(x)$ exists bilateral almost uniformly for all $x \in L_p(M, \tau)$. □.

Acknowledgment: This article is supported by ND EPSCOR through NFS Grant OSR-9452892 The author is very grateful to Professor Dŏgan Çömez for useful discussions.

Bibliography

[1] Bellow, A. (Ionescu Tulcea) *Ergodic properties of isometries in L^p-spaces,* $1 < p < \infty$; Bull. Amer. Math. Soc., 70 (1964), 366–371.
[2] Chilin, V. I., Litvinov, S. N., Skalski, A. *A few remarks in non-commutative ergodic theory*; J. Operator Theory, 53 (2005), 331–350.
[3] Goldstein, M., Litvinov, S. *Banach principle in the space of τ-measurable operators*; Studia Math., 143 (2000), 33–41.
[4] Jajte, R. *Strong limit theorems in non-commutative probability*; Lect. Notes Mathematics 1110, Springer Verlag, Berlin, 1985.
[5] Jones, R., Olsen, J. *Subsequence pointwise ergodic theorems for operators in L_p*; Israel J. Mathematics, 77 (1992), 33–54.
[6] Junge, M., Xu, Q. *Theoremes ergodiques maximaux dans les espaces L_p non commutatifs*, C.R.Math.Acad.Sci.Paris, 334 (2002), 773–778.
[7] Junge, M., Xu, Q. *Noncommutative maximal ergodic theorems*, Journal Amer. Math. Soc., 20 (2007), 385–439.
[8] Kadison, R. *Isometries of operator algebras*; Ann. Math., 54 (1951), 325–338.
[9] Krengel, U. *Ergodic theorems*; Studies in Mathematics, 6, De Gruyter, Berlin/New York, 1985.
[10] Lance, C. *Non-commutative ergodic theory*; Proc. of the meeting on C^*-algebras and their applications to theoretical physics, Roma, Gilla Universitaria, 1975, 70–79.
[11] Litvinov, S. *Weighted ergodic theorems in von Neumann algebras*; Ph.D. Thesis, North Dakota State University, Fargo, 1999.
[12] Olsen, J. H. *Dominated estimates of L_p contractions,* $1 < p < \infty$; Ph.D. Thesis, Univ. of Minnesota, Minneapolis, 1968.
[P] G. Pedersen: *C^*-algebras and their automorphism group*; Academic Press, London/New-York, 1979.
[14] Pisier, G., Xu, Q. *Non-commutative L_p-spaces*; Handbook of the geometry of Banach spaces, Volume 2, North-Holland Publ., Amsterdam, 2003, 1459–1517.
[15] Riesz, F., Sz-Nagy, B. *Functional Analysis*; Ungar Publ., New York, 1972.
[16] Sakai, S. *C^*-algebras and W^*-algebras*; Ergeb.Math., 60, Springer, 1971.
[17] Segal, I. E. *A non-commutative extension of abstract integration*; Ann. of Math. 57 (1953), 401–457.
[18] Sherman, D. *On the structure of isometries between noncommutative Lp-spaces*; Publ. RIMS, Kyoto Universty, 42 (2006), no. 1, 45–82.
[19] Takesaki, M. *Theory of Operator Algebras I*; Encyclopaedia of Mathematical Sciences, 124, Springer Verlag, Berlin, 2002.
[20] Takesaki, M. *Theory of Operator Algebras II*; Encyclopaedia of Mathematical Sciences, 125, Springer Verlag, Berlin, 2002.

[21] Yeadon, F. J. *Isometries on noncommutative L_p-spaces*; Math. Proc. Cambridge Phil. Soc. 90 (1981), 41–50.
[22] Yeadon, F. J. *Ergodic theorems for semifinite von Neumann algebras. I.*; J. London Math. Soc. 16 (1977), 326–332.
[23] Yeadon, F. J. *Ergodic theorems for semifinite von Neumann algebras. II.*; Math. Proc. Cambridge Phil. Soc. 77 (1975), 91–102.

Vesko Valov

Ranks and the approximate n-th root property of C^*-algebras

Abstract: This paper is a survey of some results on ranks of C^*-algebras obtained by the author and his co-authors. The main topics under consideration are: real and bounded ranks, weakly (strongly) infinite real and bounded ranks and the n-th root approximation property of C^*-algebras. A number of theorems included in the survey are accompanied by proofs.

Keywords: bounded and real ranks of C^*-algebras, n-th root approximation property, universal elements of C^*-algebras, weakly infinite real rank

MSC: Primary 46L05, 46L85; Secondary 54C40, 54F45

1 Introduction and preliminaries

The goal of this paper is to provide an overview of some results about ranks and approximate root properties of C^*-algebras obtained by the author and his coauthors, see [7], [8], [9]. We did not aim at the full coverage of all achievement in this very broad field. Instead, we included here the formulation and proofs of certain theorems, thus hoping to convey the key ideas to the reader.

The concept of the real rank plays an important role in a variety of problems related to general classification problems of C^*-algebras. Despite of this widely recognized fact we still see continuing attempts of defining "right dimension" (such as stable [18], analytic [14], tracial [13], exponential [17], completely positive [19] ranks) for unital C^*-algebras. Definition of the real rank [3] (as well as of its prototype – topological stable rank) of unital C^*-algebras is based on a well known characterization of the covering dimension of compact spaces. But still it is not known if the real rank of any product of unital C^*-algebras each of real rank $\leq n$ is also $\leq n$. At the same time the corresponding topological fact is true. The main advantage of the bounded rank, introduced in [8], is that the analog of the above question has a positive answer. This fact is used in the construction of a separable unital universal C^*-algebra in the class of all separable unital C^*-algebras of bounded rank $\leq n$ (see Theorem 14). A motivation for such a result lies, once again, in dimension theory. It is well known that the Menger cube μ^n contains a topological copy of any at most n-dimensional compact metric space. The definition and the main properties of the bounded rank, as well as, the relation between the real and bounded rank, are discussed in Section 2.

Vesko Valov, Department of Computer Science and Mathematics, Nipissing University, 100 College Drive, P.O. Box 5002, North Bay, ON, P1B 8L7, Canada

https://doi.org/10.1515/9783110413557-019

Section 3 is devoted to the definition and some properties of infinite real and bounded ranks defined in [9]. It is clear that some C^*-algebras of infinite real rank have infinite rank in a very strong sense. Distinguishing these types of infinite ranks is the main reason to introduce the weakly(strongly) infinite real and bounded ranks. As expected, every C^*-algebra of finite real rank has weakly infinite rank. On the other hand there exists non-commutative C^*-algebras of weakly infinite but not finite real rank (Corollary 8). The commutative case is completely settle by establishing (Theorem 19) that the function algebra $C(X)$, where X is compact, has weakly infinite real rank if and only if X is a weakly infinite-dimensional.

In the last Section 4 we consider the so called approximate n-root property of C^*-algebras. This property was introduced in [7] by modifying a similar property of the function spaces $C(X)$ with X a compactum. The commutative and non-commutative cases are considered. One of the main results is that the class of all unital separable C^*-algebras with the approximate n-th root property has universal elements which are non-commutative.

All C^*-algebras below are assumed to be unital. When we refer to a unital C^*-subalgebra of a unital C^*-algebra we implicitly assume that the inclusion is a unital $*$-homomorphism. The set of all self-adjoint elements of a C^*-algebra X is denoted by X_{sa}. The product in the category of (unital) C^*-algebras, i.e. the ℓ^∞-direct sum, is denoted by $\prod\{X_t : t \in T\}$. For a given set Y and a cardinal number τ the symbol $\exp_\tau Y$ denotes the partially ordered (by inclusion) set of all subsets of Y of cardinality not exceeding τ.

Recall that a direct system $\mathcal{S} = \{X_\alpha, i_\alpha^\beta, A\}$ of unital C^*-algebras consists of a partially ordered directed indexing set A, unital C^*-algebras X_α, $\alpha \in A$, and unital $*$-homomorphisms $i_\alpha^\beta : X_\alpha \to X_\beta$, defined for each pair of indexes $\alpha, \beta \in A$ with $\alpha \le \beta$, and satisfying the condition $i_\alpha^\gamma = i_\beta^\gamma \circ i_\alpha^\beta$ for each triple of indexes $\alpha, \beta, \gamma \in A$ with $\alpha \le \beta \le \gamma$. The (inductive) limit of the above direct system is a unital C^*-algebra which is denoted by $\varinjlim \mathcal{S}$. For each $\alpha \in A$ there exists a unital $*$-homomorphism $i_\alpha : X_\alpha \to \varinjlim \mathcal{S}$ which will be called the α-th limit homomorphism of $\mathcal{S}$.

If A' is a directed subset of the indexing set A, then the subsystem $\{X_\alpha, i_\alpha^\beta, A'\}$ of $\mathcal{S}$ is denoted $\mathcal{S}|A'$.

Definition 1. *Let $\tau \ge \omega$ be a cardinal number. A direct system $\mathcal{S} = \{X_\alpha, i_\alpha^\beta, A\}$ of unital C^*-algebras X_α and unital $*$-homomorphisms $i_\alpha^\beta : X_\alpha \to X_\beta$ is called a* direct C_τ^*-system *[4] if the following conditions are satisfied:*

(a) A is a τ-complete set (this means that for each chain C of elements of the directed set A with $|C| \le \tau$, there exists an element $\sup C$ *in A; see [6] for details).*

(b) Density of X_α is at most τ (i.e. $d(X_\alpha) \le \tau$), $\alpha \in A$.

(c) The α-th limit homomorphism $i_\alpha : X_\alpha \to \varinjlim \mathcal{S}$ is an injective $*$-homomorphism for each $\alpha \in A$.

(d) *If* $B = \{\alpha_t : t \in T\}$ *is a chain of elements of* A *with* $|T| \le \tau$ *and* $\alpha = \sup B$, *then the limit homomorphism* $\varinjlim\{i^{\alpha}_{\alpha_t} : t \in T\} : \varinjlim (\mathcal{S}|B) \to X_\alpha$ *is an isomorphism.*

We recall the concept of a $\mathcal{C}$-invertibility introduced in [5]. For a given class $\mathcal{C}$ of unital C^*-algebras the $\mathcal{C}$-invertibility of a unital $*$-homomorphism $p\colon Y \to Z$ means that for any unital $*$-homomorphism $g\colon Y \to X$, with $X \in \mathcal{C}$, there exists a unital $*$-homomorphism $h\colon Z \to X$ such that $g = h \circ p$. It is easy to see that if $\mathcal{C}$ consists of separable C^*-algebras and there exists a $\mathcal{C}$-invertible unital $*$-homomorphism $p\colon C^*(\mathbf{F}_\infty) \to Z_C$ such that $Z_C \in \mathcal{C}$, where $C^*(\mathbf{F}_\infty)$ denotes the group C^*-algebra of the free group on countable number of generators, then Z_C is an universal element in the class $\mathcal{C}$. Indeed, since every element X of $\mathcal{C}$ can be represented as the image of $C^*(\mathbf{F}_\infty)$ under a surjective $*$-homomorphism $g\colon C^*(\mathbf{F}_\infty) \to X$, the $\mathcal{C}$-invertibility of p guarantees that there exists a surjective $*$-homomorphism $h\colon Z_C \to X$ such that $g = h \circ p$.

2 Bounded and real ranks

2.1 Real rank - simple characterizations

The definition of the real rank [3] (as well as of its prototype – topological stable rank [18]) of unital C^*-algebras is based on the following standard result from classical dimension theory (see [11], [15], [16], [10]) characterizing the Lebesgue dimension dim of compact spaces: *The Lebesgue dimension* $\dim X$ *of a compact space* X *is the least integer* n *such that the set* $\{f : C(X, \mathbf{R}^{n+1}) : \mathbf{0} \notin f(X)\}$ *is dense in the space*[1] $C(X, \mathbf{R}^{n+1})$ *of all continuous maps of* X *in to the Euclidean space* $\mathbf{R}^{n+1}$.

A map $f \in C(X, \mathbf{R}^{n+1})$ may be identified with the $(n+1)$-tuple $(\pi_1 \circ f, \ldots, \pi_{n+1} \circ f)$, where $\pi_k\colon \mathbf{R}^{n+1} \to \mathbf{R}$ is a projection onto the k-th coordinate, $k = 1, \ldots, n+1$. The condition $\mathbf{0} = (0, \ldots, 0) \notin f(X)$ may be equivalently expressed as the condition $\sum_{k=1}^{n+1} f_k^2(x) \neq 0$ for any $x \in X$. This, in turn, can be equivalently rephrased as the invertibility of the element $\sum_{k=1}^{n+1} f_k^2$. These two observations lead us to the definition of the real rank:

Definition 2. *[3] The real rank* $rr(X)$ *of a unital* C^**-algebra is the least integer* n *such that each* $(n+1)$*-tuple* $(x_1, \ldots, x_{n+1})$ *of elements from* X_{sa} *can be arbitrarily closely approximated by another* $(n+1)$*-tuple* $(y_1, \ldots, y_{n+1})$ *of self-adjoint elements so that the element* $\sum_{k=1}^{n+1} y_k^2$ *is invertible.*

We have the following description of the real rank:

1 Compact-open topology is being considered.

Proposition 4. *[8] The following conditions are equivalent for a unital C^*-algebra X;*
(1) $rr(X) \leq n$;
(2) for each $(n+1)$-tuple $(x_1, \dots, x_{n+1})$ in X_{sa} and for each $\epsilon > 0$, there exists an $(n+1)$-tuple $(y_1, \dots, y_{n+1})$ in X_{sa} such that $\sum_{k=1}^{n+1} y_k^2$ is invertible and $\|x_k - y_k\| < \epsilon$ for each $k = 1, 2, \dots, n+1$.
(3) for each $(n+1)$-tuple $(x_1, \dots, x_{n+1})$ in X_{sa}, with $\|x_k\| = 1$ for each $k = 1, \dots, n+1$, and for each $\epsilon > 0$, there exists an $(n+1)$-tuple $(y_1, \dots, y_{n+1})$ in X_{sa} such that $\sum_{k=1}^{n+1} y_k^2$ is invertible and $\|x_k - y_k\| < \epsilon$ for each $k = 1, \dots, n+1$.

Let $\mathcal{K}(\mathbf{C}^m)$ denote the collection of compact subsets of $\mathbf{C}^m$ and $M \in \mathbf{N} \cup \{\infty\}$. A generalized joint M-spectrum (or simply spectrum) on an unital C^*-algebra X is a collection $\{\widetilde{\sigma}_m : m \leq M\}$ of maps $\widetilde{\sigma}_m : X^m \to \mathcal{K}(\mathbf{C}^m)$ satisfying conditions (I)–(III) below:
(I) $\widetilde{\sigma}_m(x_1, \dots, x_m)$ is a non-empty compact subset of $\mathbf{C}^m$.
(II) If $x \in X$, then $\widetilde{\sigma}_1(x) = \sigma(x)$, where $\sigma(x)$ denotes the usual spectrum of the element x.
(III) $p(\widetilde{\sigma}_m(x_1, \dots, x_m)) = \widetilde{\sigma}_n(p(x_1, \dots, x_m))$ for every polynomial mapping $p = (p_1, \dots, p_n)\colon \mathbf{C}^m \to \mathbf{C}^n$.

We also consider the following two properties:
(IV) There exists a constant $K > 0$ (called a polynomial spectral constant) such that for any rational $\delta > 0$, for any m-tuple $(x_1, \dots, x_m)$ in X (resp., in X_{sa}) and for any polynomial mapping $p = (p_1, \dots, p_m)\colon \mathbf{C}^m \to \mathbf{C}^m$ (resp., having real coefficients) with $\|p - id_{\mathbf{C}^m}\| \leq K \cdot \delta$, there exists an m-tuple $(y_1, \dots, y_m)$ in X (resp., in X_{sa}) such that $\widetilde{\sigma}(y_1, \dots, y_m) = p(\widetilde{\sigma}_m(x_1, \dots, x_m))$ and $\|x_k - y_k\| \leq \delta$ for each $k = 1, \dots, m$.
(V) There exists $K > 0$ (called a general spectral constant) such that for any rational $\delta > 0$, for any m-tuple $(x_1, \dots, x_m)$ in X (resp., in X_{sa}) and for any map f from $\widetilde{\sigma}_m(x_1, \dots, x_m)$ into $\mathbf{C}^m$ (resp., into $\mathbf{R}^m \subseteq \mathbf{C}^m$) with $\|f - id_{\widetilde{\sigma}_m(x_1,\dots,x_m)}\| \leq K \cdot \delta$, there exists an m-tuple $(y_1, \dots, y_m)$ in X (resp., in X_{sa}) such that $\widetilde{\sigma}(y_1, \dots, y_m) = f(\widetilde{\sigma}_m(x_1, \dots, x_m))$ and $\|x_k - y_k\| \leq \delta$ for each $k = 1, \dots, m$.

Few observations are in order:
(A) Concerning property (I), it is not usually requested (in a much general setting though) the spectrum of every tuple of non-commuting elements to be non-empty.
(B) Properties (I)–(III) are standard axioms [20] of joint spectra (for commuting tuples) in Banach algebras. Property (III) is known as the spectral mapping property of $\widetilde{\sigma}$.
(C) It is easy to see [20] that properties (II) and (III) imply the inclusion $\widetilde{\sigma}_m(x_1, \dots, x_m) \subseteq \prod\{\sigma(x_k)\colon k = 1, \dots, m\}$ for any m-tuple $(x_1, \dots, x_m)$ in X.
(D) Note that if, in property (III), the m-tuple $(x_1, \dots, x_m)$ consists of self-adjoint elements and the polynomials p_k, $k = 1, \dots, m$, have real coefficients, then the m-tuple

$$p(x_1, \dots, x_m) = (p_1(x_1, \dots, x_m), \dots, p_m(x_1, \dots, x_m))$$

also consists of self-adjoint elements.

(E) We do not specify whether the m-tuple $(y_1, \dots, y_m)$ in property (IV) is obtained as the image of the m-tuple $(x_1, \dots, x_m)$ under the polynomial p. Nevertheless, in light of (D), we require that all y_k's are self-adjoint provided that all x_k's are self-adjoint and all the coefficients of the polynomials p_k are real.

(F) Similar comment with respect to property (V). If all x_k's are self-adjoint and $f(\widetilde{\sigma}_m(x_1, \dots, x_m)) \subseteq \mathbf{R}^m \subseteq \mathbf{C}^m$, then all y_k's are also assumed to be self-adjoint.

(G) If T is a compact space and $(x_1, \dots, x_m)$ is an m-tuple in $C(T)$, let define $\widetilde{\sigma}_m(x_1, \dots, x_m) = \triangle\{x_k \colon k = 1, \dots, m\}(T)$, where $\triangle\{x_k \colon k = 1, \dots, m\}(t) = (x_1(t), \dots, x_m(t))$ for each $t \in T$. So obtained generalized joint spectrum has all the properties (I)–(V) with $K = 1$.

(H) For any C^*-algebra X the usual spectrum $\sigma(x)$, $x \in X$, provides an example of a generalized joint 1-spectrum. This follows from spectral mapping theorem and considerations related to the functional calculus.

The real rank can be characterized in terms of generalized joint spectra:

Proposition 5. *[8] If there exists a generalized joint $(n+1)$-spectrum on a unital C^*-algebra X, then the following conditions are equivalent:*

(a) $rr(X) \le n$;

(b) For every $(n+1)$-tuple $(x_1, \dots, x_{n+1})$ of self-adjoint elements in X and for every $\epsilon > 0$ there exists an $(n+1)$-tuple $(y_1, \dots, y_{n+1})$ of self-adjoint elements such that

(i) $\|x_k - y_k\| < \epsilon$ for each $k = 1, \dots, n+1$;

(ii) $\mathbf{0} \notin \widetilde{\sigma}_{n+1}(y_1, \dots, y_{n+1})$.

2.2 Bounded rank

The analogy between the definition of the dimension dim and the real rank is quite formal and does not go far beyond the obvious observation that $rr(X) = \dim \Omega(X)$, where $\Omega(X)$ is the spectrum of the commutative unital C^*-algebra X, and a few other straightforward extensions of certain basic facts from classical dimension theory to a non-commutative situation. But there are difficulties in finding proper algebraic interpretations in terms of the concept of the real rank of a non-commutative unital C^*-algebra X of such an immediate geometric consequence of the condition $\dim \Omega(X) \le n$ as the possibility not only to remove the image of any $f \colon \Omega(X) \to \mathbf{R}^{n+1}$ from $\mathbf{0}$, but even to push this image by an ϵ-move outside the open ball $O(\mathbf{0}, \epsilon)$ of radius $\epsilon > 0$. As it turns out in the presence of some form of functional calculus (as in the commutative or the real rank zero cases – see Corollaries 3 and 4) there exists a satisfactory analog of the above mentioned geometric fact. As for the general case, the situation remains unclear and we, as a consequence, are unable to answer the question:

Question 6. Is it true that $rr\,(\prod\{X_t\colon t \in T) \le n$ for any collection of unital C^*-algebras X_t such that $rr(X_t) \le n$ for each $t \in T$?

The corresponding topological fact – $\dim \beta\,(\oplus\{\Omega(X_t)\colon t \in T\}) \le n$ – is easy to establish[2]. Perhaps the easiest way of proving the desired inequality in terms of the above given characterization of dim is first to approximate (as close as we wish) each of the restrictions $f_t\colon \Omega(X_t) \to \mathbf{R}^{n+1}$, $t \in T$, of an arbitrarily given map $f\colon \beta\,(\oplus\{\Omega(X_t)\colon t \in T\}) \to \mathbf{R}^{n+1}$, by maps $g_t\colon \Omega(X_t) \to \mathbf{R}^{n+1} \setminus \{\mathbf{0}\}$ and then to move each of the images $g_t\,(\Omega(X_t))$ outside of the open ball O (centered at $\mathbf{0}$) of the appropriate radius by small moves (independent of $t \in T$) . This way we get a map $g\colon \oplus\,\{\Omega(X_t)\colon t \in T\} \to \mathbf{R}^{n+1} \setminus O$. Since the image $g(\oplus\{\Omega(X_t)\colon t \in T\})$ has a compact closure, g can be extended to a map $\tilde{g}\colon \beta\,(\oplus\{\Omega(X_t)\colon t \in T\}) \to \mathbf{R}^{n+1} \setminus O$ which would be the required approximation of f whose image misses $\mathbf{0}$.

Analysis of this elementary reasoning leads us to the concept of the *bounded* rank (to be more specific – bounded rank with respect to a given positive constant).

Definition 3. *[8] Let $K > 0$. We say that an m-tuple $(y_1, \ldots, y_m)$ of self-adjoint elements of a unital C^*-algebra X is K-*unessential *if for every rational $\delta > 0$ there exists an m-tuple $(z_1, \ldots, z_m)$ of self-adjoint elements of X satisfying the following conditions:*
(a) $\|y_k - z_k\| \le \delta$ for each $k = 1, \ldots, m$,
(b) The element $\sum_{k=1}^m z_k^2$ is invertible and $\left\|\left(\sum_{k=1}^m z_k^2\right)^{-1}\right\| \le \frac{1}{K\cdot\delta^2}$.
1-unessential tuples are referred as unessential.

Remark 2.1. *Obviously if $K_1 \le K_2$, then every K_2-unessential m-tuple is K_1-unessential.*

Definition 4. *[8] Let $K > 0$. We say that the* bounded rank *of a unital C^*-algebra X with respect to K does not exceed n (notation: $br_K(X) \le n$) if for any $(n+1)$-tuple $(x_1, \ldots, x_{n+1})$ of self-adjoint elements of X and for any $\epsilon > 0$ there exists a K-unessential $(n+1)$-tuple $(y_1, \ldots, y_{n+1})$ in X such that $\|x_k - y_k\| < \epsilon$ for each $k = 1, \ldots, n+1$. For simplicity, $br_1(X)$ is denoted by $br(X)$ and it is called a* bounded rank.

We record the following statement for future references.

Proposition 7. *[8] $rr(X) \le br_K(X)$ for any unital C^*-algebra X and for any $K > 0$.*

Proof. Let $br_K(X) = n$ and $(x_1, \ldots, x_{n+1})$ be an $(n+1)$-tuple of self-adjoint elements in X. Let also $\epsilon > 0$. Since $br_K(X) = n$, there exists a K-unessential $(n+1)$-tuple $(y_1, \ldots, y_{n+1})$ such that $\|x_k - y_k\| < \frac{\epsilon}{2}$ for each $k = 1, \ldots, n+1$. This in turn means

2 βK stands for the Stone-Čech compactification of a space K.

that for a rational number δ with $\delta \leq \frac{\epsilon}{2}$ there is an $(n+1)$-tuple $(z_1, \ldots, z_{n+1})$ of self-adjoint elements such that
(a) $\|y_k - z_k\| \leq \delta$ for each $k = 1, \ldots, n+1$,
(b) The element $\sum_{k=1}^{n+1} z_k^2$ is invertible and $\left\|\left(\sum_{k=1}^{n+1} z_k^2\right)^{-1}\right\| \leq \frac{1}{K \cdot \delta^2}$.

Clearly $\|x_k - z_k\| \leq \|x_k - y_k\| + \|y_k - z_k\| < \frac{\epsilon}{2} + \delta \leq \epsilon$, $k = 1, \ldots, n+1$. According to (b), $\sum_{k=1}^{n+1} z_k^2$ is invertible which shows that $rr(X) \leq n$. □

The next two propositions provide a relation between the real and bounded ranks in the presence of a generalized joint spectrum.

Proposition 8. *[8] If there exists a generalized joint $(n+1)$-spectrum on a unital C^*-algebra X satisfying conditions (IV), then the following are equivalent:*
(a) $rr(X) \leq n$;
(b) $br_{\frac{K^2}{4}}(X) \leq n$, where K is a polynomial spectral constant of the joint spectrum.

Proposition 9. *[8] If there exists a generalized joint $(n+1)$-spectrum on the unital C^*-algebra X satisfying condition (V), then the following are equivalent:*
(a) $rr(X) \leq n$;
(b) $br_{K^2}(X) \leq n$, where K is a general spectral constant of the joint spectrum.

Proposition 10. *[8] Let $(y_1, \ldots, y_m)$ be a commuting m-tuple of self-adjoint elements of the unital C^*-algebra X. Then $(y_1, \ldots, y_m)$ is K-unessential for any positive $K \leq 1$ provided $\sum_{i=1}^{m} y_i^2$ is invertible.*

Corollary 3. *[8] Let X be a commutative unital C^*-algebra and $0 < K \leq 1$. Then $br_K(X) = rr(X) = \dim \Omega(X)$, where $\Omega(X)$ is the spectrum of X.*

Proof. By Proposition 7, $rr(X) \leq br_K(X)$. The opposite inequality $br_K(X) \leq rr(X)$ follows from Proposition 10. The remaining part is well known (see [3, Proposition 1.1]. □

Corollary 4. *[8] Let X be a unital C^*-algebra and $0 < K \leq 1$. Then $br_K(X) = 0$ if and only if $rr(X) = 0$.*

Proof. According to Proposition 7, $br_K(X) = 0$ yields $rr(X) = 0$. Conversely, if $rr(X) = 0$, then, by Proposition 10, $br_K(X) = 0$. □

We also have the following property of the bounded rank:

Proposition 11. *[8] Let $K > 0$ and $p\colon X \to Y$ be a surjective $*$-homomorphism of unital C^*-algebras. Then $br_K(Y) \leq br_K(X)$.*

Comparing to the real rank, the bounded rank has the advantage that the corresponding analog of Question 6 has a positive answer. This is actually one of the main reasons of introducing the concept of bounded rank.

Proposition 12. *[8] Let $\{X_t\colon t \in T\}$ be a family of unital C^*-algebras such that $br_K(X_t) \le n$ for each $t \in T$. Then for every $K > 0$ we have $br_K\left(\prod\{X_t\colon t \in T\}\right) \le n$.*

Proof. Let $(x_1, \dots, x_{n+1})$ be an $(n+1)$-tuple of self-adjoint elements of the product $X = \prod\{X_t\colon t \in T\}$, where $x_k = \{x_k^t\colon t \in T\}$ for each $k = 1, \dots, n+1$, and let $\epsilon > 0$. Our goal is to find a K-unessential $(n+1)$-tuple $(y_1, \dots, y_{n+1})$ in X such that $\|x_k - y_k\| < \epsilon$. For a given $t \in T$ consider the $(n+1)$-tuple $(x_1^t, \dots, x_{n+1}^t)$ of self-adjoint elements in X_t. Since $br_K(X_t) \le n$, there exists a K-unessential $(n+1)$-tuple $(y_1^t, \dots, y_{n+1}^t)$ in X_t such that $\|x_k^t - y_k^t\|_{X_t} < \frac{\epsilon}{2}$ for each $k = 1, \dots, n+1$. Consider the $(n+1)$-tuple $(y_1, \dots, y_{n+1})$, where $y_k = \{y_k^t\colon t \in T\}$ for each $k = 1, \dots, n+1$. Note that $y_k \in X$ for each $k = 1, \dots, n+1$. Indeed $\|y_k^t\|_{X_t} \le \|y_k^t - x_k^t\|_{X_t} + \|x_k\|_{X_t} \le \frac{\epsilon}{2} + \sup\{\|x_k^t\|_{X_t}\colon t \in T\}$ and $\sup\{\|y_k^t\|_{X_t}\colon t \in T\} \le \frac{\epsilon}{2} + \sup\{\|x_k^t\|_{X_t}\colon t \in T\} < \infty$. Also note that $\|x_k - y_k\| = \sup\{\|x_k^t - y_k^t\|_{X_t}\colon t \in T\} \le \frac{\epsilon}{2} < \epsilon$. It only remains to show that the $(n+1)$-tuple $(y_1, \dots, y_{n+1})$ is K-unessential in X. Indeed, let $\delta > 0$ be rational. Since the $(n+1)$-tuple $(y_1^t, \dots, y_{n+1}^t)$ is K-unessential in X_t, there exists an $(n+1)$-tuple $(z_1^t, \dots, z_{n+1}^t)$ in X_t such that
$(a)_t$ $\|y_k^t - z_k^t\|_{X_t} \le \delta$ for each $k = 1, \dots, n+1$.
$(b)_t$ $\left\|\left(\sum_{k=1}^{n+1}\left(z_k^t\right)^2\right)^{-1}\right\| \le \frac{1}{K\cdot\delta^2}$.

Next, consider the $(n+1)$-tuple $(z_1, \dots, z_{n+1})$, where $z_k = \{z_k^t\colon t \in T\}$ for each $k = 1, \dots, n+1$. As above, $z_k \in X$ and obviously

$$\|y_k - z_k\| = \sup\{\|y_k^t - z_k^t\|_{X_t}\colon t \in T\} \le \delta,\ k = 1, \dots, n+1.$$

Finally, note that $\left\|\left(\sum_{k=1}^{n+1} z_k^2\right)^{-1}\right\| = \left\|\left\{\left(\sum_{k=1}^{n+1}\left(z_k^t\right)^2\right)^{-1} : t \in T\right\}\right\| = \sup\left\{\left\|\left(\sum_{k=1}^{n+1}\left(z_k^t\right)^2\right)^{-1}\right\|_{X_t} : t \in T\right\} \le \frac{1}{K\cdot\delta^2}$ as required. □

2.3 Spectral decompositions of unital C^*-algebras of bounded rank n

In this subsection we investigate the behaviour of the bounded rank with respect to direct systems and prove the existence of universal elements in $\mathcal{BR}_n^K$, where $\mathcal{BR}_n^K$ denotes the class of all separable unital C^*-algebras of bounded rank with respect to K at most n. A motivation for such a result lies, once again, in the classical dimension theory. It is a well-known observation (see, for instance, [10, Theorem 1.3.15]) that the Menger cube μ^n contains a topological copy of any at most n-dimensional metrizable compact space. This means that every commutative separable unital C^*-algebra of real rank at most n is a quotient of the C^*-algebra $C(\mu^n)$. We extend this result to the non-commutative case.

The following decomposition theorem was initially established for $n = 0$ in [5].

Proposition 13. *[8] Let $K > 0$. The following conditions are equivalent for any unital C^*-algebra X:*

(a) $br_K(X) \le n$.

(b) *X can be represented as the direct limit of a direct C^*_ω-system $\{X_\alpha, i^\beta_\alpha, A\}$ satisfying the following properties:*

  (a) *The indexing set A is cofinal and ω-closed in the ω-complete set $\exp_\omega Y$ for some (any) countable dense subset Y of X.*

  (b) *X_α is a C^*-subalgebra of X such that $br_K(X_\alpha) \le n$, $\alpha \in A$.*

The proof of Proposition 13 provided in [8] works for the real rank as well.

Corollary 5. *[8] Let $K > 0$. Every countable subset of a unital C^*-algebra X with $br_K(X) \le n$ is contained in a unital separable C^*-subalgebra X_0 such that $br_K(X_0) \le n$.*

Next, for any $K > 0$, we construct a universal separable unital C^*-algebra Z^K_n of bounded rank n. Universal in the sense that any other separable unital C^*-algebra with bounded rank $\le n$ is its quotient. Recall that $C^*(\mathbf{F}_\infty)$ denotes the group C^*-algebra of the free group on countable number of generators.

Theorem 14. *[8] Let $K > 0$. The class $\mathcal{BR}^K_n$ of all separable unital C^*-algebras with $br_K \le n$ contains an universal element Z^K_n. More precisely, there is a $\mathcal{BR}^K_n$-invertible unital $*$-homomorphism $p\colon C^*(\mathbf{F}_\infty) \to Z^K_n$, where Z^K_n is a separable unital C^*-algebra such that $br_K\left(Z^K_n\right) = n$.*

Proof. Let $\mathcal{A} = \{f_t\colon C^*(\mathbf{F}_\infty) \to X_t, t \in T\}$ denote the set of all unital $*$-homomorphisms, defined on $C^*(\mathbf{F}_\infty)$, such that $br_K(X_t) \le n$. Next, consider the product $\prod\{X_t\colon t \in T\}$. Since $br_K(X_t) \le n$ for each $t \in T$, it follows from Proposition 12 that $br_K(\prod\{X_t\colon t \in T\}) \le n$. The $*$-homomorphisms f_t, $t \in T$, define the unital $*$-homomorphism $f\colon C^*(\mathbf{F}_\infty) \to \prod\{X_t\colon t \in T\}$ such that $\pi_t \circ f = f_t$ for each $t \in T$ (here $\pi_t\colon \prod\{X_t\colon t \in T\} \to X_t$ denotes the corresponding canonical projection $*$-homomorphism). By Proposition 13, $\prod\{X_t\colon t \in T\}$ can be represented as the limit of the C^*_ω-system $\mathcal{S} = \{C_\alpha, i^\beta_\alpha, A\}$ such that C_α is a separable unital C^*-algebra with $br_K(C_\alpha) \le n$ for each $\alpha \in A$. Suppressing injective unital $*$-homomorphisms $i^\beta_\alpha\colon C_\alpha \to C_\beta$, we can, for notational simplicity, assume that C_α's are unital C^*-subalgebras of $\prod\{X_t\colon t \in T\}$. Let $\{a_k\colon k \in \omega\}$ be a countable dense subset of $C^*(\mathbf{F}_\infty)$. It is easily seen that for each $k \in \omega$ there exists an index $\alpha_k \in A$ such that $f(a_k) \in C_{\alpha_k}$. By [6, Corollary 1.1.28], there exists an index $\alpha_0 \in A$ such that $\alpha_0 \ge \alpha_k$ for each $k \in \omega$. Then $f(a_k) \in C_{\alpha_k} \subseteq C_{\alpha_0}$

for each $k \in \omega$ (see also Corollary 5). This observation coupled with the continuity of f guarantees that $f(C^*(\mathbf{F}_\infty)) = f(cl\{a_k : k \in \omega\}) \subseteq cl\{f(\{a_k : k \in \omega\})\} \subseteq clC_{\alpha_0} = C_{\alpha_0}$.

Let $Z_n^K = C_{\alpha_0}$ and p denote the unital $*$-homomorphism f considered as the homomorphism of $C^*(\mathbf{F}_\infty)$ into Z_n^K. Note that $f = i \circ p$, where $i : Z_n^K = C_{\alpha_0} \hookrightarrow \prod\{X_t : t \in T\}$ stands for the inclusion.

By construction, $br_K(Z_n^K) \leq n$. Let us show that $p : C^*(\mathbf{F}_\infty) \to Z_n^K$ is $\mathcal{BR}_n^K$-invertible in the sense of Introduction. In our situation, for any unital $*$-homomorphism $g : C^*(\mathbf{F}_\infty) \to X$, where X is a separable unital C^*-algebra with $br_K(X) \leq n$, we need to establish the existence of a unital $*$-homomorphism $h : Z_n^K \to X$ such that $g = h \circ p$. Indeed, by definition of the set $\mathcal{A}$, we conclude that $g = f_t$ for some index $t \in T$ (in particular, $X = X_t$ for the same index $t \in T$). Next observe that $g = f_t = \pi_t \circ f = \pi_t \circ i \circ p$. This allows us to define the required unital $*$-homomorphism $h : Z_n^K \to X$ as the composition $h = \pi_t \circ i$. Hence, p is $\mathcal{BR}_n^K$-invertible which yields the universality of Z_n^K. □

It is interesting to note that not only Z_n^K is universal in the above sense, but for every unital $*$-homomorphism $g : C^*(\mathbf{F}_\infty) \to \mathbf{C}$ the pushout $Z_n^K \star_{\mathbf{C}} \mathbf{C}$, generated by p and g, is also $\mathcal{BR}_n^K$-universal. To see this take any separable unital C^*-algebra X such that $br_K(X) \leq n$ and consider the unital $*$-homomorphism $h : \mathbf{C} \to X$. Since p is $\mathcal{BR}_n^K$-invertible, there exists a unital $*$-homomorphism $\tilde{h} : Z_n^K \to X$ such that $\tilde{h} \circ p = h \circ g$. The homomorphisms h and $\tilde{h}$ uniquely determine the unital $*$-homomorphism $\varphi : Z_n^K \star_{\mathbf{C}} \mathbf{C} \to X$ as required.

3 Infinite ranks

As is dimension theory, some C^*-algebras of infinite real rank have infinite rank in a very strong sense of this word, while others do not. In order to distinguish these types of infinite ranks we introduce the concept of weakly (strongly) infinite real rank. First, let us provide an alternative characterization of real rank in terms of infinite sequences.

Proposition 15. *[9] Let X be a unital C^*-algebra. Then the following conditions are equivalent:*

(i) $rr(X) \leq n$.

(ii) for each $(n+1)$-tuple $(x_1, \dots, x_{n+1})$ in X_{sa} and for each $\epsilon > 0$, there exists an $(n+1)$-tuple $(y_1, \dots, y_{n+1})$ in X_{sa} such that $\sum_{k=1}^{n+1} y_k^2$ is invertible and $\|x_k - y_k\| < \epsilon$ for each $k = 1, 2, \dots, n+1$.

(iii) for any sequence of self-adjoint elements $\{x_i : i \in N\} \subseteq X_{sa}$ and for any sequence of positive real numbers $\{\epsilon_i : i \in \mathbf{N}\}$ there exists a sequence $\{y_i : i \in \mathbf{N}\} \subseteq X_{sa}$ such that

(a) $\|x_i - y_i\| < \epsilon_i$, for each $i \in \mathbf{N}$,

(b) for any subset $D \subseteq \mathbf{N}$, with $|D| = n + 1$, the element $\sum_{i \in D} y_i^2$ is invertible.

(iv) *for any sequence of self-adjoint elements* $\{x_i : i \in \mathbf{N}\} \subseteq X_{sa}$ *and for any* $\epsilon > 0$ *there exists a sequence* $\{y_i : i \in \mathbf{N}\} \subseteq X_{sa}$ *such that*
   (a) $\|x_i - y_i\| < \epsilon$, *for each* $i \in \mathbf{N}$,
   (b) *for any subset* $D \subseteq \mathbf{N}$, *with* $|D| = n+1$, *the element* $\sum_{i\in D} y_i^2$ *is invertible.*

(v) *for any sequence of self-adjoint elements* $\{x_i : i \in \mathbf{N}\} \subseteq X_{sa}$ *such that* $\|x_i\| = 1$ *for each* $i \in \mathbf{N}$ *and for any* $\epsilon > 0$ *there exists a sequence* $\{y_i : i \in \mathbf{N}\} \subseteq X_{sa}$ *such that*
   (a) $\|x_i - y_i\| < \epsilon$, *for each* $i \in \mathbf{N}$,
   (b) *for any subset* $D \subseteq \mathbf{N}$, *with* $|D| = n+1$, *the element* $\sum_{i\in D} y_i^2$ *is invertible.*

Proposition 15 provides a basis for the next definition.

Definition 5. *[9] We say that a unital* C^* *algebra* X *has a weakly infinite real rank if for any sequence of self-adjoint elements* $\{x_i : i \in \mathbf{N}\} \subset X_{sa}$ *and any* $\epsilon > 0$ *there is a sequence* $\{y_i : i \in \mathbf{N}\} \subset X_{sa}$ *such that* $||x_i - y_i|| < \epsilon$ *for every* $i \in \mathbf{N}$ *and the element* $\sum_{i\in D} y_i^2$ *is invertible for some finite set* D *of indices. If* X *does not have weakly infinite real rank, then we say that* X *has strongly infinite real rank.*

The bounded version can be defined similarly. If $K > 0$, we say that a sequence of self-adjoint elements of a unital C^*-algebra is K-*unessential* provided it contains a finite K-unessential (in the sense of Definition 3) subset.

Definition 6. *[9] Let* $K > 0$. *We say that a unital* C^* *algebra* X *has a weakly infinite bounded rank with respect to* K *if for any sequence of self-adjoint elements* $\{x_i : i \in \mathbf{N}\} \subset X_{sa}$ *and any* $\epsilon > 0$ *there is a* K*-unessential sequence* $\{y_i : i \in \mathbf{N}\} \subset X_{sa}$ *such that* $||x_i - y_i|| < \epsilon$ *for every* $i \in \mathbf{N}$. *If* X *does not have weakly infinite bounded rank, then we say that* X *has strongly infinite bounded rank.*

Here are some properties of infinite ranks:

Proposition 16. *[9] Every unital* C^**-algebra of a finite real rank has weakly infinite real rank.*

We will see below that there exists a C^*-algebra of infinite real rank, but not finite real rank.

Proposition 17. *[9] Let* $f : X \to Y$ *be a surjective* $*$*-homomorphism of unital* C^**-algebras.*
- *If* X *has weakly infinite real rank, then so does* Y;
- *If* X *has weakly infinite bounded rank with respect to some* $K > 0$, *then so does* Y.

The next proposition provides a relation between both infinite ranks.

Proposition 18. *[9] Let $K > 0$. If the unital C^*-algebra X has weakly infinite bounded rank with respect to K, then it has weakly infinite real rank.*

Corollary 6. *[9] If a unital C^*-algebra has strongly infinite real rank, then it has strongly infinite bounded rank with respect to any positive constant.*

We are going to characterize the infinite ranks of $C(X)$, where X is a compact Hausdorff space, in terms of the dimension of X. We already observed that if X is a finite-dimensional compact space, then, according to Corollary 3, $rr(C(X)) = br_1(C(X)) = \dim X$ for any positive $K \le 1$. Our next goal is to extend this result to the infinite-dimensional situation.

First, recall that a compact Hausdorff space X is called *weakly infinite-dimensional* [1] if for any sequence $\{(F_i, H_i) : i \in \mathbf{N}\}$ of pairs of closed disjoint subsets of X there are partitions L_i between F_i and H_i such that $\bigcap_{i=1}^{\infty} L_i = \emptyset$. Here, $L_i \subset X$ is called a partition between F_i and H_i if L_i is closed in X and $X \backslash L_i$ is decomposed as the union $U_i \cup V_i$ of disjoint open sets with $F_i \subset U_i$ and $H_i \subset V_i$. Since X is compact, $\bigcap_{i=1}^{\infty} L_i = \emptyset$ is equivalent to $\bigcap_{i=1}^{k} L_i = \emptyset$ for some $k \in \mathbf{N}$. If X is not weakly infinite-dimensional, then it is *strongly infinite-dimensional.*

A standard example of a weakly infinite dimensional, but not finite-dimensional, metrizable compactum can be obtained by taking the one-point compactification $\alpha(\oplus\{\mathbf{I}^n : n \in \mathbf{N}\})$ of the discrete union of increasing dimensional cubes. The Hilbert cube $\mathbf{Q}$ is, of course, strongly infinite-dimensional.

Theorem 19. *[9] Let X be a compact Hausdorff space and $0 < K \le 1$. Then the following conditions are equivalent:*
(a) $C(X)$ has weakly infinite bounded rank with respect to K;
(b) $C(X)$ has weakly infinite real rank;
(c) X is weakly infinite-dimensional.

Proof. (a)$\Longrightarrow$(b). This implication follows from Proposition 18 (which is valid for any – not necessarily commutative – unital C^*-algebras).

(b)$\Longrightarrow$(c). Suppose that $C(X)$ has a weakly infinite real rank. Take an arbitrary sequence $\{(B_i, K_i) : i \in \mathbf{N}\}$ of pairs of disjoint closed subsets of X and define functions $f_i : X \to [-1, 1]$ such that $f_i(B_i) = -1$ and $f_i(K_i) = 1$ for every $i \in \mathbf{N}$. Then, according to our hypothesis, there is a sequence $\{g_i : i \in \mathbf{N}\} \subset C(X)$ of real-valued functions and an integer k with $\|f_i - g_i\| < 1$, $i \in \mathbf{N}$, and $\sum_{i=1}^{k} g_i^2(x) > 0$ for each $x \in X$. If C_i denotes the set $g_i^{-1}(0)$, the last inequality means that $\bigcap_{i=1}^{k} C_i = \emptyset$. Therefore, in order to prove that X is weakly infinite-dimensional, it only remains to show each C_i is a separator between B_i and K_i. To this end, we fix $i \in \mathbf{N}$ and observe that $\|f_i - g_i\| < 1$ implies the following inclusions: $g_i(B_i) \subseteq [-2, 0)$, $g_i(K_i) \subseteq (0, 2]$ and $g_i(X) \subseteq [-2, 2]$. So, $X \backslash C_i = U_i \cup V_i$, where $U_i = g_i^{-1}([-2, 0))$ and $V_i = g_i^{-1}((0, 2])$. Moreover, $B_i \subseteq U_i$ and $K_i \subseteq V_i$, i.e. C_i separates B_i and K_i.

(c)$\Longrightarrow$(a). Let us show that the weak infinite-dimensionality of X forces $C(X)$ to have a weakly infinite bounded rank with respect to K. To this end, take any sequence $\{f_i : i \in \mathbf{N}\} \subset C(X)$ of real-valued functions and any positive number ϵ. It suffices to find another sequence $\{g_i : i \in \mathbf{N}\}$ of real-valued functions in $C(X)$ such that $\|f_i - g_i\| \leq \epsilon$ for every $i \in \mathbf{N}$ and $\sum_{i=1}^m g_i^2(x) > 0$ for every $x \in X$ and some $m \in \mathbf{N}$. Indeed, if $\sum_{i=1}^m g_i^2(x) > 0$ for every $x \in X$, then the function $\sum_{i=1}^m g_i^2$ is invertible. This, according to Proposition 10, is equivalent to the K-unessentiality of the m-tuple $(g_1, \ldots, g_m)$. On the other hand, $\sum_{i=1}^m g_i^2(x) > 0$ for each $x \in X$ if and only if $\bigcap_{i=1}^m g_i^{-1}(0) = \emptyset$. Further, since X is compact, the existence of $m \in \mathbf{N}$ with $\bigcap_{i=1}^m g_i^{-1}(0) = \emptyset$ is equivalent to $\bigcap_{i=1}^\infty g_i^{-1}(0) = \emptyset$. Therefore, our proof is reduced to constructing, for each $i \in \mathbf{N}$, a function g_i which is ϵ-close to f_i and such that the intersection of all $g_i^{-1}(0)$'s, $i \in \mathbf{N}$, is empty.

For every $i \in \mathbf{N}$ let $c_i = \inf\{f_i(x) : x \in X\}$ and $d_i = \sup\{f_i(x) : x \in X\}$. We can suppose, without loss of generality, that each interval (c_i, d_i) is not empty and contains 0. For every i we choose $\eta_i > 0$ such that $\eta_i < \frac{\epsilon}{2}$ and $L_i = [-\eta_i, \eta_i] \subset (c_i, d_i)$, $i \in \mathbf{N}$. Let $Q = \prod_{i=1}^\infty [c_i, d_i]$, $Q_0 = \prod_{i=1}^\infty L_i$ be the topological products of all $[c_i, d_i]$'s and L_i's, respectively. Consider the diagonal product $f = \triangle\{f_i : i \in \mathbf{N}\} : X \to Q$ and note that $H = \bigcap_{i=1}^\infty H_i$, where $H = f^{-1}(Q_0)$ and $H_i = f_i^{-1}(L_i)$ for each $i \in \mathbf{N}$. We also consider the sets

$$F_i^- = f_i^{-1}([c_i, -\eta_i]) \text{ and } F_i^+ = f_i^{-1}([\eta_i, d_i]),\ i \in \mathbf{N}.$$

Since H is weakly infinite-dimensional (as a closed subset of X), by [1, Theorem 19, §10.4], there is a continuous map $p = (p_1, p_2, \ldots) : H \to Q_0$ and a pseudointerior point $b = \{b_i : i \in \mathbf{N}\} \in Q_0$ (i.e. each b_i lies in the interior of the interval L_i) such that

$$b \notin p(H),\ \ F_i^- \cap H \subset p_i^{-1}(\{-\eta_i\}),\ \text{ and } F_i^+ \cap H \subset p_i^{-1}(\{\eta_i\}),\ i \in \mathbf{N}.$$

Since each b_i is an interior point of $L_i = [-\eta_i, \eta_i]$, there exists homeomorphisms $s_i : L_i \to L_i$ which leaves the endpoints $-\eta_i$ and η_i fixed and such that $s_i(b_i) = 0$. Let $s = \triangle\{s_i : i \in \mathbf{N}\} : Q_0 \to Q_0$ and $q = s \circ p$. Obviously $s(b) = \mathbf{0}$ and $\mathbf{0} \notin q(H)$, where $\mathbf{0}$ denotes the point of Q_0 having all coordinates 0. Further observe that if $q_i = \pi_i \circ q$, where $q_i : Q_0 \to L_i$ denotes the natural projection onto the i-th coordinate, then

$$F_i^- \cap H \subset q_i^{-1}(\{-\eta_i\}) \text{ and } F_i^+ \cap H \subset q_i^{-1}(\{\eta_i\}),\ \ i \in \mathbf{N}.$$

Therefore, each q_i, $i \in \mathbf{N}$, is a function from H into L_i satisfying the following condition: $q_i(F_i^- \cap H) = f_i(F_i^- \cap H_i) = -\eta_i$ and $q_i(F_i^+ \cap H) = f_i(F_i^+ \cap H_i) = \eta_i$. Let $h_i : H_i \to L_i$ be an extension of q_i, $i \in \mathbf{N}$. Note that the restrictions of h_i and f_i onto the sets $F_i^- \cap H_i$ and $F_i^+ \cap H_i$ coincide. Finally, define $g_i : X \to [c_i, d_i]$ by letting

$$g_i(x) = \begin{cases} h_i(x), & \text{if } x \in H_i; \\ f_i(x), & \text{if } x \in X - H_i. \end{cases}$$

To finish the proof of the "if" part, we need to show that $g_i(x)$ is ϵ-close to $f_i(x)$ for each $i \in \mathbf{N}$ and $x \in X$, and that $\bigcap_{i=1}^{\infty} g_i^{-1}(0) = \emptyset$. Since g_i and f_i are identical outside H_i, the first condition is satisfied for $x \notin H_i$. If $x \in H_i$, then both $f_i(x)$ and $g_i(x)$ belong to L_i, so again $|f_i(x) - g_i(x)| < \epsilon$. To prove the second condition, observe first that $x \notin H$ implies $x \notin H_j$ for some j. Hence, $g_j(x) = f_j(x) \notin L_j$, so $g_j(x) \neq 0$. If $x \in H$, then $g_i(x) = q_i(x)$ for all i and, because $\mathbf{0} \notin q(H)$, at least one $g_i(x)$ must be different from 0. Thus, $\bigcap_{i=1}^{\infty} g_i^{-1}(0) = \emptyset$. □

It follows from Theorem 14 that $rr\,(C^*\,(\mathbf{F}_\infty)) > n$ for each n. Our results imply much stronger observation.

Corollary 7. *[9] The group C^*-algebra $C^*\,(\mathbf{F}_\infty)$ of the free group on countable number of generators has strongly infinite real rank.*

Proof. It is well known that every separable unital C^*-algebra is an image of $C^*\,(\mathbf{F}_\infty)$ under a surjective $*$-homomorphism. In particular, there exists a surjective $*$-homomorphism $f\colon C^*\,(\mathbf{F}_\infty) \to C(Q)$, where Q denotes the Hilbert cube. It is well known (see, for instance, [1, §10.5]) that the Hilbert cube Q is strongly infinite dimensional. By Proposition 19, $C(Q)$ has strongly infinite real rank. Finally, by Proposition 17, real rank of $C^*\,(\mathbf{F}_\infty)$ must also be strongly infinite. □

Next statement provides a formal example of a unital C^*-algebra of weakly infinite, but not finite real rank.

Proposition 20. *[9] Let $X = \alpha\,(\oplus\{I^n : n \in \mathbf{N}\})$ be the one-point compactification of the discrete topological sum of increasing-dimensional cubes. In other words, $C(X) = \prod\{C(I^n) : n \in \mathbf{N}\}$ (here $\prod$ stands for the direct product of indicated C^*-algebras). Then $C(X)$ has weakly infinite, but not finite real rank.*

Proof. Obviously X is countably dimensional and hence, by [1, Corollary 1, §10.5], it is weakly infinite dimensional. By Proposition 19, $C(X)$ has weakly infinite real rank. It only remains to note that $rr(X) > n$ for any $n \in \mathbf{N}$. □

In conclusion let us note that there exist non-commutative C^*-algebras with similar properties (compare with Corollary 7).

Corollary 8. *[9] There exist non-commutative unital C^*-algebras of weakly infinite, but not finite real rank.*

Proof. Let X be as in Proposition 20 and A be a non-commutative unital C^*-algebra of a finite real rank. Then the product $C(X) \oplus A$ has weakly infinite real rank. It is clear that $C(X) \oplus A$ is non-commutative and does not have a finite real rank. □

4 C^*-algebras with the approximate n-th root property

All topological spaces in this section are assumed to be (at least) completely regular. For a space X and an integer $n \geq 2$, we consider the following conditions ($\|\cdot\|$ denotes the supremum norm):

$(*)_n$ For each bounded continuous function $f\colon X \to \mathbf{C}$ and each $\varepsilon > 0$, there exists a continuous function $g\colon X \to \mathbf{C}$ such that $\|f - g^n\| < \varepsilon$.

$(**)_n$ For each bounded continuous function $f\colon X \to \mathbf{C}$ and each $\varepsilon > 0$, there exist bounded continuous functions $g_1, .., g_n\colon X \to \mathbf{C}$ such that $f = \prod_{i=1}^{i=n} g_i$ and $\|g_i - g_j\| < \varepsilon$ for each i, j.

Definition 7. *We say that the space $C^*(X)$ of all bounded complex-valued functions on X has the approximate n-th root property if X satisfies condition $(*)_n$.*

We have the following result established in [12] (see [2] for another characterization of function algebras with the approximate n-th root property):

Theorem 21. *Let X be a compactum with $\dim X \leq 1$ and n a positive integer. Then the following conditions are equivalent:*
(1) $C(X)$ has the approximate n-th root property.
(2) X satisfies condition $(**)_n$.
(3) the first Čech cohomology $\check{H}^1(X;\mathbf{Z})$ is n-divisible, that is, each element of $\check{H}^1(X;\mathbf{Z})$ is divided by n.

Let $\mathcal{A}(n)$ denote the class of all completely regular spaces satisfying condition $(*)_n$ and $\mathcal{A}_1(n)$ is the subclass of $\mathcal{A}(n)$ consisting of spaces X with $\dim X \leq 1$.

4.1 Properties of the classes $\mathcal{A}(n)$ and $\mathcal{A}_1(n)$

Everywhere in this subsection $\mathcal{K}_n$ will denote one of the classes $\mathcal{A}(n)$ and $\mathcal{A}_1(n)$, $n \geq 2$.

Proposition 22. *[7] A compactum X belongs to $\mathcal{K}_n$ if and only if X can be represented as the limit space of an inverse system $\{X_\alpha, p_\alpha^\beta : \alpha, \beta \in A\}$ of metrizable compacta with each $X_\alpha \in \mathcal{K}_n$.*

We say that a class of spaces $\mathcal{P}$ is *factorizable* if, for every map $f\colon X \to Y$ of a compactum $X \in \mathcal{P}$, there exists a compactum $Z \in \mathcal{P}$ of weight $w(Z) \leq w(Y)$ and maps $\pi\colon X \to Z$ and $p\colon Z \to Y$ such that $f = p \circ \pi$.

Proposition 23. *[7] Any one of the classes $\mathcal{A}(n)$ and $\mathcal{A}_1(n)$, $n \geq 2$, is factorizable.*

Corollary 9. *[7] Every space $X \in \mathcal{K}_n$ has a compactification $Z \in \mathcal{K}_n$ with $w(Z) = w(X)$.*

Proof. Obviously, $X \in \mathcal{K}_n$ implies $\beta X \in \mathcal{K}_n$. Let Y be an arbitrary compactification of X with $w(Y) = w(X)$ and let $f\colon \beta X \to Y$ be the extension of the identity on X. Then, by Proposition 23, there exists a compactum $Z \in \mathcal{K}_n$ and maps $g\colon \beta X \to Z$ and $h\colon Z \to Y$ with $h \circ g = f$ and $w(Z) = w(X)$. It remains only to observe that Z is a compactification of X. □

The final result in this subsection implies that each $\mathcal{K}_n$, $n \geq 2$, has an universal element. Recall that a map $h\colon X \to Y$ is said to be *invertible* for a given class $\mathcal{P}$ (or simply, $\mathcal{P}$*-invertible*) if for every map $g\colon Z \to Y$ with $Z \in \mathcal{P}$ there exists a map $\overline{g}\colon Z \to X$ such that $g = h \circ \overline{g}$.

Theorem 24. *[7] Let n be a positive integer and let $\mathcal{K}_n$ denote one of the classes $\mathcal{A}(n)$ and $\mathcal{A}_1(n)$. Then for every cardinal $\tau \geq \omega$ there exists a compactum $X_\tau \in \mathcal{K}_n$ of weight $\leq \tau$ and a $\mathcal{K}_n$-invertible map $f_{\mathcal{K}}\colon X_\tau \to \mathbf{I}^\tau$.*

Proof. We consider the family of all maps $\{h_\alpha\colon Y_\alpha \to \mathbf{I}^\tau\}_{\alpha \in \Lambda}$ such that each Y_α is a closed subset of $\mathbf{I}^\tau$ with $Y_\alpha \in \mathcal{K}_n$. Let Y be the disjoint sum of all Y_α and the map $h\colon Y \to \mathbf{I}^\tau$ coincides with h_α on every Y_α. We extend h to a map $\overline{h}\colon \beta Y \to \mathbf{I}^\tau$. Since $\beta Y \in \mathcal{K}_n$, by Proposition 23, there exists a compactum X of weight $\leq \tau$ and maps $p\colon \beta Y \to X$ and $f\colon X \to \mathbf{I}^\tau$ such that $X \in \mathcal{K}_n$ and $f \circ p = \overline{h}$.

Let us show that f is $\mathcal{K}_n$-invertible. Take a space $Z \in \mathcal{K}_n$ and a map $g\colon Z \to \mathbf{I}^\tau$. Considering βZ and the extension $\overline{g}\colon \beta Z \to \mathbf{I}^\tau$ of g, we can assume that Z is compact. We also can assume that the weight of Z is $\leq \tau$ (otherwise we apply again Proposition 23 to find a compact space $T \in \mathcal{K}_n$ of weight $\leq \tau$ and maps $g_1\colon Z \to T$ and $g_2\colon T \to \mathbf{I}^\tau$ with $g_2 \circ g_1 = g$, and then consider the space T and the map g_2 instead, respectively, of Z and g). Therefore, without loss of generality, we can assume that Z is a closed subset of $\mathbf{I}^\tau$. According to the definition of Y and the map h, there is an index $\alpha \in \Lambda$ such that $Z = Y_\alpha$ and $g = h_\alpha$. The restriction $p|Z\colon Z \to X$ is a lifting of g, i.e. $f \circ (p|Z) = g$. □

Corollary 10. *[7] Let $\mathcal{K}_n$ be one of the classes $\mathcal{A}(n)$ and $\mathcal{A}_1(n)$. Then for every $\tau \geq \omega$ there exists a compactum $X \in \mathcal{K}_n$ of weight τ which contains every space from $\mathcal{K}_n$ of weight $\leq \tau$.*

4.2 The approximate *n*-root property of C^*-algebras

It is easily seen that the modification of condition $(*)_n$, obtained by requiring both f and g to be of norm ≤ 1, is equivalent to $(*)_n$. This observation leads us to consider the following classes of general (non-commutative) C^*-algebras.

Definition 8. *[7] We say that a C^*-algebra X satisfies the approximation n-th root property if for every $a \in X$ with $\|a\| \le 1$ and every $\varepsilon > 0$ there exists $b \in X$ such that $\|b\| \le 1$ and $\|a - b^n\| < \varepsilon$.*

The class of all C^*-algebras with the approximate n-th root property is denoted by $\mathcal{AP}(n)$. Let $\mathcal{AP}_1(n)$ be the subclass of $\mathcal{AP}(n)$ consisting of C^*-algebras of bounded rank ≤ 1 (see Definition 4). We also consider the class $\mathcal{HP}(n)$ of C^*-algebras X with the following property: for every *invertible* element $a \in X$ with $\|a\| \le 1$ and every $\varepsilon > 0$ there exists $b \in X$ such that $\|b\| \le 1$ and $\|a - b^n\| < \varepsilon$. In the sequel, $\mathcal{AP}(n)_S$ denotes the class of all separable C^*-algebras from $\mathcal{AP}(n)$. The notations $\mathcal{AP}_1(n)_S$ and $\mathcal{HP}(n)_S$ have the same meaning.

In this subsection we investigate the behavior of the classes $\mathcal{AP}(n)$, $\mathcal{AP}_1(n)$ and $\mathcal{HP}(n)$ with respect to direct systems and then use the result to prove the existence of universal elements in the classes $\mathcal{AP}(n)_S$, $\mathcal{AP}_1(n)_S$ and $\mathcal{HP}(n)_S$. The results from the present subsection are actually a non-commutative analogues of the corresponding result from the previous subsection.

Proposition 25. *[7] Let $\mathcal{K}$ be one of the classes $\mathcal{AP}(n)$, $\mathcal{AP}_1(n)$ and $\mathcal{HP}(n)$. If X is the limit of a direct system $\mathcal{S} = \{X_\alpha, i_\alpha^\beta, A\}$ consisting of unital C^*-algebras and unital $*$-inclusions with $X_\alpha \in \mathcal{K}$ for each α, then $X \in \mathcal{K}$.*

As in the commutative case (see Proposition 22), we can establish a decomposition theorem for the classes $\mathcal{AP}(n)$, $\mathcal{AP}_1(n)$ and $\mathcal{HP}(n)$.

Proposition 26. *[7] Let $\mathcal{K}$ be one of the classes $\mathcal{AP}(n)$, $\mathcal{AP}_1(n)$ and $\mathcal{HP}(n)$. The following conditions are equivalent for any unital C^*-algebra X:*

(1) $X \in \mathcal{K}$.

(2) X can be represented as the direct limit of a direct C^*_ω-system $\{X_\alpha, i_\alpha^\beta, A\}$ satisfying the following properties:

(a) The indexing set A is cofinal and ω-closed in the ω-complete set $\exp_\omega Y$ for some (any) dense subset Y of X such that $|Y| = d(X)$.

(b) X_α is a (separable) C^*-subalgebra of X with $X_\alpha \in \mathcal{K}$, $\alpha \in A$.

Finally, we can establish the existence of universal elements in the classes $\mathcal{AP}(n)_S$, $\mathcal{AP}_1(n)_S$ and $\mathcal{HP}(n)_S$.

Theorem 27. *[7] Let $n \ge 2$ be a positive integer and let $\mathcal{K}$ be one of the classes $\mathcal{AP}(n)$, $\mathcal{AP}_1(n)$ and $\mathcal{HP}(n)$. Then there exists a $\mathcal{K}$-invertible unital $*$-homomorphism $p\colon C^*(\mathbf{F}_\infty) \to Z_{\mathcal{K}}$ of $C^*(F_\infty)$ to a separable unital C^*- algebra $Z_{\mathcal{K}} \in \mathcal{K}$.*

Proof. Let $\mathcal{B} = \{f_t\colon C^*(\mathbf{F}_\infty) \to X_t\colon t \in T\}$ denote the set of all unital $*$-homomorphisms on $C^*(\mathbf{F}_\infty)$ such that $X_t \in \mathcal{K}$. We claim that the product $\prod\{X_t\colon t \in T\}$ belongs to $\mathcal{K}$. This is obviously true if $\mathcal{K}$ is either $\mathcal{AP}(n)$ or $\mathcal{HP}(n)$. Since the bounded rank of this product is ≤ 1 provided each X_t is of bounded rank ≤ 1 (see Proposition 12), the claim holds for the class $\mathcal{AP}_1(n)$ as well. The $*$-homomorphisms f_t, $t \in T$, define the unital $*$-homomorphism $f\colon C^*(\mathbf{F}_\infty) \to \prod\{X_t\colon t \in T\}$ such that $\pi_t \circ f = f_t$ for each $t \in T$, where $\pi_t\colon \prod\{X_t\colon t \in T\} \to X_t$ denotes the canonical projection $*$-homomorphism onto X_t. By Proposition 26, $\prod\{X_t\colon t \in T\}$ can be represented as the limit of the C^*_ω-system $\mathcal{S} = \{C_\alpha, i^\beta_\alpha, A\}$ such that C_α is a separable unital C^*-algebra with $C_\alpha \in \mathcal{K}$ for each $\alpha \in A$. Suppressing the injective unital $*$-homomorphisms $i^\beta_\alpha\colon C_\alpha \to C_\beta$, we may assume, for notational simplicity, that C_α's are unital C^*-subalgebras of $\prod\{X_t\colon t \in T\}$. Let $\{a_k\colon k \in \omega\}$ be a countable dense subset of $C^*(\mathbf{F}_\infty)$. For each $k \in \omega$ there exists an index $\alpha_k \in A$ such that $f(a_k) \in C_{\alpha_k}$. Since A is ω-complete, there exists an index $\alpha_0 \in A$ such that $\alpha_0 \geq \alpha_k$ for each $k \in \omega$. Then $f(a_k) \in C_{\alpha_k} \subseteq C_{\alpha_0}$ for each $k \in \omega$. This observation coupled with the continuity of f guarantees that $f(C^*(\mathbf{F}_\infty)) = f(cl\{a_k\colon k \in \omega\}) \subseteq cl\{f(\{a_k\colon k \in \omega\})\} \subseteq clC_{\alpha_0} = C_{\alpha_0}$.

Let $Z_{\mathcal{K}} = C_{\alpha_0}$ and define the unital $*$-homomorphism $p\colon C^*(\mathbf{F}_\infty) \to Z_{\mathcal{K}}$ as f, regarded as a homomorphism of $C^*(\mathbf{F}_\infty)$ into $Z_{\mathcal{K}}$. Note that $f = i \circ p$, where $i\colon Z_{\mathcal{K}} = C_{\alpha_0} \hookrightarrow \prod\{X_t\colon t \in T\}$ stands for the inclusion.

By construction, we see $Z_{\mathcal{K}} \in \mathcal{K}$. Let us show that $p\colon C^*(\mathbf{F}_\infty) \to Z_{\mathcal{K}}$ is $\mathcal{K}$-invertible. For a given unital $*$-homomorphism $g\colon C^*(\mathbf{F}_\infty) \to X$, where X is a unital C^*-algebra with $X \in \mathcal{K}$, we need to establish the existence of a unital $*$-homomorphism $h\colon Z_{\mathcal{K}} \to X$ such that $g = h \circ p$. Indeed, by definition of the set $\mathcal{B}$, we conclude that $g = f_t\colon C^*(\mathbf{F}_\infty) \to X_t = X$ for some index $t \in T$. Observe that $g = f_t = \pi_t \circ f = \pi_t \circ i \circ p$. This allows us to define the required unital $*$-homomorphism $h\colon Z_{\mathcal{K}} \to X$ as the composition $h = \pi_t \circ i$. Hence, p is $\mathcal{K}$-invertible. □

It is well-known that every separable C^*-algebra is a surjective image of $C^*(\mathbf{F}_\infty)$. Therefore, if $\mathbb{R}$ is a class of separable C^*-algebras and $p\colon C^*(F_\infty) \to Y_{\mathbb{R}}$ is a $\mathbb{R}$-invertible homomorphism with $Y_{\mathbb{R}} \in \mathbb{R}$, then $Y_{\mathbb{R}}$ is universal for the class $\mathbb{R}$. Hence, Theorem 27 implies the following

Corollary 11. *[7] Each of the classes $\mathcal{AP}(n)_s$, $\mathcal{AP}_1(n)_s$ and $\mathcal{HP}(n)_s$ has a universal element.*

Let us note that there exists a non-commutative C^*-algebra which belongs to any one of the classes $\mathcal{AP}(n)_s$, $\mathcal{AP}_1(n)_s$ and $\mathcal{HP}(n)_s$. Indeed, let $X = M(m)$ be the algebra of all $m \times m$ complex matrixes, where $m \geq 2$ is a fixed integer. By [3], the bounded rank of X is 0. Moreover, using the canonical Jordan form representation, one can show that if $A \in X$ and $n \geq 2$, then A can be approximated by a matrix $B \in X$ with $C^n = B$ for some $C \in X$. Hence, X is a common element of $\mathcal{AP}(n)_s$, $\mathcal{AP}_1(n)_s$ and $\mathcal{HP}(n)_s$. This implies that the universal elements of $\mathcal{AP}(n)_s$, $\mathcal{AP}_1(n)_s$ and $\mathcal{HP}(n)_s$ are also non-commutative.

Bibliography

[1] Alexandroff, P. and Pasynkov, B. *Introduction to Dimension Theory*, Nauka, Moscow, 1973 (in Russian).

[2] Brodskiy, N., Dydak, J., Karasev, A., and Kawamura, K. *Root closed function algebras on compacta of large dimension*, Proc. Amer. Math. Soc. 135 (2007), 587–596.

[3] Brown, L.G. and Pedersen, G.K. *C^*-algebra of real rank zero*, J. Functional Anal. 99 (1991), 131–149.

[4] Chigogidze, A. *Uncountable direct systems and a characterization of non-separable projective C^*-algebras*, Mat. Stud. 12, # 2 (1999), 171–204.

[5] Chigogidze, A. *Universal C^*-algebra of real rank zero*, Infinite Dimensional Analysis, Quantum Probability and Related Topics, 3 (2000), 445–452.

[6] Chigogidze, A. *Inverse Spectra*, North Holland, Amsterdam, 1996.

[7] Chigogidze, A., Karasev, A., Kawamura, K., and Valov, V. *On C^*-algebras with the approximate n-th root property*, Bull. Austral. Math. Soc. 72 (2005), 197–212.

[8] Chigogidze, A., Valov, V. *Bounded rank of C^*-algebras*, Topology Apll. 140 (2004), 163–180.

[9] Chigogidze, A., Valov, V. *C^*-algebras of infinite real rank*, Bull. Austral. Math. Soc. 66 (2002), 487–496.

[10] Engelking, R. *Dimension Theory*, PWN, Warsaw, 1978.

[11] Hurewitz, W., Wallman, H. *Dimension Theory*, Princeton Univ. Press, 1941.

[12] Kawamura, K. and Miura, T. *On the existence of continuous (approximate) roots of algebraic equations*, preprint.

[13] Lin, H. *The tracial topological rank of C^*-algebras*, Proc. London Math. Soc. 83 (2001), 199–234.

[14] Murphy, G.J. *The analytic rank of a C^*-algebra*, Proc. Amer. Math. Soc. 115 (1992), 741–746.

[15] Nagata, J. *Modern Dimension Theory*, North-Holland, Amsterdam, 1965.

[16] Pears, A.R. *Dimension Theory of General Spaces*, Cambridge Univ. Press, London, 1975.

[17] Phillips, N.C. *Simple C^*-algebras with the property (FU)*, Math. Scand. 69 (1991), 121–151.

[18] Rieffel, M.A. *Dimension and stable rank in the K-theory of C^*-algebras*, Proc. London Math. Soc. 46 (1983), 301–333.

[19] Winter, W. *Covering dimension for nuclear C^*-algebras*, J. Funct. Anal. 199 (2003), no. 2, 535–556.

[20] Zelazko, W. *An axiomatic approach to joint spectra I*, Studia Math. 64 (1979), 249–261.

Wiesław Żelazko

Dense ideals in topological algebras: some results and open problems

Abstract: The present paper is an extended version of the author's lecture during the conference ICTAA-2014.

Keywords: topological algebras; maximal ideals

MSC: 46L

1 Introduction

Unless otherwise stated, all algebras in this paper will be commutative, complex, unital, with the unity denoted by e.

By a topological algebra we mean a topological vector space equipped with a jointly continuous associative multiplication. We shall be always assuming completeness of the considered algebras (the jointly continuous multiplication always extends to such a multiplication in the completion of the algebra in question).

Many topological algebras contain (proper) dense ideals. One of most celebrated open problems in theory of these algebras is the Michael Problem ([10]) asking whether all multiplicative linear functionals of m-convex B_0-algebras are continuous (see Problem 1 below). It is the question whether such an algebra can possess a dense ideal of codimension one.

In this paper we shall consider the following classes of topological algebras:

– The class $\mathcal{T}$ of all topological algebras.

– The class $\mathcal{LC}$ of all locally convex algebras. The topology of such an algebra can be given by means of a family $(\|\cdot\|_\alpha)$ of seminorms such that $\|e\|_\alpha = 1$ and for each α there is an index β so that

$$\|xy\|_\alpha \leq \|x\|_\beta \|y\|_\beta \tag{1}$$

for all elements x, y of the algebra in question.

– The class $\mathcal{MLC}$ of all multiplicatively convex algebras (shortly: m-convex algebras). Here the formula (1) is replaced by

$$\|xy\|_\alpha \leq \|x\|_\alpha \|y\|_\alpha \tag{2}$$

for all indices α.

Wiesław Żelazko, Mathematical Institute, Polish Academy of Sciences, Śniadekich 8, 00-956 Warszawa, Poland, e-mail: zelazko@impan.pl

https://doi.org/10.1515/9783110413557-020

– The class $\mathcal{F}$ of all F-algebras, i.e. completely metrisable topological algebras.The topology of an F-space is given by meas of an F-norm, i.e. functional $x \mapsto \|x\|$ satisfying

(i) $\|x\| \geq 0$, and $\|x\| = 0$ if and only if $x = 0$,

(ii) $\|x + y\| \leq \|x\| + \|y\|$,

(iii) the map $(t, x) \mapsto \|tx\|$ is a jointly continuous map from $\mathbf{C} \times A$ to A.

– The class $\mathcal{B}_0 = \mathcal{LC} \cap \mathcal{F}$ of B_0-algebras, their topology is given by means of a sequence of seminorms

$$\|x\|_1 \leq \|x\|_2 \leq \ldots \tag{3}$$

satisfying

$$\|xy\|_i \leq \|x\|_{i+1}\|y\|_{i+1}. \tag{4}$$

– The class $\mathcal{MB}_0 = \mathcal{MLC} \cap \mathcal{F}$ of all m-convex B_0-algebras, where the relations (4) are replaced by (2).

In the last section we shall also consider the (defined there) class of semi-topological algebras, which contains properly all classes mentioned above.

For more specific information on these classes the reader is referred to [9], [10], [12] and [13].

In the second section of this paper we shall deal with dense maximal ideals in topological algebras (some classes of), and we shall recall some, known already, facts. In the third one, we shall deal with topologically invertible elements. We define here new concepts of the Arens Property and of finitely closed ideals, and prove that for topological algebras with Arens Property the maximal ideals must be finitely closed. We also show that some results on MLC-algebras extend to topological algebras possessing the Arens Property. Finally, in the last section, we discuss some non-commutative situations , some "topological" situations, and give a "warning" example of an inductive limit.

2 Dense Maximal Ideals

One of most challenging open problems in the Theory of Topological Algebras is the following problem, called Michael Problem, or Michael-Mazur Problem (Michael [10]).

Problem 1. *Let A be a commutative unital m-convex B_0-algebra. Is it true that all its multiplicative-linear functionals are continuous ?*

In other words: is it true that all maximal ideals of A of codimension one are closed ?

Recall that a topological algebra is said a Q-algebra, if the set (group) $G(A)$ of its invertible elements is open. We have the following result dealing with maximal ideals of codimension one (see [14]):

Theorem 28. *Let A be a (commutative) m-convex algebra. Then the following are equivalent*
(i) *All maximal ideals of A are of codimension one,*
(ii) *Each element x of A has a bounded spectrum,*
(iii) *Each element of A has a compact spectrum*
(iv) *The algebra A is a complete m-convex Q-algebra under some topology stronger than the original one.*
(v) *The algebra A is a complete m-convex Q-algebra under some topology.*
(vi) *The space $\mathcal{M}^\#$ of all multiplicative linear functionals of A is compact in the weak–star topology.*
If, moreover, A is a barrelled space (in particular, if it is an m-convex B_0-algebra), the above are equivalent to
(vii) *Every maximal ideal of A is closed.*
(vii) *A is a Q-algebra.*

Corollary 12. *If an m-convex B_0-algebra is not a Q-algebra, then it must have a dense maximal ideals of infinite codimension (this is a consequence of the classical Frobenius theorem stating that a complex division algebra is either equal to* **C**, *or is infinite dimensional).*

Thus the positive answer to the Problem 1 would be equivalent to the fact that all dense maximal ideals of A must have an infinite codimension.

We give now some examples illustrating the above results.

Example 29. *Denote by ω_1 the first uncountable ordinal. The closed segment $[1, \omega_1]$ is a compact space under the segment topology and the algebra $C[1, \omega_1]$ is a uniform Banach algebra. The half-open segment $[1, \omega_1)$ is non compact, and we can consider the algebra $A = C([1, \omega_1))$ of all its continuous functions provided with the open-compact topology. It is the topology of uniform convergence on compact subsets of this segment, and it is given by the seminorms*

$$\|x\|_\alpha = max\{|x(t)| : 1 \leq t \leq \alpha < \omega_1\}.$$

It is known that all functions in A are constant beginning from some point on. Thus, as a set, the algebra A is identical with $C[1, \omega_1]$. Consequently all ideals of A are of codimension one, but one of them is non-closed and so dense, as the kernel of the discontinuous multiplicative-linear functional $f(x) = x(\omega_1)$. A is not a Q-algebra (in Q-algebras all maximal ideal are closed), but it is a Q-algebra under the stronger topology of $C([1, \omega_1])$.

Example 30. *Consider the algebra $A = C([0, \infty))$ with the pointwise algebra operations and with the compact-open topology. It is an m-convex B_0-algebra, and it is easy to show that all its multiplicative-linear functionals are point evaluations, and so they* *are continuous. However, A has a dense ideal* $I = \{x \in A : x(t) = 0 \text{ for } t \geq c(x)\}$, *where* $c(x)$ *is an unbounded positive-valued function on* A. *Thus all maximal ideals containing* I *are dense, and so all of them are of infinite codimension.*

In the LMC-algebras and in B_0-algebras all maximal closed ideals must be of codimension one. This follows from the fact that these classes (in fact, all considered here classes of topological algebras) are closed with respect to taking quotients by closed ideals and by the Gelfand-Mazur-type theorems true for these classes ([10], [12]). However, for the F-algebras the following problem, again one of most challenging problems in Topological Algebras, is open.

Problem 2. *Is it true that a division algebra of type F is isomorphic to* **C** *?*

Note, that an LMC-algebra cannot contain the field of rational functions. This is not true for B_0-algebras as shows an example constructed by Williamson [11]. Thus the completeness is an essential assumption concerning the Problem 2.

A part of Theorem 2.1 is true also for F-algebras ([3]).

Theorem 31. *Let A be a commutative unital F-algebra, then all its maximal ideals are closed if and only if it is a Q-algebra.*

A similar result fails to be true in the non-metrisable case. The algebra of all polynomials provided with the maximal locally convex topology has all ideals (in fact all linear subspaces) closed, but it is not a Q-algebra (cf. [16]) .

It is known ([12]) that any (commutative) B_0-algebra which is a Q-algebra, must be m-convex (by [15] this fact is not true in the non-commutative case). Thus we have

Corollary 13. *Let A be a non-m-convex* B_0*-algebra. Then A has a dense maximal ideal.*

We close this section with two more problems, the first one contained in the second, and both containing Problem 1.

Problem 3. *Let A be be a commutative algebra of type* B_0. *Is it true that all its multiplicative-linear functionals are continuous ?*

Problem 4. *Let A be a commutative algebra of type F. Does it follow that all its multiplicative-linear functionals are continuous ?*

3 Topologically invertible elements and related topics

Let A be a (commutative) topological algebra. We say that an element x in A is *topologically invertible* if there is a net $(z_\alpha) \subset A$ such that

$$\lim_\alpha z_\alpha x = e. \tag{1}$$

The set (group) of all invertible elements of A will be denoted by $G(A)$ and the set of topologically invertible elements – by $G^t(A)$. In general, it is only a multiplicative semigroup and it contains $G(A)$. The *proper* topologically invertible elements are defined as elements of $G^t(A) \setminus G(A)$. An element in $G^t(A)$ is proper if and only if the principal ideal $I = xA$ is dense.

For some classes of topological algebras there are no proper topologically invertible elements. It is so for algebras in MLC, which follows from the fact that an element of such an algebra is invertible if and only if it does not belong to the kernel of any continuous multiplicative linear functional (see [10]). The relation (5) shows that a proper topologically invertible element cannot belong to such a kernel (the author does not know any example of topological algebra possessing a (necessarily discontinuous) multiplicative-linear functional with topologically invertible element in its kernel). Another example of a topological algebra possessing no proper topologically invertible element is the algebra $L_0[0, 1]$ of all Lebesgue measurable functions on the unit interval with pointwise algebra operations and with the topology of convergence in measure (it belongs to $\mathcal{F}$). This follows from the fact that a non-invertible element x of $A = L_0[0, 1]$ must vanish on a set of positive measure, but then the principal ideal xA cannot be dense. Also the Q-algebras cannot have proper topologically invertible elements.

On the other hand, there are B_0 algebras possessing proper topologically invertible elements. For instance it is so for the Arens algebra ([4])

$$A = L^\omega[0, 1]) = \bigcap_{p \geq 1} L_p[o, 1]$$

provided with the pointwise algebra operations and the L_p-norms (this topology is given by means of any increasing sequence (p_i) tending to an infinity). To see it consider the element $x(t) = t$, its inverse does not belong to A, but setting

$$z_n(t) = \begin{cases} 0, & \text{if } 0 \leq t < 1/n; \\ t^{-1} & \text{if } 1/n \leq t \leq 1, \end{cases}$$

for $n = 1, 2, \ldots$, we obtain

$$z_n x(t) = \begin{cases} 0, & \text{if } 0 \leq t < 1/n; \\ 1 & \text{if } 1/n \leq t \leq 1, \end{cases}$$

and so the relation (5) holds true.

The reason that $L^\omega[0,1]$ has proper topologically invertible elements is that its inverse is discontinuous. It is known (see [12] or [13]) that an F-algebra A has a continuous inverse if and only if the set $G(A)$ is a G_δ-set. On the other hand, we have (see [17])

$$G^t(A) = \bigcap_{n=1}^{\infty} \{x \in A : \exists z \in A \; \|e - zx\| < 1/n\},$$

where $\|\cdot\|$ is the F-norm giving the topology of A. The above formula implies that the set $G^t(A)$ is a G_δ-set for every F-algebra. Thus such an algebra with a discontinuous inverse must have proper topologically invertible elements. There is an open question whether the converse is true.

Problem 5. *Suppose that an F-algebra has a continuous inverse. Does it follow that is has no proper topologically invertible elements ?*

A positive answer to this question would give a characterization of F-algebras possessing proper topologically invertible elements.

For non-metrisable algebras such a problem has an answer in negative. There exists a complete locally convex (non-metrisable) infinite dimensional field. Such a field cannot have proper topologically invertible element and its inverse cannot be continuous. Another B_0-algebra possessing proper topologically invertible elements is mentioned above Williamson's algebra, since it also has a discontinuous inverse.

For an F-algebra with a discontinuous inverse we can indicate a construction of proper topologically invertible elements ([18]). Let A be such an algebra. There is a sequence $(x_n) \subset G(A)$ with $\lim_n x_n = e$, but with $\|x_n^{-1} - e\| > \delta$ for some positive δ. Passing, if necessary, to a subsequence, we can assume that the product

$$x = \prod_{n=1}^{\infty} x_n$$

converges in A.

Setting $z_n = \prod_{i=1}^{n} x_i^{-1}$, we have $z_n x = \prod_{i=n+1}^{\infty} x_i$, and so $\lim_n z_n x = e$.

On the other hand, the sequence (x_n^{-1}) is divergent, so it is not a Cauchy sequence. Thus there is a subsequence (z_{i_n}) with $\|z_n - z_{i_n}\| \geq \delta$ for some positive δ. Consequently the sequence $u_n = z_n - z_{i_n}$ does not converge to zero, and $\lim_n x u_n = 0$. So x is a topological divisor of zero and so is non-invertible. Note that the proper topologically invertible elements must be also topological divisors of zero. This follows from the fact that in the formula (5), the net (z_α) cannot be convergent and so it is not a Cauchy net.

R.F. Arens has shown in [5] that the MLC-algebras cannot have dense finitely generated ideals (this result generalizes the fact that m-convex algebra do not have topologically invertible elements). We shall call this result the Arens Theorem.

Problem 6. *Suppose that a topological algebra has a dense, finitely generated ideal*

$$I = x_1A + \cdots + x_nA.$$

Does it follow that A has proper topologically invertible elements ?

We can ask also a stronger question, whether the ideal I must have a topologically invertible element.

We introduce now the concept of the *Arens Property* suggested by the above mentioned Arens Theorem.

Definition 3.1. *Definition. We say that a topological algebra A has the Arens property if it has no dense finitely generated ideal.*

By the above mentioned Arens Theorem, all MLC-algebras have the Arens property. Another example is the mentioned earlier algebra $A = L_0[0,1]$ of measurable functions. The reason is that if $x_1, \ldots, x_n$ generate a proper ideal in A, then the intersection of their zero sets must have positive measure, but then, the ideal $I = I(x_1, \ldots, x_n)$ generated by these elements cannot be dense. Other examples are all Q-algebras and the free countably generated algebra provided with the maximal locally convex topology (see [16]).

Let A be a topological algebra, and let $x_1, \ldots, x_n \in A$. Denote by $\bar{I}(x_1, \ldots, x_n)$ the closure of $I(x_1, \ldots, x_n)$ – the ideal generated by $x_1, \ldots, x_n$.

Definition 3.2. *Let I be a proper ideal of a (commutative) topological algebra A. We say that I is* finitely closed, *if together with any n-tuple* $(x_1, \ldots, x_n) \subset A$ *it contains the ideal* $\bar{I}(x_1, \ldots, x_n)$. *For an ideal* $I \subset A$ *its finite-closure is defined as*

$$\bar{\bar{I}} = \bigcup_{(x_1,\ldots,x_n)\subset A} \bar{I}(x_1, \ldots, x_n). \tag{2}$$

It is clear, that an ideal $I \subset A$ is finitely closed if and only if it equals to its finite-closure.

All maximal ideals in $L_0[0,1]$ are finitely closed, as well as all maximal ideals in the algebra $C[0,\infty)$ of the example (2.4). The latter can be easily seen directly, but there is a more general fact stating that all maximal ideals in MLC algebras are finitely closed. It follows from the Arens Theorem, and from the following result which is the main theorem of this section.

Theorem 32. *Let A be a (commutative) topological algebra. Then all its maximal ideals are finitely closed if and only if A has the Arens Property.*

Proof. If A does not have the Arens Property, then it contains a dense finitely generated ideal $I = I(x_1, \ldots, x_n)$. But then every maximal ideal containing I is not finitely closed. On the other hand, suppose that some maximal ideal M of A is not finitely closed. Then M contains an ideal of the form $I(x_1, \dots, x_n)$ whose closure has an element z, which does not belong to M. Thus the smallest ideal of A containing M and z is the whole of A. This means that

$$zy + w = e \tag{3}$$

for some y in A and w in M. Consequently, the ideal $I(x_1 y, \dots, x_n y, w)$ is dense in A, and so A fails to have the Arens Property. □

Maybe the following Corollary could be useful in solving the Problem 1.

Corollary 14. *Suppose that f be a discontinuous multiplicative-linear functional of an MB_0-algebra A. Then the kernel of f is finitely closed.*

A topological algebra A is finitely generated, if there is a finite set $(g_1, \dots, g_n) \subset A$ such that the algebra of all polynomials in these elements is dense in A (the constant polynomials are treated as the multiples of unity)

The following result is well known for m-convex algebras as a consequence of Arens Theorem. We shall prove it in a more general setting.

Theorem 33. *Let A be a topological algebra with the Arens Property, and let f be its multiplicative linear functional. Then the restriction of f to every finitely generated subalgebra of A is continuous.*

Proof. Let A_o be a subalgebra of A generated by elements $g_1, \dots, g_n$ and put $u_i = g_i - f(g_i)e$, $1 = 1, 2, \dots, n$. The elements u_i also generate A_0 and $f(u_i) = 0$, $i = 1, \dots, n$. Every polynomial p_α in n variables u_i can be written as

$$p_\alpha(u_1, \dots, u_n) = u_1 q_{\alpha,1}(u_1, \dots, u_n) + \dots + u_n q_{\alpha,n}(u_1, \dots, u_n) + p(0, \dots, 0) \tag{4}$$

for suitable polynomials $q_{\alpha,i}$, and $f(p_\alpha(u_1, \dots, u_n) = p_\alpha(0, \dots, 0)$. Denote by P the algebra of all such polynomials, it is clear that it is a dense subalgebra of A_0. It is easy to see that the formula (8) implies

$$f(p) = p(0, \dots, 0) \tag{5}$$

for all p in P. Thus this functional is continuous on P because, by Theorem 3.1, the ideal $I(u_1, \dots, u_n)$, which is the kernel of f restricted to P, is non-dense in A and so in P. Thus, by the density of P in A_0, this restriction can be extended to a continuous multiplicative linear functional F on A_0. It remains to be shown that F equals to f on A_0. To this end take any element z in A and put $u = z - f()z)e$ The elements $u, u_1, \dots, u_n$ also generate A_0 and, by an analogous reasoning as above, we show that $f(z) = F(z)$. Since the element z was chosen arbitrarily, we have $F(x) = f(x)$ for all x in A_0 and the conclusion follows . □

The above result fails to be true, if we replace the kernel of a multiplicative linear functional by an arbitrary maximal ideal (in the sense that the intersection of this ideal with a finitely generated subalgebra must be closed). The algebra of example 2.4 has dense maximal ideals while it itself is singly generated.

The proof of the Theorem 3.3 would be much simpler if we had in positive the answer of following question:

Problem 7. *Let A be a topological algebra possessing the Arens property and let A_0 be its closed subalgebra. Does it follow that A_0 has the Arens Property too ?*

The difficulty here lies in the fact that an element non-invertible in A_0 can be invertible in A.

As a corollary to the Theorem 3.3 we obtain following result (known in the case of MLC-algebras).

Corollary 15. *Let A be a finitely generated topological algebra with the Arens Property. Then all multiplicative linear functionals in A are continuous.*

4 Final Remarks

In this section we shall occasionally deal with the non-commutative case. Whenever it occurs it will be clearly stated.

We note first the following non-commutative version of the Theorem 2.5 ([18]).

Theorem 34. *Let A be an F-algebra. Then the following are equivalent:*
(i) *All maximal left ideals of A are closed,*
(ii) *The set $G_l(A)$ of all left-invertible elements of A is open,*
(iii) *A is a Q-algebra,*
(iv) *The set $G_r(A)$ of all right-invertible elements of A is open,*
(v) *All maximal right ideals of A are closed.*

Also the given in the previous section construction can be carried over in the non-commutative case. Suppose that A is an F-algebra with a discontinuous inverse. As previously we take a sequence (x_i) of (not necessarily commuting) invertible elements, tending to the unity e, such that following both products converge

$$a = \lim_n x_1 x_2 \dots x_n \quad \text{and} \quad b = \lim_n x_n x_{n-1} \dots x_1, \tag{1}$$

but the inverses x_i^{-1} do not tend to the unity. Similarly as in the commutative case, we can show that the element a is topologically left invertible and the element b is topologically right invertible. We do not know, however, whether for an element with is both topologically right and left invertible, these inverses (nets) can be equal. Even in the commutative case a net (u_α) which is a topological inverse of an element x is not determined in a unique way (under natural identification of two such nets by the relation $\lim_\alpha(u_\alpha - v_\alpha) = 0$. The reason is that a proper topologically invertible element is at the same time a topological divisor of zero, and adding such a divisor to a topological inverse, we obtain another topological inverse of the element in question. Some topological algebra problems can be posed and studied by replacing invertibility by topological invertibility, so there are TQ-algebras (see e.g. [1] and [2]), i.e. algebras with an open set containing the unity e and consisting of topologically invertible elements, one can study an "TQ-spectrum", or an TQ-joint spectrum. As a sample result concerning the latter we quote the main result of the paper [20]. This result concern still more general class $\mathcal{ST}$ of semi-topological algebras (algebras with only separately continuous multiplication). Here we cannot assume completeness because the multiplication of an $\mathcal{ST}$-algebra usually does not extend to the multiplication of its completion.

The spectrum $\sigma(x)$ of an element $x \in A$ is given by

$$\sigma(x) = \{\lambda \in \mathbf{C} : x - \lambda e \notin G(A)\},$$

or equivalently, by

$$\sigma(x) = \{\lambda \in \mathbf{C} : I(x - \lambda e) \neq A\}.$$

If A is in $\mathcal{ST}$, then the topological spectrum of an element x in A is

$$\sigma^t(x) = \{\lambda \in \mathbf{C} : \bar{I}(x - \lambda e) \neq A\}.$$

The joint spectrum of an n-tuple of elements of an algebra is given by

$$\sigma^t(x_1, \ldots, x_n) = \{(\lambda_1, \ldots, \lambda_n) \in \mathbf{C}^n : I(x_1 - \lambda_1 e, \ldots, x_n - \lambda_n e) \neq A\}.$$

Definition 4.1. *Let A be a (commutative, unital) semi-topological algebra, and let $(x_1, \ldots, x_n) \subset A$. The topological joint spectrum of this n-tuple is defined as*

$$\sigma^t(x_1, \ldots, x_n) = \{(\lambda_1, \ldots, \lambda_n) \in \mathbf{C}^n : \bar{I}(x_1 - \lambda_1 e, \ldots, x_n - \lambda_n e) \neq A\}. \tag{2}$$

If A is a Banach algebra, then $\sigma(x_1, \ldots, x_n) = \{(f(x_1), \ldots, f(x_n)) \in \mathbf{C}^n : f \in \mathcal{M}(A)\}$, where $\mathcal{M}(A)$ is the set of all maximal ideals (identified with multiplicative linear functionals) of A. In the case when elements $x_1, \ldots, x_n$ generate A, this joint spectrum is topologically identical with $\mathcal{M}(A)$. In the case of A in $\mathcal{ST}$, we have an algebraic identification of these sets

Theorem 35. *Let $A \in \mathcal{SL}$ be finitely generated by elements $x_1, \ldots, x_n$. Then there is a one-to-one correspondence between continuous multiplicative linear functionals of A and points of the topological joint spectrum $\sigma^t(x_1, \ldots, x_n)$.*

This correspondence, restricted to the algebra $P(x_1, \ldots, x_n)$, of complex polynomials in variables x_i) is given by $f(p(x_1, \ldots, x_n)) = p(\lambda_1, \ldots, \lambda_n)$ and the conclusion is obtained by showing that the functionals given by this formula on P are continuous, and so extend onto the whole of A.

Corollary 16. *If A is a finitely generated semi-topological (commutative) algebra, then it has continuous)(non-zero) multiplicative linear functionals if and only if the topological joint spectrum ot its generators is non-void.*

The concept of Arens Property can be extended onto semi-topological algebras. In addition to Theorem 3.1 the following result gives a characterization of algebras possessing Arens Property.

Theorem 36. *Let A be a (commutative) semi-topological algebra. Then A has the Arens Property if and only if for every n-tuple* $(x_1, \ldots, x_n) \subset A$, $n = 1, 2, \ldots$ *we have*

$$\sigma(x_1, \ldots, x_n) = \sigma^t(x_1, \ldots, x_n). \tag{3}$$

Proof. Evidently we have

$$\sigma^t(x_1, \ldots, x_n) \subset \sigma^t(x_1, \ldots, x_n) \tag{4}$$

for each n-tuple $(x_1, \ldots, x_n) \subset A$. If there is an

$$(\lambda_1, \ldots, \lambda_n) \in \sigma(x_1, \ldots, x_n) \setminus \sigma^t(x_1, \ldots, x_n), \tag{5}$$

then the ideal $I(x_1 - \lambda_1 e, \ldots, x_n - \lambda_n e)$ is propper and dense, and so the algebra A fails to have the Arens Property. If A does not have the Arens Property, then there is a proper dense ideal of the form $I(x_1, \ldots, x_n)$. In this case the n-tuple $(0, \ldots, 0)$ is in $\sigma 9 x_1, \ldots, x_n)$, but not in $\sigma^t(x_1, \ldots, x_n)$. The conclusion follows. □

We can also study a "T-field" Such a "field" was given by Aharon Atzmon ([6])who constructed a locally convex algebra with all ideals dense. Consequently all its non-zero elements are topologically invertible. This example is non-metrisable, so we have

Problem 8. *Does there exist an F-algebra (or even a* B_0*-algebra) with all (proper) ideals dense ?*

For any infinite dimensional locally convex field its inversion must be discontinuous, however, we do not know whether the same concern "T-fields". In particular we do not know whether in the Atzmon's example the inverse is discontinuous. This example, given as an inductive limits of Bergman spaces, has very complicated construction. The author tried to simplify it, and "succeeded" in the sense that this example has the needed properties. The only problem was to see what is its topology, and it turned out that the construction does not work. It is explained below as a "warning" example against not careful enough use of the inductive limits.

Example 37. *Denote by* Δ_n *the disc* $\{\zeta \in CC : |\zeta - 1/n| < 1/n\}$, *and by* A_n *– the* MB_0*-algebra of all holomorphic functions in* Δ_n *provided with the pointwise algebra operation*

and the compact-open topology. The maps $h_n : A_n \mapsto A_{n+1}$ given by restriction of functions in A_n to the disc Δ_{n+1} are continuous one-to-one algebra homomorphisms. Denote by

$$A = \lim_{\rightarrow} A_n \tag{6}$$

The inductive limit of the sequence (A_n, h_n), i.e. the union $\cup A_n$ under the strongest vector space topology making the embeddings h_n continuous. In the above we identify an element x in A_n with $h_n(x)$. It was easy to see that A is not a field, and all its non-invertible elements are topologically invertible.

A subset U of an inductive limit (15) is a neighborhood of zero if and only if its intersection with any A_n is such a neighborhood in its original topology (see e.g. [8], p. 447).

Fix now an arbitrary neighborhood U and an arbitrary element x_0 of A, and find another such a neighborhood V satisfying

$$V + V \subset U. \tag{7}$$

The element x_0 belongs to some A_m, so we can find a compact $K_m \subset \Delta_m$, and a positive ε_m such that

$$\{x \in A_m : |x(\zeta)| < \varepsilon_m \text{ for } \zeta \in K_m\} \subset V \cap A_m. \tag{8}$$

We can also find an index n, so that $K_m \cap \Delta_n = \emptyset$. For this index, similarly as before, we find a positive ε_n and a compact $K_n \subset \Delta_n$, so that

$$\{x \in A_n : |x(\zeta)| < \varepsilon_n \text{ for } \zeta \in K_n\} \subset V \cap A_n. \tag{9}$$

By the Mergelyan theorem (see e.g.[7], p.47), there is a polynomial p with

$$|p(\zeta)| < \varepsilon_n \text{ for } \zeta \in K_n \tag{10}$$

and

$$|x_0(\zeta) - p(\zeta)| < \varepsilon_m \text{ for } \zeta \in K_m. \tag{11}$$

The relations (17),(18),(19) and (20) imply $p \in V$ and $x_0 - p \in V$. Thus, by (16)

$$x_0 = x_0 - p + p \in V + V = U. \tag{12}$$

Since x_0 was chosen arbitrarily, we have $U = A$. Since U was chosen arbitrarily, the only open set in A is the whole of A! and the example does not work.

Bibliography

[1] Abel, M. and Zelazko, W. *Topologically invertible elements and topological spectrum*, Bull. Polon. Acad. Sc. Math. 54 (2006), 257–271.

[2] Abel, M. and Zelazko, W. *Properties of TQ-algebras*, Proc. Est. Acad. Sc.: Phys. Math. 60 (2011), 141–148.

[3] Akkar, M., Nacir, C. *Continuité automatique dans les limits inductives localement convexes de Q-algébres de Fréchet*, Ann. Sci. Math. Québec 19 (1995), 115–130.

[4] Arens, R.F. *The space L^{ω} and convex topological rings, Bull. Amer. Math. Soc. 52 (1946), 931–935. limit rings*, Michigan Math. J. 5 (1958).

[5] Arens, R.F. *Dense inverse limit rings*, Michigan. Math. J. 5 (1958), 169–182.

[6] Atzmon, A. *An operator without invariant subspace on a nuclear Fréchet space*, Ann. of Math. 117 (1983), 665–684.

[7] Gamelin, T.W. *Uniform Algebras*, Prentice-Hall, 1969.

[8] Kantorovich, L.V. and Akilov, G.P. *Functional Analysis in Normed Spaces*, Pergamon Press, New York, 1964.

[9] Mallios, A. *Topological Algebras. Selected Topics*, North Holland, 1986.

[10] Michael, E. *Maultiplicatively Convex Topological Algebras*, Mem. Amer. Math. Soc. 11, 1952.

[11] Williamson, J.H. *Topologizing the field $C(t)$*, Proc. Amer. Math. Soc. 5 (1954), 729–734.

[12] Zelazko, W. *Metric Generalizations of Banach Algebras*, Dissert. Math. 47 (1965).

[13] Zelazko, W. *Selected Topics in Topological Algebras*, Aarhus University Lecture Notes Series No 31, 1970.

[14] Zelazko, W. *On maximal ideals in commutative m-convex algebras*, Studia Math. 58 (1976), 291–298.

[15] Zelazko, W. *Concerning entire functions in B_0-algebras*, Studia Math. 110 (1994), 283–290.

[16] Zelazko, W. *On topologization of countably generated algebras*, Studia Math. 112 (1994), 83–88.

[17] Zelazko, W. *F-algebras: some results and open problems*, in *Functional Analysis and its Applications*, North-Holland Math. Stud. 197, 317–326, Elsevier 2004.

[18] Zelazko, W. *When a commutative F-algebra has a dense principal ideal*, in *Topological Algebras and Their Applications (Proc. ICTAA 4)*, Contemporary Mathematics vol. 341, Providence 2004, pp. 133–137.

[19] Zelazko, W. *When a unital F-algebra has all maximal left (right) ideals closed*, Studia Math. 175 (2006), 279–284.

[20] Zelazko, W. *Continuous characters and joint topological spectrum*, Control and Cybernetics 36 (2007), 859–864.

De Gruyter Proceedings in Mathematics

J. Pintz, A. Biró, K. Győry, G. Harcos, M. Simonovits, J. Szabados (Eds.)
Number Theory, Analysis, and Combinatorics: Proceedings of the Paul Turán Memorial Conference held August 22–26, 2011 in Budapest, 2013
ISBN 978-3-11-028237-5, e-ISBN 978-3-11-028242-9

I. Assani (Ed.)
Ergodic Theory and Dynamical Systems: Proceedings of the Ergodic Theory workshops at University of North Carolina at Chapel Hill, 2011–2012, 2013
ISBN 978-3-11-029813-0, e-ISBN 978-3-11-029820-8

B. Landman, M. B. Nathanson, J. Nešetril, R. J. Nowakowski, C. Pomerance, A. Robertson (Eds.)
Combinatorial Number Theory: Proceedings of the „Integers Conference 2011", Carrollton, Georgia, USA, October 26–29, 2011, 2013
ISBN 978-3-11-028048-7, e-ISBN 978-3-11-028061-6

C. Francisco, L. C. Klingler, S. Sather-Wagstaff, J. C. Vassilev (Eds.)
Progress in Commutative Algebra 1: Combinatorics and Homology, 2012
ISBN 978-3-11-025034-3, e-ISBN 978-3-11-025040-4

C. Francisco, L. C. Klingler, S. Sather-Wagstaff, J. C. Vassilev (Eds.)
Progress in Commutative Algebra 2: Closures, Finiteness and Factorization, 2012
ISBN 978-3-11-027859-0, e-ISBN 978-3-11-027860-6

K. K. Sabelfeld, I. Dimov (Eds.)
Monte Carlo Methods and Applications: Proceedings of the 8th IMACS Seminar on Monte Carlo Methods, August 29–September 2, 2011, Borovets, Bulgaria, 2012
ISBN 978-3-11-029347-0, e-ISBN 978-3-11-029358-6

A. D. Bruno, A. B. Batkhin (Eds.)
Painlevé Equations and Related Topics: Proceedings of the International Conference, Saint Petersburg, Russia, June 17-23, 2011, 2012
ISBN 978-3-11-027558-2, e-ISBN 978-3-11-027566-7

Bei Fragen zur Produktsicherheit wenden Sie sich bitte an:
If you have any questions regarding product safety,
please contact:

Walter de Gruyter GmbH
Genthiner Straße 13
10785 Berlin
productsafety@degruyterbrill.com